Marine Polysaccharides

Food Applications

Marine Polysaccharides

Food Applications

Vazhiyil Venugopal

CRC Press
Taylor & Francis Group
Boca Raton   London   New York

CRC Press is an imprint of the
Taylor & Francis Group, an **informa** business

CRC Press
Taylor & Francis Group
6000 Broken Sound Parkway NW, Suite 300
Boca Raton, FL 33487-2742

First issued in paperback 2016

© 2011 by Taylor and Francis Group, LLC
CRC Press is an imprint of Taylor & Francis Group, an Informa business

No claim to original U.S. Government works

ISBN 13: 978-1-138-19844-9 (pbk)
ISBN 13: 978-1-4398-1526-7 (hbk)

Visit the Taylor & Francis Web site at
http://www.taylorandfrancis.com

and the CRC Press Web site at
http://www.crcpress.com

Dedicated to

Late Prof. K. P. Antony, Department of Chemistry,

St. Thomas' College, Thrissur, India

Contents

Section II. Food Applications

Section III. Biomedical Applications

Preface

Increasing public awareness of the importance of healthy living is presenting new challenges for the commercial food processing sector. Consumer perceptions of processed foods are changing, and processed foods are being recognized as convenient vehicles for the delivery of bioactive compounds and nutraceuticals. This changing scenario has had a profound effect on the global food processing industry, which must consider nutritional quality, cost of production, added value, consumer safety, and convenience. A major contributor to successfully addressing these challenges is the availability of novel and safe additives with fascinating functional properties that can be used to impart appealing properties to foods, such as modified texture, stability, foam and emulsion capacities, water retention, fat replacement, microbial protection, control of rancidity, and enhancement of fiber content, among others. Polysaccharides are water-soluble biopolymers (also referred to as *hydrocolloids* or *gums*), derived from diverse renewable sources such as seeds, fruits, vegetables, plant exudates, microorganisms, and animals, that can meet most of these requirements for food additives. Polysaccharides were introduced to food processing as early as the 1940s, initially through the use of modified specialty starches and high fructose sweeteners. Explosive growth in the use of other carbohydrate ingredients, such as maltodextrins and microcrystalline cellulose, occurred from the 1960s to the 1980s. These compounds are of natural origin and are considered safe and edible. Further, their chemical structures allow modification of their functional properties and expand their potential applications in food, medicine, biotechnology, and other fields.

Marine resources, apart from providing a wealth of protein foods, also provide several polysaccharides with numerous possible uses in diverse fields. Because marine environments are extremely diverse in terms of, for example, available nutrients, temperature, and pressure, organisms inhabiting the oceans have adapted themselves to these varying habitats by evolving several unique compounds. It is likely that marine polysaccharides possess some unique properties that may be useful in food technology and other fields. Although some marine polysaccharides, such as carrageenans, and agar, have found wide use as hydrocolloids due to their ability to form gels and function as thickeners and stabilizers in a variety of foods, the potential of many of these hydrocolloids has yet to be explored. This book attempts to compile recent data on the food applications of marine polysaccharides from such diverse sources as fishery products, seaweeds, microalgae, microorganisms, and corals.

The chapters of the book are grouped into three sections. The chapters in the first section are devoted to discussions on the isolation of polysaccharides from marine sources and their general properties, particularly those important from a food technology point of view. The second section of the book focuses on the actual food applications of these compounds, and the chapter in the third section provides a brief discussion of biomedical applications. Chapter 1 discusses major sources of marine polysaccharides, including crustacean shellfish, macroalgae (seaweed), microalgae, and marine microorganisms, as well as coral. Chapter 2 provides an overview of the general functional properties of polysaccharides, such as their structure; their hydration, gelation, emulsification, and rheological properties; and interactions among themselves and with other food components such as proteins that are relevant to food processing. Chapters 3, 4, and 5 further discuss the isolation and food-related properties of various marine polysaccharides.

The second section covers the use of these polysaccharides for food product and biopackaging development. Recent developments in composite films and nanotechnology have greatly contributed to this field, as discussed in Chapter 9. The safety and regulatory aspects of food ingredients are very important factors to consider during product development, and Chapter 10 addresses these aspects with respect to polysaccharides. The book concludes with an overview of recent developments in the biomedical applications of marine polysaccharides.

I thank Stephen Zollo, Chief Editor, and Patricia Roberson, Project Coordinator, Taylor & Francis, Boca Raton, FL, for their editorial support.

Vazhiyil Venugopal
venugopalmenon@hotmail.com
vvenugopalmenon@gmail.com

Author

Vazhiyil Venugopal received his MSc in chemistry from the University of Kerala and his PhD in biochemistry from the University of Bombay, India. He began his career at the Central Institute of Fisheries Technology, Cochin, India, and later moved to the Bhabha Atomic Research Center, Mumbai, where he was the head of the Seafood Technology Section of the Food Technology Division. He has been a postdoctoral Research Fellow at the National Institutes of Health, Bethesda, Maryland, and a visiting scientist at the Memorial University of Newfoundland, St. John's, Newfoundland, Canada. His main interests are the value addition of fishery products, radiation processing of seafood, and marine proteins. His more than 120 publications in these areas include research papers, review articles, and book chapters. He has previously published two books, *Seafood Processing: Adding Value Through Quick Freezing, Retortable Packaging, and Cook-Chilling* (CRC Press, 2006) and *Marine Products for Healthcare: Functional Compounds and Bioactive Nutraceuticals from the Ocean* (CRC Press, 2008). He is a Fellow of the National Academy of Agricultural Sciences, New Delhi, India.

Section I

Isolation and Properties of Marine Polysaccharides

1

Polysaccharides: Their Characteristics and Marine Sources

1.1 Introduction

Commercial food processing is becoming more sophisticated in response to growing consumer demands for processed foods, of both animal and vegetable origin, as a result of increasing globalization and changes in lifestyles of the general public all over the world. Among the diverse food processing sectors, the field of fabricated foods is witnessing considerable ingenuity and creativity, as well as rapid growth. In developing such foods, the use of hydrocolloids is crucial to impart specific sensory properties. Hydrocolloids are used as thickeners, emulsifiers, syneresis inhibitors, gel and film matrices, and water retention and texture enhancers.[1-3] Polysaccharides are the major hydrocolloids invariably used for these applications. They are also referred to as *gums*, in view of their ability to thicken and bind various components in a food product. Currently, the food industry uses polysaccharides such as agar, alginates, gum arabic, carrageenan, konjac flour, locust bean gum, methyl cellulose and hydroxylpropylmethylcellulose, microcrystalline cellulose, xanthan, curdlan, gellan, pullulan, dextran, and pectin. These compounds are isolated from plant and microbial sources. The rising demand for processed foods has created a need for novel food hydrocolloids with interesting functional properties.

Marine habitats, inhabited with diverse animals, are capable of providing such interesting polysaccharides. The sea has long been considered a treasure trove of a variety of protein foods, including not only numerous species of finfish but also several species of crustaceans and mollusks. The high fecundity (in some cases, over 1 million eggs per female) of fish and shellfish has no parallel among other animal protein sources, such as livestock and poultry, and has been supporting life on Earth for centuries. The oceans, however, are also home for diverse macroalgae (seaweed), corals, and microorganisms. Not only do many of these oceanic creatures serve as a source of food, but many of them are also sources of various life-supporting compounds, including a variety of nutraceuticals and bioactive compounds such as vitamins, carotenoids, proteins, and polyunsaturated fatty acids (omega-3 fatty acids), among others. Finfish, shellfish, seaweeds, microalgae (including cyanobacteria), and marine microorganisms are all sources of these compounds in varying proportions.[4-6]

A major class of compounds that can be isolated from these oceanic creatures is polysaccharides. These high-molecular-weight marine carbohydrates have found a variety of applications in the food technology, biotechnology, pharmacy, and chemical industries. They have the potential to enhance the quality, shelf life, and acceptability of diverse food products. The voluminous information made available through recent research has made it necessary to consolidate our current understanding on marine polysaccharides, particularly with respect to their food applications. This chapter begins with a brief discussion on the characteristics of carbohydrates in general, which is essential to understanding the role of marine polysaccharides in food product development. Toward the end of the chapter, the discussion focuses on identifying major sources of polysaccharides from the marine environment.

1.2 Carbohydrates

Carbohydrates are defined as polyhydroxy aldehydes, ketones, alcohols, and acids, or their simple derivatives and polymers having polymeric linkages of the acetal type. They are classified according to their degree of polymerization: sugars (mono- and disaccharides), oligosaccharides (containing three to ten monosaccharide units), and polysaccharides (containing ten or more monosaccharide units). Monosaccharides are commonly referred as *simple sugars* (e.g., glucose and fructose), which cannot be broken down by hydrolysis. Oligosaccharides are carbohydrate chains that yield three to ten monosaccharide molecules upon hydrolysis. These compounds may also be linked to non-sugar organic compounds. The ring structures (pyranose or furanose rings) of carbohydrate molecules are hemiacetals that can react with an alcohol to form glycosides, giving acetals of sugars. Hydrolysis of a glycoside in an acidic solution releases the monosasccharide and the alcohol. Glycosides, particularly of phenolic compounds, are widely distributed in plant tissues.[7]

In terms of their nutritional role, carbohydrates are often classified as available or unavailable. Available carbohydrates are hydrolyzed by enzymes of the human gastrointestinal system to monosaccharides, which are then absorbed in the small intestine to enter the pathways of carbohydrate metabolism. Oligosaccharides in the human body include those derived from the hydrolysis of starch and other oligosaccharides, such as fructooligosaccharides. Unavailable carbohydrates are not hydrolyzed by endogenous human enzymes, although they may be fermented to varying extents in the large intestine. Consumption of carbohydrates through food provides 40 to 80% of the body's energy needs. Following digestion and absorption, the available carbohydrates from food may be used to meet the immediate energy needs of tissue cells, converted to glycogen and stored in the liver and muscle, or converted to fat as a reserve of energy. Besides providing energy, carbohydrates are integrally involved in a multitude of biological functions such as regulation of the immune system, cellular signaling (communication), cell malignancy, antiinfection responses, and host–pathogen interactions.[8,9] Many carbohydrates, particularly polysaccharides, also function as fiber, offering a wide range of beneficial physiological functions (such as increased transit time and increased satiety). They can also contribute to the prevention of cancer, heart disease, and weight gain, thereby contributing to health and wellbeing.[10]

Specific dietary carbohydrates also function as prebiotics to increase the numbers of beneficial bacteria (probiotics) within the colon, in a selective manner. Prebiotics are nondigested carbohydrates that selectively modify the composition of the colonic flora in favor of those that contribute to good health. This results in a wide range of physiological benefits for the host, including reduced gut infections and constipation, improved lipid metabolism, higher mineral absorption, enhanced immunomodulation, and reduced risk of carcinogenesis. The concept of prebiotics emerged when it was discovered that mammalian digestive enzymes could not hydrolyze the β-glucosidic bonds found in certain carbohydrates. These oligosaccharides and polysaccharides, therefore, escape metabolism in the upper intestine and become available to bacteria in the colon via fermentation. Prebiotics can be easily incorporated into food products, where they confer technological and nutritional advantages.[11]

Food processing has a direct effect on dietary carbohydrates. During wet heat treatment, as in blanching, boiling, and canning of vegetables and fruits, a considerable loss of low-molecular-weight carbohydrates into the processing water occurs. The loss of glucose and fructose during boiling is higher than that of sucrose. The loss of low-molecular-weight carbohydrates also depends on various cultivars and harvest and storage conditions.[11,12]

1.3 Polysaccharides

Polysaccharides are relatively complex carbohydrates and are the most abundant of organic compounds, constituting about half of the organic carbon on Earth. Polysaccharides are macromolecules made up of many monosaccharides joined together by glycosidic bonds, hence they are very large and often branched. Polysaccharides differ not only in the nature of their component monosaccharides but also in the length of the chains and in the amount of chain branching. Depending on its chemical composition, a polysaccharide can be a *homopolysaccharide*, where all of the monosaccharides in the polysaccharide are of the same type, or a *heteropolysaccharide*, where more than one type of monosaccharide is present. Homopolysaccharides are often named for the sugar unit they contain, so glucose homopolysaccharides are called *glucans*, and mannose homopolysaccharides are *mannans*. Generally, polysaccharides are also called *glycans*, which differ in the type of glucosidic linkages, degree and type of branching, length of glucan chains, molecular mass, and conformation of polymers. Glucans strongly contribute to specific polysaccharide characteristics such as solubility, rheology, and other physical characteristics.

Polysaccharides have a general formula of $C_x(H_2O)_y$, where x is usually a large number between 200 and 2500. Considering that the repeating units in the polymer backbone are often 6-carbon monosaccharides, the general formula can also be represented as $(C_6H_{10}O_5)_n$, where $n = 40$ to 3000. The most common constituent of polysaccharides is D-glucose, but D-fructose, D-galactose, D-galactose, D-mannose, L-arabinose, and D-xylose are also found. Some monosaccharide derivatives found in polysaccharides include the amino sugars (D-glucosamine and D-galactosamine), as well as their derivatives (N-acetylneuraminic acid and N-acetylmuramic acid) and simple sugar acids (glucuronic and uronic acids). Derived compounds from polysaccharides include carboxymethylcellulose, butyrates, cellulose nitrates, hydroxyalkylcellulose, and methylcellulose from cellulose; acetate, adipates, phosphates, succinates, and carboxymethyl, hydroxylethyl, hydroxylpropyl, and cationic salts from starches; and carboxymethyl, hydroxylpropyl, and cationic salts from guar gum; as well as several other compounds.[7] The largest functional group is the hydroxyls. A given sugar residue can form only one glycosidic linkage with a hydroxyl group of another molecule.

Polysaccharides and oligosaccharides have reducing and nonreducing ends. The reducing end of an oligo- or polysaccharide is the one end not involved in a glycosidic linkage. The sugar units constituting all of the other ends are attached through glycosidic linkage (acetal) and therefore are nonreducing ends. In a polysaccharide, the hydroxyl groups may sometimes be methylated or converted to sulfate esters or ketals, formed with pyruvic acid. The ability to form branched structures distinguishes polysaccharides from proteins and nucleic acids, which occur only as linear polymers. Despite the variety of different monomer units and the kinds of linkages present, the conformational positions of carbohydrate chains are limited, unlike proteins.

Polysaccharides are produced by renewable resources, including plants, animals, microorganisms, seeds, fruits, vegetables, plant exudates, seaweeds, and microalgae. Carbohydrates constitute about 75% of the dry weight of all land and marine plants; most are found in cell walls or structural compounds. Cellulose, hemicellulose, and starch represent the main constituents of plant biomass; they accumulate through the photosynthetic fixation of carbon dioxide and occur as cell wall and storage carbohydrates, transportation carbohydrates, and glycol conjugates. As a component of connective tissues, they provide structure to the cell; chains of specific polysaccharides are covalently linked to a polypeptide chain. They are also implicated in such functions as nutrition for growing pollen tubes, cell–cell adhesion, cell malignancy, immune responses, and host–pathogen interactions.

Major sources of starch are the polysaccharides of plant cell walls (e.g., cellulose, hemicellulose, pectin). Starches are generally regarded as the most important constituent of cereals. Milling of cereals transforms the whole grain into forms suitable for human consumption. These grains contain two polysaccharide molecules, amylose and amylopectin, in varying proportions. Cereal starches can serve as fat replacements in food, and maltodextrins and other modified starches provide high viscosity and creaminess to low-fat foods with little or no gelling.[10,13] The term *complex carbohydrate* is often used in the United States to describe starch, dietary fiber, and nondigestible oligosaccharides, all of which exhibit a large variety of unique physiological functions and a wide range of potential applications, particularly for foodstuffs. Universally recognized as safe for human consumption, plant polysaccharides such as starch have been used as additives in foods for a very long time. Starch provides most of the calories in the average human diet. Starch, although considered fully digestible, sometimes remains partly undigested in the human gastrointestinal tract. This fraction of starch-resisting digestion *in vivo* is known as *resistant starch*. Upon fermentation in the gut, the resistant starch can yield butyric acid.[14]

Agricultural waste is a good source of carbohydrates and other nutrients for the fermentation and chemical industries. Production of pulp from wood cellulose, applications of starch for papermaking, and the use of glucose and saccharine for fermentation are the most important chemical uses of carbohydrates. Their importance is also steadily growing in biotechnological processes.[14]

Microbial polysaccharides are primarily the exogenous metabolites. Newer biotechnological tools such as genetic manipulation and protein engineering are useful in developing novel polysaccharides.[15] The β-glucans (polysaccharides characterized by β-(1,4) linkages; see Section 1.3.1) are abundant in microorganisms. They are also found in higher plants as structural entities of cell walls, as cytoplasmic and vacuolar reserve materials, and as extracellular substances. Some microorganisms, such as yeast and mushrooms, and such cereals as oats and barley are of interest because they contain large amounts of β-glucans. These compounds can stimulate the immune system and modulate cellular immunity, in addition to offering humoral and anticytotoxic, antimutagenic, and antitumorogenic functions. Also, they exhibit hypocholesterolemic and anticoagulant properties, making them promising candidates as pharmacological promoters of health.[16,17]

1.3.1 Isolation and Identification

The composition and molecular weights of polysaccharides can vary with source and with the conditions used for isolation and preparation, thus determining their physical and functional properties. Various types of a given polysaccharide can be produced by controlling the source and isolation procedure. Polysaccharides are isolated from biological extracts derived from either plant or animal tissues or from the cultures of microorganisms. Extracellular polysaccharides (EPSs) from microorganisms are isolated from cultures obtained by fermentation of the required organism. The crude extracts consist of mixtures of proteins, polysaccharides, and secondary metabolites. Purification is necessary to obtain products that are suitable for human or animal consumption or for use in pharmaceuticals.

Analytical ultracentrifugation has provided fundamental physical information about polysaccharides in solution, mainly because of the technique's diversity, absolute nature, and inherent fractionation ability without the need for separation columns and membranes. Size-exclusion chromatography, high pH anion-exchange chromatography, gas–liquid chromatography, high-performance liquid chromatography, and gel electrophoresis are also useful for isolation and characterization.[18–21] Analytical methods to determine total carbohydrates in environmental and food samples usually require a preliminary chemical hydrolytic procedure to convert polysaccharides into monosaccharides prior to detection by colorimetric or chromatographic techniques. A method for hydrolysis based on the application of ultrasound at room temperature provides better accuracy and reduced time required for analysis (4 to 5 hours), which is useful for the determination of total carbohydrates in samples including seawater and marine mucilage.[22] To reveal the structures of polysaccharides, it is necessary to determine not only the monosaccharides present and their linkage positions and sequence but also the anomeric configuration of the linkages, the ring size (furanose or pyranose), and the absolute configuration (D or L), in addition to determining the presence of any other substituents. Component analysis, methylation analysis, glycosidic hydrolysis, mass spectrometry (MS) methods, and nuclear magnetic resonance (NMR) spectroscopy are some of the relevant analytical methods useful in this regard.[7,23–25] Table 1.1 lists the various techniques used to characterize the biophysical properties of polysaccharides in solution.

Polysaccharides adopt different types of secondary structures in the solid state, each exhibiting a distinctive set of helix parameters. The type-A helix is a ribbon structure formed by structural polysaccharides such as cellulose, hemicellulose, or pectin. Alginate and carrageenan also adopt a type-A secondary structure, which is formed by polysaccharides having the β-(1,4) linkage; these polysaccharides exhibit relatively strong interchain hydrogen bonding and exclusion of water. The type-B helix is formed by storage polysaccharides (amylase, amylopectin, and glycogen) and is relatively less compact, with a large number of residues ($n = 8$) per turn. The hollow helix

TABLE 1.1

Techniques for Characterizing Polysaccharides

Methodology	Information
Analytical centrifugation	Molecular weight, polydispersibility, sedimentation coefficient and distribution, solution conformation, flexibility, interaction, and complex formation
Static (multi-angle laser) light scattering	Molecular weight, radius of gyration, solution conformation, and flexibility
Size-exclusion chromatography	Molecular weight, molecular weight distribution, polydispersibility, radius of gyration and distribution, solution conformation, and flexibility
Dynamic light scattering	Hydrodynamic or Stoke's radius
Small-angle x-ray scattering	Chain contour length and flexibility
Viscometry	Intrinsic viscosity [η], conformation, and flexibility

Source: Adapted from Harding, S.E., *Adv. Polym. Sci.*, 186, 211, 2005.

is relatively water soluble and unstable in solution unless present as a double helix. The type-C helix, or flexible coil, is formed by an expanse of monomers joined by β-(1,2) linkages. This structure is expected to show substantial steric hindrance and hence a low probability of occurrence. Table 1.2 shows features of the primary structure of polysaccharides, and Table 1.3 shows the types of glycosidic linkages in the compounds.

The shape of a biopolymer in a solution gel or solid substrate is called its *tertiary structure*. Dry polymers have a monocrystalline and amorphous tertiary structure. Heating polysaccharides results in a glassy state. In the case of a glassy polymer, no rotation around bonds adjoining adjacent monomers is observed in the polymer backbone. The backbone remains rigid, and the viscosity of the system is very high. The glass transition temperature (T_g) when the backbone mobility begins to occur can be measured by a number of techniques (e.g., rheology, differential scanning calorimetry). Glass transition is restricted to only selected polysaccharides, such as amylopectin, pullulan, and guar gum. Water is a very effective plasticizer for biopolymers, and the addition of water lowers the glass transition temperature. The glass transition

TABLE 1.2

Features of Primary Structure of Polysaccharides

Presence of different sugars (monomers)

Substitute groups (e.g., sulfate, phosphate) and substitution points

Sequence of sugars

Glycosidic linkages: (1→4), (1→3), (1→2)

Anomeric configuration (α or β configuration)

Ring sizes (pyranose or furanose rings)

Absolute configuration (D or L)

TABLE 1.3

Type of Glycosidic Linkages in Some Polysaccharides and Their Occurrence

Polysaccharide	Glucosidic Linkage	Occurrence
Amylose	(1→4) (linear)	Plant
Amylopectin	(1→4), (1→6) (branched)	Plant
Glycogen	(1→4), (1→6) (branched)	Mammals
Cellulose	(1→4) (linear)	Plant cell wall
Laminarin	(1→3) (linear)	Seaweed
Alginic acid	(1→4) (linear)	Seaweed
Carrageenan	(1→3), (1→4)	Seaweed
Xanthan	(1→4), with branching at C_3	Microorganism (*Xanthomonas* spp.)
Dextran	(1→6), with branching at (1→3) and occasionally at (1→4) or (1→2)	Microorganism (*Streptococcus* and *Leuconostoc* spp.)
Unnamed	(1→3), (1→6) (branched)	Fungi, mushroom

Source: Adapted from Misaki, A., in *Food Hydrocolloids: Structure, Properties and Functions*, Nishinari, K. and Doi, E., Eds., Plenum Press, New York, 1994, pp. 1–20; Shimizu, Y. and Kamiya, H., in *Marine Natural Products*, Vol. 5, Sheuber, P.J., Ed., Academic Press, New York, 1985, p. 403; Welman, A.D. and Maddox, I.S., *Trends Biotechnol.*, 21, 269–274, 2003.

temperature has a significant influence on such processes as collapse, caking, and crystallization and on such operations as drying, extrusion, and flaking.[25,26] In the dissolved state, polymer–polymer contacts are replaced by polymer–solvent interactions, leading to a random conformation.[7,26]

1.3.2 Properties and Food Uses

Polysaccharides are odorless, colorless, and nontoxic. They are amorphous and insoluble in water, and they have no sweet taste. Polysaccharides have been widely used in food technology as additives to modify the rheology and texture of aqueous suspensions. The presence of polysaccharides in food dictates its texture and gives the product desirable consumer appeal. In food technology, polysaccharides are usually used at levels as low as 1 to 3% of formulation weight. Their use arises from the fact that these biopolymers at low concentrations dissolve or disperse in water to give highly viscous dispersions or gels with various gelling, thickening, stabilizing, binding, and emulsifying effects. Their gelling ability has allowed the development of a wide variety of processed food products, such as jams, jellies, salad dressings, and sauces. These compounds also slow down the retrogradation of starch, increase moisture retention, enhance fiber content, and improve overall quality of the product. Polysaccharides control ice crystal formation in frozen food products and confer stability to products undergoing successive freeze–thaw cycles. They have also shown good properties as fat replacers in different products.[27] Functional properties of polysaccharides relevant to food product development are discussed in detail in Chapter 2. It must be pointed out that, although their versatility makes them useful in many fields of application, currently the use of polysaccharides in food product development probably accounts for about 10% of their total applications. The majority of polysaccharides currently employed in the food industry are derived from plants and seaweeds. Starch and modified starch are the most widely used polysaccharides in foodstuffs.[2,10,28–30] Regardless of the purpose of the polymer used in food product development, it must conform to the food additive regulations applying to that particular product (see Chapter 10).

1.4 Marine Sources of Polysaccharides

Life on Earth owes much to the oceans for its survival. Covering more than 70% of the Earth's surface, the oceans help maintain an environment suitable for life by functioning as the largest sink for atmospheric carbon dioxide. The oceans receive a greater proportion of sunshine than land does, particularly in the tropical and subtropical belt, and this sunshine is efficiently utilized through photosynthesis by hundreds of species of microalgae. Algal photosynthesis provides food for higher organisms, including fish, shellfish, and numerous other living organisms of both plant and animal origin, which, in turn, contribute to sustenance of life on Earth. More than 80% of Earth's living organisms are found in marine ecosystems; of these, only 5% are utilized by humans. Marine organisms contain a variety of potentially bioactive substances associated with their secondary metabolites, which differ considerably from those of terrestrial organisms. Various marine biotechnology tools have been used to identify and investigate bioactive substances with antitumor, antiinflammation, antiallergy, antioxidant, antifungal, anti-HIV, and antihypertensive activities.[4,31]

1.4.1 The Marine Environment

The marine ecosystem, the largest on the planet, can be divided into photic, pelagic, benthic, epipelagic, and aphotic zones, the depths of which vary from 200 to 10,000 meters. The more than 40,000 different species of phytoplankton are divided into the major classes of Cyanobacteria, Chlorophyta, Cryptophyta, Rhodophyta, Heterokontophyta, Dinophyta, Haptophyta, and Euglenophyta.[32] Marine microbiological communities have significant influence on the marine food chain. Marine bays and inlets receiving sewage or industrial waste can have very high phytoplankton and bacterial populations. This, in turn, supports higher densities of chemotrophic bacteria and aquatic animals such as fish and shellfish.[33]

The dominant autotrophs (living organisms capable of producing energy) are single-celled microscopic plants of various groups of algae, which comprise the first stage of the marine food chain. Much of the primary productivity in the open oceans, even at significant depths, is due to the photosynthetic activities of these microalgae, aerobic prokaryotes that contain chlorophylls.[34] The photosynthetic fixation of carbon dioxide is limited by the nutrients available; therefore, inshore ocean areas, which are nutritionally richer than open waters, support more dense populations of phototropic microorganisms. Nevertheless, significant numbers of prokaryotic cells (in the range of 10^5 to 10^6 cells per milliliter) are suspended in the open ocean, in addition to very small eukaryotic organisms (about 10^4 cells per milliliter).

In tropical and subtropical oceans, the planktonic filamentous marine cyanobacterium *Trichodesmium* represents a significant fraction of the biomass suspended in these waters. This organism is capable of nitrogen fixation and is thought to be a major link in the nitrogen cycle in the marine environment. The very small phototropic algae *Ostreococcus*, measuring only about 0.7 μm in diameter (smaller than a cell of *Escherichia coli*), are also involved in nitrogen fixation. Many prokaryotes in the photozone of the ocean (down to around 300 m) contain a form of the visual pigment rhodopsin, which the cells use to convert light energy into adenosine triphosphate (ATP).

Organisms that inhabit the deep sea are faced with three types of extreme environmental conditions: low temperature, high pressure, and low nutrient levels. Below depths of about 100 m, ocean water stays at a constant temperature of 2 to 3°C. Pressure increases by 1 atmosphere every 10 m; thus, organisms growing at 5000 m must be able to withstand pressures as high as 500 MPa. These extreme conditions result in a reduction in microbial levels with increasing depth. Compared with about 10^5 to 10^6 cells per milliliter of surface waters, 2000 m down the cell counts may be as low as 3×10^3 per milliliter. Whereas temperatures below 0°C are found in the Arctic and Antarctic oceans, temperatures exceeding 100°C are found in the hydrothermal vents in the ocean bottoms. Salinities as high as 6 N have been found in salt marshes and mines. These diverse environmental conditions have resulted in an immense biodiversity of marine organisms. Each organism has a metabolism adapted to its particular conditions. Cold-adapted enzymes from fish living at temperatures near the freezing point of seawater and thermoresistant enzymes from organisms, including crustaceans, living in hydrothermal vents have been reported.[35]

Dense, thriving animal communities supported by the presence of microorganisms cluster about thermal springs in deep-sea waters. Several inedible species, including sponges, crustaceans, and other animals, also live in these deep marine environments. Seasonal environmental characteristics such as temperature, chlorophyll content, salinity, microbial water quality, and algal lipid composition have significant influence on marine organisms with respect to their meat content, shell size, and microbial and lipid composition.

Despite obvious differences, marine biodiversity patterns in environmental conditions of the various oceanographic regions show a consistency worldwide.[36,37]

Marine organisms have developed unique properties and bioactive compounds that, in some cases, are unparalleled by their terrestrial counterparts. The marine world, therefore, can be considered to be a large reservoir of bioactive ingredients that can be isolated for varied applications, including food processing, storage, and fortification. Enzymes extracted from fish and other marine microorganisms, for example, offer numerous advantages over traditional enzymes used in food processing due to their ability to function at extreme temperatures and pH levels. Fish proteins such as collagens and their gelatin derivatives function at relatively low temperatures and can be used in such heat-sensitive processes as gelling and clarifying. A number of marine-derived compounds, such as omega-3 polyunsaturated fatty acids and photosynthetic pigments, are also important to the nutraceutical industry. These bioactive ingredients provide myriad health benefits, including reduction of coronary heart disease and their anticarcinogenic and antiinflammatory activity.[4] Despite the vast potential for the use of marine organisms in the food industry, tools of biotechnology are required for successful cultivation and isolation of these unique bioactive compounds.[38]

Marine organisms are a virtually untapped reservoir of polysaccharides. Marine life is adapted to widely different environments, so polysaccharides from these organisms differ greatly in their properties, more so than their terrestrial counterparts. Innumerable applications related to food processing, storage, and food fortification are possible, in addition to other uses in healthcare, cosmetics, biotechnology, etc. Some polysaccharides derived from marine sources such as seaweeds are currently being utilized for their ability to form gels and act as thickeners and stabilizers in a variety of foods, but there is ample potential for tapping other marine organisms. The major classes of marine organisms serving as sources of polysaccharides are described below.

1.4.2 Marine Fisheries

Seafood, in general, is comprised of groups of biologically divergent animals, including not only fish but also shellfish such as crustaceans and mollusks. Crustaceans include crayfish, crab, shrimp, and lobster. Mollusks include bivalves such as mussel, oyster, and scallop; univalves such as abalone, snail, and conch; and cephalopods such as squid, cuttlefish, and octopus. According to the *The State of World Fisheries and Aquaculture*, published by the Food and Agriculture Organization (FAO) of the United Nations, 95 mt of marine fishes were harvested in 2008. Marine crustaceans as a whole totaled 5.7 mt in 2006.[39] Capture fisheries together with aquaculture supplied the world with 106 mt of food fish in 2004, an apparent per capita availability of 16.6 kg (live weight equivalent).[39] The provisional total world marine

TABLE 1.4

World Fish Production (mt) and Utilization from 2000 to 2006

Products	2000	2002	2004	2005	2006
Marine capture	86.8	84.5	85.9	84.2	81.9
Inland capture	8.8	8.8	9.2	9.6	10.1
Inland aquaculture	21.1	23.9	27.2	28.9	31.6
Marine aquaculture	14.3	16.5	18.3	18.9	20.1
Total marine capture	101.1	101.0	104.1	103.1	102.0
Total aquaculture	35.5	40.4	45.5	47.8	51.7
Human consumption	96.9	100.2	105.6	107.2	110.4

Source: Adapted from FAO, *The State of World Fisheries and Aquaculture*, Food and Agriculture Organization of the United Nations, Rome, 2008.

capture in 2006 was 102 mt, while marine and inland aquaculture operations contributing 20.1 and 31.6 mt, respectively. The total amount of fish utilized for human consumption in 2006 was 110 mt, representing a per capita consumption of 16.6 kg, as in 2004.

The global seafood trade has grown dramatically. With a net flow of seafood from developing countries to richer nations, the current international trade in fish products is US$71.5 billion. The United States imported 50,000 t of shrimp worth US$3.7 billion in 2004. India exported over 600,000 t of seafood, worth US$1.85 billion, during the period 2008 to 2009 to Japan, the United States, and the European Union.[40] Table 1.4 shows total world fish production from 2000 to 2006. Globally, about 1 billion people rely on fish as their main source of animal proteins, and the dependence on fish is usually highest in coastal areas. About 20% of the world's population derives at least 20% of their animal protein from fish, although some small island nations depend on fish almost exclusively. Seafood products are considered inexpensive compared with those from land animals. Several factors point to continued growth in the demand for seafood, including increasing awareness of the health benefits of seafood and improvements in aquaculture methods.[41] Although many wild fish stocks have become depleted and total global fisheries catches have reached a plateau, the aquaculture production of fish and shellfish has risen significantly. Many farmed shellfish and fish are preferred in industrialized countries. The Asia–Pacific region accounts for 88% of the total production of aquacultured shrimps and prawns.[39] As much as 30% of the annual global fish catch is used for animal feed inputs, the majority going toward aquaculture.[42]

The seafood processing industry generates tremendous amounts of wastes comprised of heads, bones, and skin from finfish and the exoskeleton, cephalothorax, and carapace from crab, shrimp, and lobster. Recent estimates suggest that annual discards from the world fisheries exceed 20 mt, equivalent to 25% of the total production of marine capture fisheries. Trawl fishing for shrimp and demersal finfish accounts for over 50% of the total estimated discards, representing approximately 22% of total landings.[39] Shrimp is the most popular internationally traded commodity and accounted for 16.5% of the total value of internationally traded fishery products in 2004.

The rise in aquaculture, particularly shrimp aquaculture, has further contributed to the problem of biowastes. Large producers of farmed shrimp in Asian countries operate centralized processing plants to serve the needs of importing countries. These facilities generate considerable processing wastes, especially crustacean shell; however, shell waste from shrimp, lobster, crab, squid, cuttlefish, and prawn is rich in the polysaccharide chitin. Another promising source of chitin is the Antarctic krill (*Euphausia superba*); the potential annual catch of this crustacean is as high as 100 mt, which can yield about 2.0 mt of chitin. The economy of krill fisheries depends not only on the efficiency of utilizing the edible parts as rich sources of nutrients but also on the full utilization of all inedible constituents, such as chitin and carotenoids. Atlantic regions of Canada harvest about 50,000 t of crab (*Chionoecetes opilio*); as much as 80% of the catch is waste that offers another good source of chitin. There are significant opportunities for the marine bioprocess industry to convert these seafood wastes to valuable polysaccharides (and other nutraceuticals) using marine biotechnology. In addition, legal restrictions, high costs, and environmental concerns regarding the disposal of marine processing wastes have amplified interest in extracting useful byproducts from marine wastes.[43]

1.4.3 Seaweed

Seaweed is a colloquial term encompassing about 9000 species of macroscopic, multicellular, benthic marine algae—the plants of the sea. Growth of these algae is dependent on the presence of seawater (or at least brackish water) and the availability of sunlight sufficient to drive photosynthesis. Seaweed grows more frequently on rocky shores than on sand or shingle. Luxuriant growth of seaweed occurs in both shallow and deep waters up to 150 m deep, as well as in estuaries and backwaters. Seaweeds are classified into four main groups on the basis of their color: red algae (Rhodophyceae), brown algae (Phaeophyceae), blue–green algae (Cyanophyceae), and green algae (Chlorophyceae). Red algae is the most abundant group (6000 species), followed by brown (2000) and green (1200). Brown seaweed (also known as kelp) is one of the most abundant seaweed groups of economic importance. Kelp is usually large, often as long as 2 to 4 m, and can resemble thick forests in the sea. It attaches to a substrate by a structure called a *holdfast* rather than by true roots. The stem arises from the holdfast, which ends in one or more broad, flat blades. The major kelps, which include such genera as *Macrocystis*, *Laminaria*, *Sargassum*, *Pterygophora*, and *Nereocystis*, grow upward like trees and spread their blades at the surface of the water, where they obtain the maximum amount of light. The Pacific coast of both North and South America is dominated by *Macrocystis*, whereas *Laminaria* is dominant in Atlantic waters and in Japan. Kelp is most recognized for its health benefits due its high levels of iodine; the nutritional composition of kelp can assist the healing process. Brown algae belonging to the genus *Sargassum* are widely distributed in tropical and subtropical regions. Free-floating brown algae depend on gas-filled sacs to maintain an acceptable depth, while others have adapted to living in tidal pools. Some brown algae, such as members of the order Fucales (the rockweeds), are commonly found along rocky seashores. Most brown algae contain the pigments fucoxanthin and chlorophyll *a*, which are responsible for the distinctive greenish-brown color. Brown algae concentrate a number of chemicals from the sea and consequently are used as fertilizer by coastal farmers.

Red algae are a large, morphologically diverse group of algae representing more than 700 genera and 6000 species that can be found at a maximum depth of 200 m. As compared with brown macroalgae, red seaweed is usually smaller, generally ranging from a few centimeters to about a meter in length. The popular red alga *Chondrus crispus* (commonly known as Irish moss) is a husky, rather rigid seaweed composed of flattened, forked branches about half an inch wide and 2 to 3 inches tall. Its color ranges from greenish in shallow, clear water to a blackish-purple in deep or turbid water. The plant grows in tidal pools and from the lowest parts of intertidal zones to a depth of 50 feet or more. It is found along the Atlantic coast of North America extending from New Jersey to Newfoundland.

As its name implies, red algae are usually red to violet in color due to the pigment phycoerythrin; however, rhodophytes can also be black, brownish, violet, yellow, or green. Some species contain the blue pigment phycocyanin, and all species contain chlorophyll *a* but not chlorophyll *b*. Red algae are important sources of commercial colloids, including agar. Among the red algae, the genus *Gracilaria*, consisting of more than 40 species, contributes about 70% of the raw materials required for the production of hydrocolloid agar.[6] The plants have been thought to have medicinal value in the Orient since the time of Shen Nung (3000 B.C.), who is considered to be the father of medicine.[44]

The green algae are more common in freshwater and on land, although they are also found in the marine habitats. They form symbiotic relationships with protozoa, sponges, and coelenterates. Green algae are an early link in the food chain, as they serve as food for fish and also for the natives of Africa and of Mexico. They contain chlorophylls *a* and *b* and store food as starch in their plastids. Green seaweed tends to be found toward the top of the water; brown seaweed species are found from the top to deeper waters, as they are adapted to carrying on photosynthesis at lower lights levels; and red seaweeds tend to dominate the deeper, darker waters.[44,45] The Southern Ocean has immense potential for the development and exploitation of seaweed resources.[46] Although seaweeds are of high economic importance, some blooms of some macroalgae may be considered harmful because of their direct and indirect impacts on living resources due to the depletion of oxygen. Also, decomposing mats of macroalgae that accumulate along shorelines can be an odorous nuisance to local residents.

As noted earlier, seaweed is highly valued for its high contents of valuable nutraceuticals.[4] Seaweed polysaccharides include agar, alginates, carrageenans, and fucoidans, as well as others collectively known as phycocolloids (*phyco* in Greek means "seaweed"), hydrocolloids, or gums. Because of their commercial importance, selected seaweed species are farmed in several countries using simple and cost-effective cultivation methods, select germ plasma as seed stock, and good farm management practices. The FAO has developed guidelines on the proper management of marine living resources, including seaweeds.[47] These include improved methods of assessing and managing wild seaweed resources, understanding the impact of seaweed cultivation and harvesting on other commercial marine resources, and determining the contribution of macroalgae to marine food chains, particularly commercial fish populations, in addition to the processing, marketing, and trade of seaweed resources.

World aquatic plant production in 2006 reached 151 mt (worth US$7.2 billion) and has experienced an average annual growth rate of 8% since 1970, most of it coming from China, the Philippines, Republic of Korea, and Japan.[39] The commercial output of the Japanese kelp *Laminaria japonica* was 4.5 mt in 2006, followed by 2.5 mt of wakame (*Undaria pinnatifida*) and 1.3 mt of nori (*Porphyra tenera*).[39] *Kappaphycus alvarezii* and *Eucheuma denticulatum* are red seaweeds currently farmed in the Philippines for carrageenans. The *Kappaphycus* species are of particular interest in the Philippine seaweed industry because of their improved resistance to disease, fast growing characteristics, and content of κ-carrageenan, whereas *E. denticulatum* is grown for its ι-carrageenan content.[48,49]

Tissue culture techniques facilitate development and propagation of seaweed genotypes of commercial importance. Tissue-cultured plants have a higher growth rate than cultivated strains. A typical technique for the tissue culture of *Gelidiella acerosa* (Gelidiales, Rhodophyta), an economically important red alga, involves preparation of axenic material, culture of explants, subculture of excised callus, and regeneration of *de novo* plants from the callus. Sequential treatments of explants with sterile seawater containing small amounts of liquid detergent, 2% betadine containing 0.5% w/v iodine, and a 3.5% broad-spectrum antibiotic mixture with nystatin for 2 days can yield viable explants as high as 90%. A prolific and rapid growth of filamentous callus on explants has been observed on cut surfaces during the first month of culture. The highest level of callus induction occurs in Provasoli's enriched seawater (PES) medium solidified with 1.5% agar and incubated at 20 to 22°C. The callus mass with bud or shoot developments continues to grow when transferred to semisolid PES medium (0.2% agar w/v). In 4 months, these shoots give rise to 2- to 3-cm long plantlets of *Gelidiella acerosa*. The tissue-cultured *Gelidiella* germlings successfully grow into full plants in the field on coral stones in 6 months.[50]

Protoplasts are living plant cells without cell walls which offer a unique, uniform single-cell system that facilitates several aspects of modern biotechnology, including genetic transformation and metabolic engineering. Reliable procedures are now available to isolate and culture protoplasts from diverse groups of seaweeds. Of the total species studied for protoplasts, most belong to Rhodophyta (with 13 genera and 41 species), followed by Chlorophyta (5 genera and 24 species) and Phaeophyta (18 genera and 24 species). Regeneration of protoplast-to-plant systems is available for a large number of species, with extensive literature regarding their culture methods and morphogenesis.[51]

In the context of plant genetic manipulation, somatic hybridization by protoplast fusion has been accomplished in a number of economically important species with various levels of success. Isolated protoplasts are also used to understand membrane function, cell structure, and the biochemical synthesis of cell walls.[51] *Polyculture* and *integrated aquaculture* are terms used to describe the farming of one or more species simultaneously in the same area. There is renewed interest in incorporating seaweed into integrated aquaculture systems, where the macroalgae primarily serve as nutrient scrubbers of seawater used in the mariculture of fish and crustaceans, either before or after the water is used by these animals. The cultured seaweed and crustaceans utilize the nutrients in the waters surrounding the sea pens and provide the fish farmer with an additional cash crop. When two species of agarophytes, *Gelidium sesquipedale* and *Gracilaria tenuistipitata*, were cultured in chemostat systems under different light qualities to study the production of polysaccharides, their yields of methoxyl groups and sulfates were influenced by the light quality.[52]

1.4.4 Microalgae

Microalgae is the largest primary biomass, covering almost three quarters of the Earth's surface to a depth up to 200 m and forming the base of the marine food web through their photosynthetic activity. Marine microalgae protect the environment through their photosynthetic activity by absorbing nearly half of the carbon dioxide being emitted by fossil fuels. They are a virtually untapped resource of more than 25,000 species. Diatoms (Bacillariophyceae) are a major group of microalgae. Diatoms from the oceans are estimated to contribute up to 45% of total oceanic primary production. Most diatoms are unicellular, although some form chains or simple colonies and include both autotrophs and heterotrophs. Diatoms such as *Chaetoceros calcitrans*, *C. gracilis*, *C. muelleri*, and *Skeletonema costatum* are commonly used as live feeds for bivalve mollusks, crustacean larvae, and zooplankton.

Dinoflagellates (Dinophyceae) represent a significant portion of primary planktonic production in both oceans and lakes. They are microscopic, usually unicellular, and flagellated; they are commonly regarded as algae. Approximately 130 genera and more than 2000 species have been described, most of them belonging to a marine habitat. Cryptophyte species, another major group of phytoplankton, are unicellular flagellates with more than 20 genera comprised of 200 species. They are distributed both in freshwater and marine environments. Most possess various colored plastids with chlorophylls, carotenoids, and phycobiliprotein. The phylum Euglenophyta encompasses unicellular flagellate organisms and is comprised of 40 genera and 900 species. The phylum Haptophyta is a group of unicellular flagellates having a brownish or yellowish-green color due to chlorophylls a and c_1/c_2 and carotenoids such as β-carotene, fucoxanthin, and others. About 70 genera and 300 species have been isolated to date, most being tropical marine species providing food for aquatic communities.[32,33,53]

Cyanobacteria are often referred to as blue–green algae, although they are not really algae; the description is primarily used to reflect their appearance and ecological role rather than their evolutionary lineage. They are oxygenic phytosynthetic prokaryotes that show large diversity in their morphology, physiology, ecology, biochemistry, and other characteristics. More than 2000 species of cyanobacteria have been recognized. They are efficient in fixing carbon dioxide in the form of sugars. Although distributed widely not only in saltwater but also in freshwater, brackish water, polar areas, and hot springs, cyanobacteria are generally associated with marine plants and animals. The cyanobacteria of the genus *Prochlorococcus* are the smallest (0.6 μm diameter) and most numerous of the photosynthetic marine organisms. It has been estimated that a drop of seawater contains up to 20,000 cells of organisms belonging to *Prochlorococcus*. *Prochlorococcus* and another marine pelagic, *Synechococcus*, contribute significantly to global oxygen production. Some also exist in symbiotic association with sponges, ascidians, echiuroid worms, planktonic diatoms, and dinoflagellates in marine environments. These associations have helped these organisms survive under highly stressful growth conditions such as high salinity, high and low temperatures, and limited nutrient availability.

Microalgae have the potential in the future to serve as a renewable energy source through the commercial production of hydrocarbons. For most of these applications, the market is still developing but undoubtedly the biotechnological use of microalgae will extend into new areas.[32,33,54]

As noted earlier, microalgae are also capable of producing several secondary metabolites, including polysaccharides,[55,56] and there is great interest in the culture of microalgae to produce vitamins, proteins, cosmetics, and health foods. The inherent characteristics of microalgae make them amenable to culture. The advantages of microalgal cultivation include: (1) the absence of complex reproductive organs, thus rendering the entire biomass available for use; (2) simple and rapid reproduction through cell division, which enables them to complete their lifecycle faster; (3) their efficiency in the utilization of solar energy for photosynthesis; (4) being able to manipulate their growth rate and biomass yield by varying such culture parameters as light, temperature, pH, nutrient status, and CO_2, nitrogen, and phosphorus concentrations; and

(5) the production of commercially valuable compounds such as polysaccharides, proteins, lipids, and pigments by these organisms. Some drawbacks of mass cultivation, however, include poor light penetration with an increase in cell density, which affects the growth rate of the organisms.[45,57,59,64]

Currently, both open and closed systems have been developed for the culture of microalgae. For large-scale cultivation, the open system is the simplest method, offering advantages in construction costs and ease of operation. A typical open system should have a shallow depth of about 12 to 15 cm for optimal light penetration, but they require a large surface area. Contamination by different algal species and other organisms is a serious problem in open culture systems. Furthermore, rain may dilute the nutrients. Closed systems (photobioreactors) overcome the disadvantages of open systems. They are similar to conventional fermenters, but the major difference is that they are driven by light instead of an organic carbon source. Closed systems have been well studied for the cultivation of *Spirulina*, *Chlorella*, and other microalgal species.[60,61] Photobioreactors offer the advantages of easier maintenance of monoalgal cultures and reduced harvest costs. These reactors, apart from preventing contamination, provide better process control of, for example, light intensity, temperature, aeration, and pH. In addition, they offer higher efficiencies with increased biomass yield (2 to 8 g per liter) within a shorter cultivation time (2 to 4 weeks) as compared to the lower yields (0.1 to 1 g per liter) and longer cultivation time times (6 to 8 weeks) for open systems.[45,62,63] Recovery of the biomass from cultures involves flocculation employing such flocculants as ferric chloride and alum. Chitosan can also be employed as a nontoxic flocculant. These flocculants increase the particle size by concentrating the microalgal cells for easy and effective sedimentation, centrifugation, filtration, dehydration, cell disruption, extraction, and purification of the targeted metabolite. Typically, the downstream recovery of microbial products accounts for 70 to 80% of the total cost of production. With the further development of sophisticated culture and screening techniques, it is expected that microalgal biotechnology will be able to meet the challenging demands of food and pharmaceutical industries.[45,59,63,64]

1.4.5 Coral Reefs and Corals

Coral reefs are massive deposits of calcium carbonate in the oceans that harbor a rich and diverse ecosystem of animals. They are produced primarily by corals with minor additions from calcareous algae and other organisms that secrete calcium carbonate. The skeletal remains of corals and plants on the reef may be considered the marine equivalent of tropical rain forests. The reefs are unique among marine associations in that they are built up entirely by biological activity. Coral reefs are widespread. They can be found in clean coastal waters of the tropics and subtropics, which offer optimal conditions, such as moderate temperature and good sunlight, favoring the growth of reef-forming organisms. It has been estimated that coral reefs occupy about 600,000 square miles of the surface of the Earth, or about 0.17% of the total area of the planet. On the continental shelves of northern and western Europe, extensive reefs can be found at depths of 60 to 2000 m. Reefs can be thousands of years old; the Great Barrier Reef of Australia is said to be more than 9000 years old. Corals, the major organisms that form the basic reef structure, are members of the phylum Cnidaria, class Anthozoa, and order Madreporaria. The phylum Cnidaria includes such diverse forms as jellyfish, hydroids, the freshwater *Hydra*, and sea anemones.

A bewildering array of other organisms is associated with reefs. The rate of growth of different tiny corals varies widely; for example, members of the genera *Acropora* (Stag's horn coral) and *Pocillopora* (stony coral) grow rapidly and represent a considerable proportion of tropical coral reefs. Stony corals are the foundation of coral reef ecosystems. Coralline algae (algae that also secrete calcium carbonate and often resemble corals) contribute to the calcification of many reefs. Shallow-water corals owe their beautiful colors in part to symbiotic algae that live inside the coral cells. About 27 species of sponges are abundant on reefs, but they have little to do with reef construction. The important genera are *Callyspongia*, *Oceanapia*, *Haliclona*, *Axinella*, and *Sigmadocia*. Marine sponges are the most primitive multicellular animals and contain many metabolites yet to be fully understood, many of which have been shown to possess diverse biological activities. A comprehensive taxonomy has been developed by Hooper et al.[65]

1.4.6 Marine Microorganisms

The ocean depths are home to myriad species of microorganisms that vary with the environment. Many parts of the ocean are characterized by unfavorable survival conditions, such as extreme high or low temperatures, acidity, alkalinity, salt concentrations, and pressure. Marine environments include polar regions, hot springs, acid and alkaline springs, and the cold, pressurized depths of the oceans. Deep-sea hydrothermal vent environments, discovered in 1977, are characterized by high pressure and temperature gradients and sometimes high levels of toxic elements such as sulfides or heavy metals. It is now recognized that these environments, once thought to be too hostile to allow life, are the natural habitats of certain microorganisms known as extremophiles. Thermophiles are one example; they grow optimally at temperatures between 60 and 80°C and are widely distributed among the genera *Bacillus*, *Clostridium*, *Thermoanaerobacters*, *Thermus*, *Fervidobacter*, *Thermotoga*, and *Aquifex*. Psychrophiles survive and proliferate at low temperatures and have successfully adapted to challenges including reduced enzyme activity; decreased membrane fluidity; altered transport of nutrients and waste products; decreased rates of transcription, translation, and cell division; protein cold denaturation; inappropriate protein folding; and intracellular ice formation. Recently, functional genomics, especially proteome analyses, has opened up revolutionary insights into the adaptation strategies of marine organisms in response to the challenges of their habitat.[66] The role of bacteria in the marine food webs is twofold in that they serve as primary food sources and as components of the microbial communities of marine animals. Generally, these organisms, including fungi, are involved in a symbiotic association with hosts such as algae and corals.[67,68]

1.5 Summary

The sea is a rich reserve of a multitude of resources. Numerous marine organisms, including shellfish, seaweed, microalgae, and corals, can be good sources of polysaccharides having interesting functional properties (see Chapter 2). Marine polysaccharides, like their counterparts from terrestrial animals, have the potential to be used for food processing, storage, and fortification, as well as in medicine as carriers of drugs and as nutraceuticals. With the exception of some seaweeds, the potential of many of these organisms in this regard has not been fully exploited.[4,5] It is possible that developments in marine biotechnology will contribute to the isolation and characterization of novel polysaccharides from diverse marine organisms.

Marine microorganisms are sources of several important nutraceuticals. In recent times, the marine environment has been recognized as a rich and largely untapped source of microbial extracellular polysaccharides that can be harnessed and developed for potential biotechnological applications.[69,70] The majority of these marine microbial organisms, however, cannot be cultured under artificial laboratory conditions because of specific growth requirements, so detailed taxonomical and physiological characterizations are difficult to achieve. Nevertheless, advanced molecular techniques have altered our perspectives on the naturally occurring diversity and distribution of such marine microorganisms; for example, it might be possible to use the gene pools of marine bacteria for recombinant DNA technology to increase polysaccharide yield.[67,71–74]

References

1. Cui, S. W., Ed., *Food Carbohydrates: Chemistry, Physical Properties, and Applications*, CRC Press, Boca Raton, FL, 2005.
2. Smith, J. and Hong-Shum, L., Eds., *Food Additives Databook*, Blackwell Science, Boston, MA, 2003.
3. Norton, I. T. and Foster, T. J., Hydrocolloids in real food systems, in *Gums and Stabilizers for the Food Industry 11*, Williams, P. A. and Philips, G. O., Eds., The Royal Society of Chemistry, Cambridge, U.K., 2002, p. 187.

4. Venugopal, V., *Marine Products for Healthcare: Functional and Bioactive Nutraceuticals from the Ocean*, CRC Press, Boca Raton, FL, 2008, chap. 2.
5. Kim, S.-K. et al., Prospective of the cosmeceuticals derived from marine organisms, *Biotechnol. Bioprocess Eng.*, 13, 511, 2008.
6. Tressler, D. K. and Lemon, J. M. W., *Marine Products of Commerce: Their Acquisition, Handling, Biological Aspects, and the Science and Technology of Their Preparation and Preservation*, Reinhold, New York, 1951, chap. 1.
7. Bemiller, J. N., *Carbohydrate Chemistry for Food Scientists*, 2nd ed., American Association of Cereal Chemists, St. Paul, MN, 2007; Eliasson, M., Ed., *Carbohydrates in Food*, Marcel Dekker, New York, 1996.
8. Tharanathan, R. N., Food-derived carbohydrates: structural complexity and functional diversity, *Crit. Rev. Biotechnol.*, 22, 65, 2002.
9. FAO, *Carbohydrates in Human Nutrition*, FAO Food and Nutrition Paper 66, Food and Agriculture Organization of the United Nations, Rome, 1998.
10. Anon., Carbohydrates and fibre: a review of functionality in health and well-being, *FST Bull.*, September, 2005 (http://www.foodsciencecentral.com/fsc/ixid14071).
11. Anon., Prebiotics: a nutritional concept gaining momentum in modern nutrition, *FST Bull.*, January, 2006 (http://www.foodsciencecentral.com/fsc/ixid13932).
12. FAO, Effects of food processing on dietary carbohydrates, *Asia Pacific Food Indust.*, 20(4), 64–69, 2008.
13. Khatkar, B. S., Panghal, A., and Singh, U., Applications of cereal starches in food processing, *Indian Food Indust.*, 28, 37, 2009.
14. Peters, D., Carbohydrates for fermentation, *Biotechnology. J.*, 1, 806, 2006.
15. Steinbuchel, A. and Rhee, S. K., Eds., *Polysaccharides and Polyamides in the Food Industry*, Vol. 1., John Wiley & Sons, New York, 2005, p. 275.
16. Mantovani, S. et al., β-Glucans in promoting health: prevention against mutation and cancer, *Rev. Mut. Res.*, 658, 154, 2008.
17. Vargas-Albores, I. and Yepiz-Plascencia, G., Beta glucan binding protein and its role in shrimp immune response, *Aquaculture*, 191, 13, 2000.
18. Harding, S. E., Analysis of polysaccharides by ultra centrifugation, size, conformation and interactions in solutions, *Adv. Polym. Sci.*, 186, 211, 2005.
19. Francisco, M., Goycoolea, F., and Chronakis, L. S., Specific methods for the analysis of identity and purity of functional food polysaccharides, *Dev. Food Sci.*, 39, 99, 1998.
20. Khan, T., Park, J. K., and Kwon, J.-H., Functional biopolymers produced by biochemical technology considering applications in food engineering, *Korean J. Chem. Eng.*, 24, 816, 2007.
21. Muffler, K. and Ulber, R., Downstream processing in marine biotechnology, *Adv. Biochem. Eng./Biotechnol.*, 96, 85, 2005.
22. Mecozzi, M. et al., Ultrasound-assisted analysis of total carbohydrates in environmental and food samples, *Ultrasonics Sonochem.*, 6, 133, 1999.
23. Blackwood, A. D. and Chaplin, M. F., Disaccharide, oligosaccharide and polysaccharide analysis, in *Encyclopedia of Analytical Chemistry*, Meyers, R. A., Ed., Springer-Verlag, 2000, pp. 741–765.
24. Montgomery, R., Development of biobased products [review paper], *Bioresource Technol.*, 91, 1, 2004.
25. Dumitriu, S., Ed., *Polysaccharides: Structural Diversity and Functional Versatility*, 2nd ed., CRC Press, Boca Raton, FL, 2004.

26. Chinachoti, P., Carbohydrates: functionality in foods, *Am. J. Clin. Nutr.*, 61, 922S, 1995.
27. Bárcenas, M. E. and Rosell, C. M., Different approaches for increasing the shelf life of partially baked bread: low temperatures and hydrocolloid addition, *Food Chem.*, 100, 1594, 2007.
28. Williams, P. A. and Phillips, G. O., The use of hydrocolloids to improve food texture, in *Texture in Food*. Vol. 1. *Semi-Solid Foods*, McKenna, B. M., Ed., Aspen Publishers, New York, 2003, chap. 11.
29. Biliaderis, C. G. and Izydorczyk, M. S., Eds., *Functional Food Carbohydrates*, CRC Press, Boca Raton, FL, 2006.
30. Karovicová, J., Application of hydrocolloids as baking improvers, *Chem. Papers,* 63, 26, 2009.
31. Kohajdová, Z. et al., Marine biotechnology for production of food ingredients, *Adv. Food Nutr. Res.*, 52, 237, 2007.
32. Nybakken, J. W., *Marine Biology: An Ecological Approach*, 4th ed., HarperCollins College Publishers, New York, 1997.
33. Madigan, M. T. and Martinko, J. M., *Biology of Microorganisms*, 11th ed., Pearson Education, London, 2005.
34. Matsunaga, T. et al., Marine microalgae, *Adv. Biochem. Eng./Biotechnol.*, 96, 165, 2005.
35. Haard, N. F., Specialty enzymes from marine organisms, *Food Technol.*, 52(7), 64, 1998.
36. Irigoien, X. et al., Global biodiversity patterns of marine phytoplankton and zooplankton, *Nature*, 429, 863, 2004.
37. Steele, R. L., Comparison of marine and terrestrial ecosystems, *Nature*, 313, 355, 1985.
38. Rasmussen, R. S. and Morrissey, M. T., Marine biotechnology for production of food ingredients, *Adv. Food Nutr. Res.*, 52, 237, 2007.
39. FAO, *The State of World Fisheries and Aquaculture*, Food and Agriculture Organization of the United Nations, Rome, 2008.
40. MPEDA, *Annual Report 2006–2007*, The Marine Products Export Development Authority, Ministry of Commerce and Industry, Government of India (http://www.mpeda.com).
41. Anon., *U.S. Market for Fish and Seafood, with a Focus on Fresh*, Packaged Facts, Division of Market Research Group, LLC, Rockville, MD, 2009 (http://www.packagedfacts.com/prod-toc/Seafood-Focus-Fresh-1737415/).
42. Goldberg, R. J., Aquaculture, trade, and fisheries linkages: unexpected synergies, *Globalization*, 5, 183, 2008.
43. Hayes, M. et al., Mining marine shellfish wastes for bioactive molecules: chitin and chitosan. Part A. Extraction methods, *Biotechnol. J.*, 3, 871, 2008.
44. Chapman, V. J. and Chapman, D. J., Sea vegetables (algae as food for man), in *Seaweeds and Their Uses*, 3rd ed., Chapman & Hall, London, 1980, p. 95.
45. Lele, S. S. and Kumar, J., *Algal Bioprocess Technology*, New Age Publications, New Delhi, India, 2008, p. 75.
46. Dhargalkar, V. K. and Verlecar, X. N., Southern ocean seaweeds: a resource for exploration in food and drugs, *Aquaculture*, 287, 229, 2009.
47. Caddy, J. A. and Fisher, W. A., FAO interests in promoting understanding of world seaweed resources, their optimal harvesting, and fishery and ecological interactions, *Hydrobiologia*, 124, 111, 1985.

48. Aguilan, J. T., Structural analysis of carrageenan from farmed varieties of Philippine seaweed, *Bot. Mar.*, 46, 179, 2003.
49. Fruedkabderm, J., Advances in the cultivation of Gelidales, *J. Appl. Phycol.*, 20, 451, 2008.
50. Kumar, G. R. et al., Tissue culture and regeneration of thallus from callus of *Gelidiella acerosa* (Gelidiales, Rhodophyta), *Phycologia*, 43, 596, 2004.
51. Reddy, C. R. K. et al., Seaweed protoplasts: status, biotechnological perspectives and needs, *J. Appl. Phycol.*, 20, 619, 2008.
52. Carmona, R. et al., Effect of light quality on polysaccharide yield and composition of two red algae, *J. Phycol.*, 36, 10, 2000.
53. Kurano, N. and Miyachi, S., Microalgal studies for the 21st century, *Hydrobiologia*, 512, 27, 2004.
54. Raja, A. et al., Perspective on the biotechnological potential of microalgae, *Crit. Rev. Microbiol.*, 34, 77, 2008.
55. Shah, V. et al., Characterization of the extracellular polysaccharide produced by a marine cyanobacterium and its exploitation towards metal removal from solutions, *Curr. Microbiol.*, 40, 274, 2000.
56. Kawaguchi, T. and Decho, A. W., Biochemical characterization of cyanobacterial EPS from modern marine stromatolites (Bahamas), *Prep. Biochem. Biotechnol.*, 30, 321, 2000.
57. Olaizola, M., Commercial development of microalgal biotechnology: from the test tube to the marketplace, *Biomol. Eng.*, 20, 459, 2003.
58. Borowitzka, M. A., Commercial production of microalgae: ponds, tanks, tubes and fermenters, *J. Biotechnol.*, 70, 313, 1999.
59. Cysewski, G.R. and Lorenz, T.R., Industrial production of microalgal cell-mass and secondary products. Species of high potential—*Haematococcus*, in *Handbook of Microalgal Culture: Biotechnology and Applied Phycology*, Richmond, A., Ed., Blackwell, Oxford, 2004, pp. 281–288.
60. Dodd, J. C., *Handbook of Mass Culture*, CRC Press, Boca Raton, FL, 1986, p. 265.
61. Bhattacharya, S. and Shivaprakash, M. K., Evaluation of three *Spirulina* species grown under similar conditions for their growth and biochemicals, *J. Sci. Food Agric.*, 85, 333, 2005.
62. Torzillo, G. et al., A two-plane tubular photobioreactor for outdoor culture of *Spirulina*, *Biotechnol. Bioeng.*, 42, 891, 1992.
63. Lee, Y. K., Commercial production of microalgae in the Asia Pacific rim, *J. Appl. Phycol.*, 9, 403, 1997.
64. Richmond, A., Microalgae of economic potential, in *Handbook of Microalgal Mass Culture*, Richmond, A., Ed., CRC Press, Boca Raton, FL, 1986, pp. 199–243.
65. Hooper, J. N. A. and Van Soest, R. W. M., Eds., *Systema Porifera: A Guide to the Classification of Sponges*, Kluwer Academic/Plenum Press, New York, 2002.
66. Schweder, T. et al., Proteomics of marine bacteria, *Electrophoresis*, 29, 2603, 2008.
67. Nichols, C.M. et al., Chemical characterization of EPSs from Antarctic marine bacteria, *Microbial Ecol.*, 49, 578, 2005.
68. Raghukumar, C., Marine fungal biotechnology: an ecological perspective, *Fungal Diversity*, 31, 19, 2008.
69. Mancuso Nichols, C. A., Guezennec J., and Bowman, J.P., Bacterial exopolysaccharides from extreme marine environments with special consideration of the Southern Ocean, sea ice, and deep-sea hydrothermal vents: a review, *Mar. Biotechnol.*, 7, 253, 2005.

70. Guezennec, J., Deep-sea hydrothermal vents: a new source of innovative bacteria EPSs of biotechnological interest?, *J. Indian Microbiol. Biotechnol.*, 29, 204, 2002.
71. Zhenming, C. and Yan, F., EPSs from marine bacteria, *J. Ocean Univ. China*, 4, 67, 2005.
72. Weiner, R. M. et al., Applications of biotechnology to the production, recovery and use of marine polysaccharides, *Biotechnology*, 3, 899, 1985.
73. Misaki, A., Structural aspects of some functional polysaccharides, in *Food Hydrocolloids: Structure, Properties and Functions*, Nishinari, K. and Doi, E., Eds., Plenum Press, New York, 1994, pp. 1–20.
74. Welman, A. D. and Maddox, I. S., Exopolysaccharides from lactic acid bacteria: perspectives and challenges, *Trends Biotechnol.*, 21, 269–274, 2003.

2

Functional Properties Relevant
to Food Product Development

2.1 Introduction

The modern food processing industry is aimed at the manufacture of tailor-made foods to satisfy consumer needs. The development of such food products depends on appropriate uses of ingredients at the levels necessary to impart proper texture and flavor as well as storage stability to the processed products.[1-3] Polysaccharides contribute immensely to food product development through such functional properties as their ability to bind water and undergo gelation at defined conditions, their interaction with other food components such as proteins, and their capacity to emulsify oil. Also, the presence of some polysaccharides enhances the fiber content of foods. This chapter discusses the general functional properties of polysaccharides in food products and provides some general guidelines on the use of polysaccharides in food formulations.

2.2 Major Functions of Polysaccharides in a Food System

2.2.1 Water-Binding Capacity

Polysaccharides are able to bind large amounts of water and disperse it in the food. Differences in the water-binding capacity displayed by polysaccharides, from such diverse sources as seaweeds, crustacean shellfish, and microorganisms, are due to functional group differences in their structures, which strongly interact with water. Alginates, for example, contain large amounts of carboxylic groups; carrageenans, depending on the type, contain varying amounts of sulfonic groups; and chitosan has amino groups in its structure. These groups facilitate the binding of water as much as 98 to 100 times their weight. Xanthan can bind as much as 230 mL water per gram, while guar gum and alginate are able to bind only 40 and 25 mL water per gram, respectively. Propylene glycol alginate (PGA) and locust bean gum (LBG) are the poorest water binders. Nuclear (^{1}H) magnetic methods have been used to measure the degree of water binding in polysaccharide gels.[12]

The water-binding properties of polysaccharides, and thus their performance in various foods, are related to their rheological properties.[1–3] The ability of polysaccharides to undergo gelation (see Section 2.2.2) under mild conditions of temperature and pH and the presence of sodium or potassium ions offers various benefits regarding the modification of food texture, matrix stabilization, and many other functions useful for adding value to various food commodities. In addition, the ability to bind exceptionally high amounts of water makes polysaccharides ideal materials as cellular scaffolds, biodegradable packaging, and coatings for biomedical waste disposal, as well as carriers for controlled-release drugs.

2.2.2 Gelation

The ability to undergo gelation is probably the most important functional property of polysaccharides. A gel is an intermediate state between a solution and solid. The word "gel" is derived from "gelatin," and the terms "gel" and "jelly" can both be traced back to the Latin *gelu* for "frost" and *gelare*, meaning "freeze" or "congeal." This indicates a solid-like material that does not flow but is elastic and retains some fluid characteristics. A gel is defined as a solid composed of at least two components, one of which (polymer) forms a three-dimensional network in the medium of the other component (liquid). Gels can be classified as covalently cross-linked gels, entangled networks, or physical gels.[4] Food gels are predominantly physical gels. Usually, physical gels are formed by cooling heated solutions of polymers; the gels melt upon heating, indicating their thermoreversible character. Such gels are formed through noncovalent interactions, such as hydrogen bonding and hydrophobic and ionic interactions, which fluctuate with time and temperature.

During these interactions, junction zones are formed among the polymer molecules; these junction zones are the entanglements occurring at network junctions resulting from hydrogen bonding and ionic interactions. In solutions, the formation of junction zones between the polysaccharide and the solvent results in an elastic gel. A minimum amount of the solvent is required to maintain the flexible and elastic properties of the gel. Polysaccharides having branched structures such as cellulose do not form junction zones and hence do not form strong elastic gels. Chemical gels, on the other hand, are thermally irreversible.[4,9] In a food system, gel networks are formed as a result of interactions between macromolecules such as proteins and polysaccharides with water under appropriate conditions of temperature, pH, and pressure. The macromolecules are capable of retaining a large amount of water in the network which stabilizes the system. Pioneering studies in this area were initiated by Ferry in 1948.[5] The field of polysaccharide (and protein) gelation has received much attention because of the roles they play in determining food texture.

Gels are metastable systems that continue to evolve after the initial gelation process has taken place.[6] Polysaccharide gel formation proceeds via a disorder-to-order transition induced by cooling. The process is reversible, and gel melting is possible by reheating. The way in which cross-links between individual chains are formed depends on the individual polysaccharide (see Chapters 3 to 5 for further discussion of the gelation of individual polysaccharides). The concentration and types of junction zones in the gel network govern the characteristics of polysaccharide gels. Gels can be rigid, flowing, brittle, firm, soft, spreadable, sliceable, rubbery, or grainy, depending on the degree of interaction of the polymers. If the junction zones are short and the chains are not held together strongly, then the polysaccharide molecules will separate under physical pressure or with a slight increase in temperature. Important characteristic of gels are reversibility, specific texture (brittle, elastic, plastic, firm, soft, or mushy), and a tendency for syneresis. Gels tend to become more brittle as concentration increases.[6,9] A polysaccharide gel usually has a high water content (98 to 99%), with the polysaccharide component being as low as 1 to 2%.[7,8]

Polysaccharides differ in many ways in their gelation behavior when compared with proteins. Polysaccharide gels, in contrast to protein gels, are reversible, cold-setting gels. The critical concentrations required for protein gelation are generally five- to tenfold higher compared with polysaccharides. Protein gels are often turbid except in the case of gels formed at low ionic strength. The stability of gels is important with regard to the storage behavior of processed foods. Some gels experience a slow, time-dependent deswelling resulting in an exudation of liquid known as *syneresis*. Syneresis and precipitation are commonly encountered in protein gels. In general, syneresis primarily increases with temperature and the ionic strength of the dispersing medium. The loss of liquid may result in shrinking of the gels, changes in texture, and reduced quality. Syneresis resulting from compression can give a feeling of juiciness in the mouth. In aerated gels, some of the exuded liquid could be trapped inside the gas bubbles and would not appear on the surface of the product during storage. Additives are frequently used to achieve the desired functionality, including rheological properties; in fact, optimization of product formulations and processing conditions involving food gels depends on understanding the influence of additives on processing and rheological properties.[9]

Polysaccharide-based hydrogels exhibit a tissotropic property; that is, they can be injected via a needle without loss of their rheological properties.[8] Hydrogels containing 1 to 2% polymer and 98 to 99% water can also be shaped as spheres or films with good mechanical stability. During gel formation, molecules such as flavor components, antioxidants, antimicrobials, and fat replacers, as well as air, can be trapped in the polysaccharide gel matrices. Fluid gels are formed by applying an appropriate flow field to a biopolymer solution while it undergoes gelation under conditions of either heat or pH. During gelation, the hydrocolloids are well mixed or subjected to a uniform, sufficiently vigorous flow. When heat is used, the temperature is kept to a minimum by mixing or cooling slowly to prevent the buildup of any unsheared gel regions in the sample. The materials formed in this manner consist of a concentrated suspension of gel particles of irregular shape in a continuous phase, usually a purely aqueous medium; such fluid or sheared gels have considerable potential for use as stabilizers in food products.[10,11,32,70] The term *weak gel* is used for differentiating a gel-like polymer dispersion from a true gel. A weak gel exhibits strong shear or thinning behavior (see Section 2.4.4), which facilitates handling these gel materials at a relatively high shear rate. An aqueous solution of xanthan is the best known weak gel used in the food industry. Weak gels of polysaccharides such as curdlan and carrageenan have been reported.[9]

In recent years, new methodologies and instruments have provided a more accurate view of the relationships between chemical structure and gelling characteristics of polysaccharides. Gelation kinetics, mechanical spectra, thermal scanning rheology, and differential scanning calorimetry (DSC) in aqueous solutions are all used to study the gelation of polymers, including polysaccharides. The sol–gel transition can be monitored by light scattering, nuclear magnetic resonance (NMR), and DSC.[7]

2.2.3 Emulsions and Emulsifiers

Emulsions and foams are important features that influence the development of colloidal systems in many products, including food items. A *colloid* is defined as a dispersion of discrete particles in a continuous medium. Emulsions and foams are fine dispersions of oil, water, or air (droplets and air bubbles) in an immiscible liquid. Generally, an emulsion consists of at least two immiscible liquids (usually oil and water), with one of the liquids being dispersed as small spherical droplets in the other. The substance that makes up the droplets in an emulsion is the *dispersed phase*, whereas the substance that makes up the surrounding liquid is the *continuous phase*. The formation of an emulsion results in a large interfacial area between two immiscible phases and therefore is associated with an increase in free energy; consequently, emulsions are thermodynamically unstable and tend to undergo phase separation over time via a variety of physicochemical mechanisms, including gravitational separation, flocculation, coalescence, and Ostwald ripening.

Gravitational separation is one of the most common causes of instability in food emulsions and may take the form of either *creaming* or *sedimentation*, depending on the relative densities of the dispersed and continuous phases. Food emulsions can be mainly of three types: (1) *oil-in-water* (O/W) emulsions, oil droplets dispersed in an aqueous phase, or *water-in-oil* (W/O) emulsions, water droplets dispersed in an oil phase; (2) *foam*, in which air (gas) bubbles are dispersed in an aqueous medium; and (3) *sol*, which is small solid particles dispersed in a liquid medium. Multilayer emulsions of oil and water such as W/O/W and O/W/W are also possible. Most food emulsions are oil-in-water emulsions, including milk, cream, mayonnaise, sauces, salad dressings, custard, fabricated meat products, and cake batter. Butter, margarine, and spreads are examples of water-in-oil emulsions. Ice cream and fabricated meat products are complex oil-in-water emulsions, where the continuous phase is semisolid or a gel. Emulsions of the type O/W, W/O/W, or O/W/W are commonly used as delivery systems for bioactive lipids in the food and other industries.[13–15]

Emulsifiers are used to facilitate the formation, stabilization, and controlled destabilization of emulsions. An emulsifier is adsorbed at the oil–water interface to form a film around the droplets that prevents coalescence by virtue of its elasticity and viscosity. An effective emulsifier favors emulsion formation by rapidly reducing interfacial tension at the freshly formed oil–water interface, binding strongly to the interface once adsorbed, and protecting the newly formed droplets against flocculation or coalescence. The final structure of an emulsion is determined by the water droplets, oil droplets, gas cells, starch granules, casein micelles, and fat crystals used. Popular emulsifiers include monomeric emulsifiers such as mono- and diglycerides, lecithins and lysolecithins, and ingredients derived from milk and eggs. The most widely used polysaccharide emulsifiers in food applications are gum arabic (*Acacia senegal*), modified starches, modified celluloses, some kinds of pectin, and some galactomannans. Synthetic surfactants and emulsifiers are also used in many foods, but their use depends on their acceptability and safety. Emulsion stability is an important consideration in food product development and depends on the intricate interactions among various structural elements within the emulsion, such as oil droplet–matrix, protein–protein, protein–carbohydrate, polymer–whey protein, protein–casein micelle, protein–lipid surfactant, and fat crystal–fat crystal interactions. Controlled destabilization of the emulsion can result in the partial coalescence necessary for the proper form and texture development of ice cream and whipped cream.[15]

To form a fine emulsion, large deformable drops must be broken down, which can be accomplished using such homogenizers as high shear mixers, high-pressure homogenizers, colloid mills, ultrasonic homogenizers, and membrane homogenizers. Various types of emulsification equipment are used to produce a range of sizes of emulsion droplets and influence the food texture. High-pressure valve homogenizers or microfluidizers generally produce emulsions with droplet diameters of less than 100 to 500 nm. Such emulsions are often referred to as *nano-emulsions*.[16–18,70]

Numerous analytical techniques have been developed to characterize the stability of food emulsions and their functional characteristics—for example, the effect of emulsifiers on droplet size, surface forces, and rheology. Optical

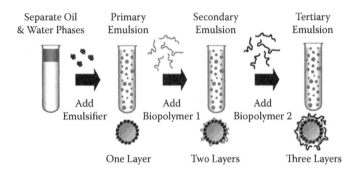

FIGURE 2.1
Schematic representation of the formation of conventional and multilayer emulsions. (From Dickinson, E., *Soft Matter*, 4, 932, 2008. With permission from Royal Society of Chemistry. Publishers.)

properties of an emulsion are determined by the relative refractive index, droplet concentration, and droplet size. The color and opacity of emulsions can be quantitatively described using tristimulus values; for example, in the $L^*a^*b^*$ system, L represents lightness, and a and b are color coordinates. A $+a$ value is in the red direction; $-a$, the green direction; $+b$, the yellow direction; and $-b$, the blue direction. A low L value is dark, and a high L is light. The opacity of an emulsion can therefore be characterized by the lightness (L), and the color intensity can be characterized by the chroma: $C = (a^2 + b^2)/2$. For an oil-in-water emulsion, the lightness increases steeply as the oil droplet concentration is increased from 0 to 5% by weight but then increases more gradually at higher droplet concentrations.[13,15,19] Figure 2.1 illustrates the formation of conventional and multilayer emulsions. Guidelines for processing emulsion-based foods have recently been developed, and the various types of emulsification equipment, the nature of flow fields, the breakup and coalescence of droplets, and predicting drop size during emulsion formation have been discussed.[12,92]

2.2.3.1 Foams

Foams are an integral component of many foods. Ideal foam is characterized by a desired volume of air (foamability) and stability. Air bubbles are structural elements usually present in the dispersed phase of solid food foams such as bread, cakes, and meringue; in semisolid foams such as whipped cream or mayonnaise; and in beverages such as milk shakes. Introducing a gas phase into the food matrix not only affects its texture and firmness, making the product lighter, but also changes its appearance, color, flavor, and mouth feel. An aerated structure facilitates mastication and enzyme accessibility to the substrate and is used to reduce the caloric density of foods and induce satiety.[20] Because aerated liquids are thermodynamically unstable, bubbles must be stabilized at their air–liquid interface, usually by surface-active agents (e.g., proteins, emulsifiers, or solid particles such as fat crystals). Factors that decrease surface charge (pH ≈ pI or high ionic strength) cause a more rapid adsorption of these compounds at the air–water interface, leading to increased viscoelasticity and increased foam yield stress. Most foam-containing foods products incorporate emulsifiers such as egg white (ovalbumin) or milk proteins (casein, sodium caseinate, calcium caseinate) at concentrations of 0.01 to 1% (w/w). These emulsifiers enhance the introduction of a gaseous phase in the form of bubbles and enhance texture and perception, in addition to serving other functions such as flavor encapsulation, delivery of bioactive molecules, satiety improvement, and creation of novel structures. Aerated gels can be produced by traditional methods such as mixing, cutting, and heating, as well as by nonconventional technologies such as the use of membrane processes or microfluidics. A mechanical whipper with a six-blade curved impeller has also been developed to make protein-stabilized foams.[11,15,22–24] Polysaccharides are also used to influence the stability of foams in foods.[21]

2.3 Food Texture

Texture is an important factor determining the acceptability of food. Texture is perceived when food materials are stirred, poured, pumped, stretched, and, finally, eaten.[25,26] The International Organization for Standardization has defined texture as "all the mechanical, geometrical, and surface attributes of a product perceptible by means of mechanical, tactile, and, where appropriate, visual and auditory receptors."[27] The sensory perception of food texture depends on the composition of food at molecular, microstructure, and macroscopic levels—that is, its geometrical and mechanical properties.[28] Whereas the molecular properties are related to the chemical composition of individual components of the food, the microstructure involves the organization of important components, including polysaccharides, proteins, and lipids, within a food product and their interactions.[11,29,31] In many food products (e.g., processed meats, cheese, yogurt, confectionary products), the desired texture is primarily achieved by the formation of a gel network of macromolecules, including proteins and polysaccharides, and their interactions among themselves. The sensory perception of texture involves one or many stimuli (visual, auditory, and kinesthetic) working in combination. Descriptive sensory analysis and instrumental measurements are both used to assess the textural properties of foods. Sensory analysis involves the use of panelists trained to detect and evaluate specific textural attributes, such as "hardness" and "stickiness."[30–33,36] Table 2.1 lists and defines terms used when evaluating the texture of foods.

TABLE 2.1

Terms Used in Texture Evaluation of Foods and Their Definitions

Term	Definition
Cohesiveness	Degree to which the chewed mass holds together
Adhesiveness	Degree to which the chewed mass sticks to mouth surfaces
Firmness	Force required to fracture sample with molars
Smoothness	Degree to which sample was perceived as smooth when evaluated with tongue
Small-strain force	Force required to cause 10% deformation
Crumbliness/fracturability	Degree to which the sample fractures into pieces
Deformability	Degree of deformation prior to fracture
Fracture force	Force required to fracture sample with molars
Smoothness of mouth coating	Smoothness felt after expectoration

Source: Adapted from Foegeding, E.A., *Curr. Opin. Colloid. Interface Sci.,* 12, 242, 2007.

Imitative and empirical instrumental tests have been used to evaluate mechanical texture properties. Texture profile analysis using a texturometer produces a force–displacement curve (see Figure 2.2) obtained from a double compression test that provides a meaningful interpretation of a number of texture features: hardness, cohesiveness, viscosity, elasticity, adhesiveness, brittleness, chewiness, and gumminess.[34] The texture profile analysis method devised by Szczesniak's group[32] is considered to be the gold standard for texture characterization. It utilizes a double compression test with flat plates attached to an Instron® universal testing machine (UTM). Much information can be extracted from the force–displacement curve generated by the test (e.g., hardness, cohesiveness, springiness, chewiness). A new bicyclical instrument for texture evaluation (BITE Master) measures crisp, crunchy, and crackly textures and correlates them with sensory analysis.[35] Oral processing, the most important factor in textural perception and appreciation, consists of a series of complex operations, including grip and first bite, first-stage transportation involving chewing and mastication, and second-stage transportation involving bolus formation and finally swallowing. Texture perception and appreciation are dynamic processes that depend on the continuous destruction and breakdown of food material in the mouth.[36]

2.3.1 Rheological Evaluation of Food Texture

Rheology is the study of the deformation and flow of materials. Rheological methods that are useful for characterizing the mechanical properties of gel networks include both small-strain rheology, which nondestructively probes the linear region, and large-strain rheology, which probes the nonlinear region and fracture. Rheological properties are usually considered to be those that are evaluated prior to fracture, and rheological characteristics are influenced

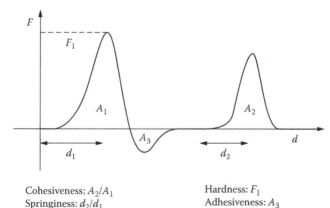

Cohesiveness: A_2/A_1 Hardness: F_1
Springiness: d_2/d_1 Adhesiveness: A_3
Gumminess: Hardness × Cohesiveness
Chewiness: Gumminess × Cohesiveness

FIGURE 2.2
A typical force–displacement curve obtained from a double compression test using the texture profile analysis approach. One single test is capable of characterizing a number of textural parameters.

by both temperature and moisture. Rheology begins with a consideration of two ideal materials—an elastic solid and a viscous liquid. An elastic solid is a material with a definite shape which, after being deformed within a certain limit by an external force, will return to its original dimensions upon removal of that force. A viscous liquid has no definite shape and will flow irreversibly upon application of an external force, including gravity. Most food materials have rheological properties somewhere between these two models and are classified as *viscoelastic*. Most solid foods are viscoelastic; that is, they behave somewhat as elastic solids but also exhibit viscous flow behavior.

Food texture is usually evaluated in terms of deformation, whereas viscosity is assessed in terms of flow.[37] The evaluation of food texture by rheology requires understanding the response of food materials to applied force (stress). Food is subjected to two different kinds of external stress. Tensile or compressive stress is applied at right angles to a surface; the resulting fractional change in length or volume is termed *strain*. The second form of stress is *shear* or tangential stress, which is applied in a direction parallel to the surface, leading to movement of the surface layer in relation to the underlying layers. The strain rate (γ^*) is the strain per unit area. The stiffness modulus relates stress and strain (deformation). For a solid, the stiffness modulus measures elasticity. Young's modulus is the constant relating to tensile stress and strain. When a solid is subjected to compressive stress, the size of the strain depends on a constant of proportionality known as the bulk modulus (K). The shear stress is proportional to the strain rate, and the constant of proportionality in this case is the shear modulus (G). The complex modulus is comprised of the elastic (storage) modulus (G'), which represents the elastic character, and the viscous (loss) modulus (G''), which represents the viscous behavior. The damping factor tan $\delta = G''/G'$ indicates whether a body is mainly elastic or viscous.[51] Dynamic rheological measurements performed at low strain in the linear domain are useful for characterizing such network properties as gelation, aging, and mechanical recovery after shearing. These measurements can be performed on strong gels; in other cases, calibrated pieces of gels are tested in compression at a very low degree of strain to avoid solvent expulsion.[38-42]

The viscosity of a fluid is its resistance to flow when subjected to an external stress. For the simplest case of an ideal viscous liquid (Newtonian), shear stress σ (force per unit area, N/m^2 or Pa) acting throughout a liquid contained between two parallel plates will result in deformation or strain (γ) at a rate of $d\gamma/dt$. The time derivative of strain is often expressed as γ' sec^{-1}. The value of σ/γ' is viscosity η and is expressed as Pascal second (Pa·s) or poise (P), where $1\,P = 0.1$ Pa·s. For most fluids, the ratio σ/γ' is not constant but changes (usually decreases) with increasing shear rate; that is, the relationship between shear stress and shear rate is nonlinear (non-Newtonian). Three broad categories of fluids may be considered as being non-Newtonian:

1. *Time-independent fluids*—For these fluids, the shear stress at any position within the fluid is independent of shear rate. The most common example of a time-independent non-Newtonian fluid is pseudoplastic flow, in which the fluid exhibits shear thinning (reduced viscosity with increasing rate of shear) over a wide range of shear rates. Time-independent non-Newtonian flow can be represented by the power law model:

$$\sigma = \sigma_y = m\gamma^n$$

where σ_y is yield stress, the minimum stress that must be applied to a material to initiate flow; m is the consistency coefficient; and n is the flow behavior index. Yield stress is characteristic of a particular product and hence its quality. A material will flow only when the yield stress is exceeded. The consistency coefficient (m) is numerically equal to shear stress or apparent viscosity at a shear rate of 1 sec^{-1}. The flow behavior index (n) provides a convenient measure of shear thinning ($n < 1$) and shear thickening ($n > 1$) types of flow. The above equation represents the relationship between force and flow rate for a spreadable food; other models have also been developed to represent the flow characteristics of these fluids.[42]

2. *Time-dependent fluids*—For these fluids, shear stress is a function of both the magnitude and direction of shear and possibly time lapse between consecutive applications of shearing treatments. These fluids will return to normal after withdrawal of shear.

3. *Viscoelastic fluids*—These fluids show partial elastic recovery upon removal of shear stress.

Some foods show mixtures of the above properties. A comparatively small group of fluids exhibits *dilatant flow*, which is an increase in resistance to flow with increasing shear rate.

Rheological properties that are important in characterizing gels include both static elastic properties and viscoelastic properties, in addition to the frequency dependence of storage modulus G' and loss modulus G''. Information on the rheological behavior of fluids has been gained through the development of controlled stress and strain rheometers capable of measurements to a very low shear rates ($<10^{-3}$ sec^{-1}). When the viscosity is high enough, a plane–cone viscometer can be used for both flow and dynamic types of experiments. It is commonly assumed that flow is laminar.

Three consecutive dynamic tests are performed on gels: (1) oscillatory time sweeps at a frequency of 1 Hz in the linear viscoelastic zone, usually with 120 sec between each measurement, using autostrain adjustment; (2) frequency sweep tests carried out between 0.01 and 15 Hz using autostrain adjustment; and (3) strain sweep tests at a frequency of 1 Hz and a strain between 0.01 and 100%. The gap angles in these measurements should be low, generally less than 4°. To avoid gel drying, a plastic cap containing damp cotton is used. Generally, a 1% polysaccharide gel is loaded and the solutions are heated to 70 to 80°C and cooled at a rate of 5°C per min. During cooling, the dynamic viscosity of the solutions is recorded at a frequency of 1 Hz and with a 0.05% strain amplitude. The samples are allowed to equilibrate at 20°C for 30 min before a frequency sweep is performed. Flow curves are then obtained from steady-stress sweep tests performed between 0.1 and 100 Pa.

Another type of study provides insight into the viscosity characteristics of polymer solutions with regard to storage modulus G' and loss modulus G''.[9,43] When a sinusoidal deformation is imposed on a solution over a wide range of frequencies, the response is a complex modulus decomposed into an in-phase response (G', reflecting the elastic character) and a cut-off phase response (G'', reflecting the viscous response). At low frequencies, G'' is larger than G', but over a critical frequency G' becomes larger than G'', corresponding to the presence of entanglements (transitory cross-links).

Emulsions exhibit flow and elastic properties. The presence of yield stress can help control deterioration in quality, such as coalescence and emulsion droplet formation. Relatively dilute emulsions are normally characterized in terms of their apparent shear viscosity, which is largely determined by the continuous phase viscosity, the droplet concentration, and the nature of the droplet–droplet interactions. Normally, the viscosity of an emulsion increases with increasing droplet concentration. At or above the droplet concentration where close packing occurs (typically around 50 to 60% for a non-flocculated O/W emulsion), the emulsion exhibits solid-like characteristics such as viscoelasticity and plasticity. In addition, shear thinning behavior is observed in flocculated emulsions due to deformation and breakdown of the flocculated structure as shear stresses increase. The impact of droplet characteristics on the overall rheology of an emulsion is an important consideration

TABLE 2.2

Typical Food Colloids, Methods of Preparation, and Stabilization Mechanisms

Food	Emulsion Type	Preparation Method	Stabilization Mechanism
Milk	Oil-in-water	Natural	Protein membrane
Cream	Fat + oil-in-water	Centrifugation	Protein membrane and particle stabilization of air
Ice cream	Fat + oil-in-water	Homogenization	Cream plus ice network
Butter and margarine	Water-in-oil	Churning and votator	Fat crystal network
Sauces	Oil-in-water	High-speed mixing and homogenization	Protein and polysaccharides
Fabricated meat products	Oil-in-water	Low-speed mixing and chopping	Gelled protein matrix
Bakery products	Oil-in-water	Mixing	Starch and protein network

when designing a delivery system for a particular food application.[15,19] Table 2.2 provides some typical food colloids, their preparation methods, and stabilization mechanisms.

2.3.2 Relationship between Rheological and Sensory Properties

Elucidation of the relationship between food texture perception and food structure is of increasing importance for food processors wishing to produce texturally attractive food products. Efforts in this direction attempt to correlate both classical rheology and mechanical data (small and large deformations) with the sensory properties of foods. Rheological methods, however, have certain limitations in assessing actual situations; for example, evaluating small deformation rheology properties is hardly applicable to sensory tests. Studies of the relationship between rheological properties and sensory perception were initiated by Sherman's group;[44] after correlation of viscosity data for a wide range of food materials (with viscosity values ranging between 1 mPa·s and 10 Pa·s), they concluded that during consumption of highly viscous foods the applied stress increased in proportion to the viscosity increase.

Rheological data relative to the flow, deformation, and breakdown of materials subjected to stress can have some relevance to the behavior of foods during oral processing and sensory perception of food. Additionally, properties associated with friction, adhesion, and lubrication of interacting surfaces can be used to explain oral perceptions of food structures. Frictional conditions in the mouth have been implicated in the perception of such important food attributes as astringency, mealiness, smoothness, roughness, and slipperiness. The perception of taste, flavor, and texture of food during consumption

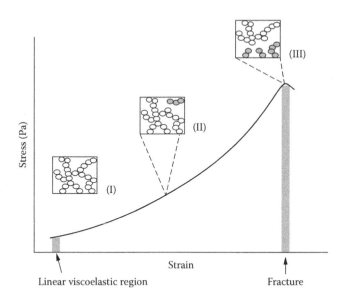

FIGURE 2.3
Hypothetical stress–strain plot for a biopolymer gel. Gray bars indicate boundary regions of linear viscoelasticity and fracture. Inserted graphics depict microstructures that are (I) intact, (II) showing microcracks, and (III) fractured. Gray-filled elements of the microstructure indicate separation from the network. (From Foegeding, E.A., *Curr. Opin. Colloid Interface Sci.*, 12, 242, 2007. With permission from Elsevier.)

is influenced by α-amylase present in saliva which initiates the digestion of starch, resulting in a decrease in the perceived thickness. Major differences in oral processing and the sensory perception of food can be related to human individuality (e.g., age, sex, health status).[30,36,45]

A hypothetical stress–strain curve of a biopolymer is given in Figure 2.3. The curve is separated into three regions: (I) linear viscoelasticity (LVE), representing properties determined at nondestructive strains; (II) nonlinear viscoelasticity, representing strain hardening or weakening followed by stress; and (III) strain at fracture. Large-strain rheological tests typically measure the forces and deformations associated with the first bite during consumption, which represents only 2 to 10% of the total normal mastication time. Combining the linear, nonlinear, and fracture properties of a specific gel network offers a complete mechanical characterization of the structure–function relationship of the gel network and aids in the development of multicomponent gel systems.[30] It is still necessary to more comprehensively define the relationship between sensory attributes and microstructural and macroscopic rheological properties, incorporating measures of food structure, mechanical properties, oral processing, and sensory texture attributes.[46]

2.3.3 Rheological Properties of Polysaccharide Solutions

The rheological properties of an aqueous polysaccharide solution are directly related to its molecular properties—composition, molecular weight, branched or linear structure, and changes occurring during isolation, purification, and concentration (e.g., drying), as well as intermolecular and intramolecular interactions and interactions with a solvent (e.g., water). Measurements of the viscoelastic and flow behavior properties can provide information on network formation and the aggregation tendencies of the polymers. Small-strain rheology nondestructively probes the linear region, and large-strain rheology probes the nonlinear region and fracture. Polysaccharide solutions normally exhibit Newtonian behavior at concentrations well below a critical concentration; however, above that critical concentration, non-Newtonian behavior is observed. In general, the rheology of polysaccharide solutions is pseudoplastic or thixotrophic (i.e., they show shear thinning). The viscosity shear rate dependency increases with increasing molecular mass, and the shear rate at which shear thinning occurs shifts to lower values.

The viscosity values of polymer solutions show a marked increase above a critical concentration, commonly referred to as C*, which corresponds to the transition from the so-called dilution region, where the polymer molecules are free to move independently in solution, to the semidilute region, where molecular crowding gives rise to the overlap of polymer coils resulting in interpenetration. C* is a function of the hydrodynamic volume of the polymer and is given by the relationship, $C^* = a/\eta$, where a is an integer that varies with different polysaccharides and η is the viscosity. A logarithmic plot of the viscosity vs. polymer concentrations below C* (dilute region) gives a straight line having a slope of about 1.4, while above C* (semidilute region), the slope is much higher—about 3.3 for random coil-type polymers. Polysaccharide solutions normally exhibit Newtonian behavior at concentrations well below C*; that is, their viscosity is independent of shear rate. Above C*, non-Newtonian behavior is usually observed.[8]

The viscosity of carrageenan is particularly dependent on temperature; that of xanthan gum, the least. The consistency coefficient (m) and flow behavior index (n) are sensitive to changes in temperature and carrageenan concentration. In one study, shear stress vs. shear rate data for this polysaccharide in the presence of 1% salt fit the Herschel–Bulkley model.[47] In another study, the rheological properties of 12 polysaccharide solutions were investigated at concentrations ranging from 0.05 to 0.5%. The viscous and elastic components of the complex viscosity (η^*), elastic yield stress, and tan δ were measured as functions of oscillatory shear. A substantial increase in tan δ was observed for most 0.5% gum solutions at shear rates beyond 10 sec^{-1}, indicating a shift from a viscoelastic regime to a purely viscous one. For gums that showed substantial viscoelasticity, peak tan δ values ranged from 5.7 to 68.3. Analysis of the data led to the conclusion that the viscosity of a hydrocolloid depends on its mass, molecular size, shape, charge, concentration, and presence of electrolytes.

TABLE 2.3

Comparison of Viscosities (mPa·s) of Aqueous Polysaccharides
at Varying Concentrations at 25°C

Concentration (%)	Carrageenan	Sodium Alginate	Methylcellulose
1	57	214	39
2	397	3760	512
3	4411	29,400	3850
4	25,356	—	12,750
5	51,425	—	17,575

Source: Adapted from Belitz, H.-D. et al., *Food Chemistry*, 3rd ed.,
Springer-Verlag, Heidelberg, 2004, p. 245.

A 1% aqueous solution of agar has been determined to have a G' value of 50,000 Pa, a G'' value of 3500 Pa, and a tan δ value of 0.07. In comparison, starch gels had much lower G' and G'' values, in the range of 16 to 141 Pa and 5.5 to 7.0 Pa, respectively. Myofibrillar protein, at a 7% concentration, has values for G', G'', and tan δ of 2500 Pa, 320 Pa, and 0.12, respectively.[7,48–50]

The characteristic of yield stress has been found to be beneficial with regard to enhancing the stability of liquid food products containing xanthan. When a xanthan solution is subjected to shear due to mixing, shaking, or chewing it will thin out, but once the shear forces are removed the gum will return to its original consistency. It remains thick at rest in a bottle that keeps the mixture fairly homogeneous, but the shear forces generated by shaking the bottle thin it so it can be easily poured. This property has practical use when incorporated in salad dressing.[50]

The elastic properties of polysaccharides are particularly significant from an industrial standpoint. As discussed earlier, when a sinusoidal deformation is imposed on a solution over a wide range of frequencies, the response is a complex modulus decomposed into an in-phase response (G', reflecting the elastic character) and a cut-off phase response (G'', reflecting the viscous response). At low frequencies, G'' is larger than G', but over a critical frequency G' becomes larger than G'', corresponding to the presence of entanglements (transitory cross-links).

An increase in molar mass or concentration will lead to a decrease in the critical shear rate at which the normal stress (elastic behavior) will replace the shear stress (viscous behavior).[51] Such findings have helped us understand the behavior of polysaccharides not only for food product development but also in other areas such as biopackaging and the entrapment and controlled release of nutraceuticals and drugs.[52] Table 2.3 compares viscosities (mPa·s) of some polysaccharides in aqueous suspensions of varying concentrations at 25°C.

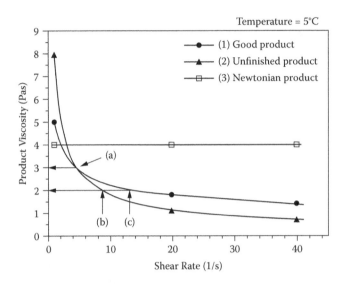

FIGURE 2.4

Flow curves for two shear-thinning products and one Newtonian product measured at a constant temperature (From Mason, S., *Challenges of In-Process Food Viscosity Measurement*, IFIS Publishers, Reading, Berkshire, U.K., February, 2007. With permission.)

2.3.3.1 In-Process Viscosity Measurement

When a food product is heated, its viscosity decreases; cooling causes the viscosity to rise. Viscosity also decreases as the shear rate increases. If a viscosity measurement is conducted at only one shear rate, it can be difficult to identify the product, particularly if the temperature varies only slightly. For a non-Newtonian food product, single-point viscosity measurement might not be completely sufficient to determine its quality. In-process viscosity measurement helps us understand the behavior of food under production conditions, particularly the behavior of food biopolymers with respect to their concentration, shape, size, and polydispersibiity in a food system. Because production temperatures may fluctuate, viscosity is often measured at a specified reference temperature.[53] In-process measurement of viscosity involves placing multiple viscosity sensors at locations along the processing route; determining the food microstructure using microscopy or laser light scattering complements the viscosity measurement. Equipment used for the in-process measurement of viscosity includes the rotational viscometer, which measures viscosity values at multiple shear rates. From the data gathered, a viscosity vs. shear rate curve can be generated. The torque necessary to generate the required rotational speed of the rotor is proportional to the viscosity. The oscillatory viscometer provides one viscosity data point at an unknown shear rate.[54,55] Figure 2.4 gives flow curves for two shear-thinning products and one Newtonian product measured at a constant temperature.

2.4 Interactions of Polysaccharides with Food Components

Interactions of polysaccharides with food components have a profound influence on the sensory quality of processed foods and are of some significance in the development of multicomponent biodegradable and edible films.

2.4.1 Protein–Polysaccharide Interactions

Proteins are an important component of many foods. Interactions among proteins and polysaccharides in aqueous media are mainly mediated through electrostatic forces, which can lead to complex formation through aggregation or phase separation, the latter being most generally observed. Physicochemical factors such as pH, ionic strength, temperature, pressure, shearing rate, mixing time, ratio of protein to polysaccharide, charges on the macromolecules, and their molecular weights all affect these interactions and the stability of the resulting products. Oppositely charged hydrocolloids (e.g., a protein below its isoelectric point and an anionic polysaccharide) are likely to associate, and the complexation results in improved thermal stability and increased resistance to external treatments (e.g., high pressure) involved in food processing. Complexation also enhances other properties such as oil emulsification capacity and rheological features compared to individual, uncomplexed components.[14,56,57] Protein–polysaccharide interactions can still be affected even when the protein is hydrolyzed, as has been observed for sunflower protein subjected to various degrees of hydrolysis.[58]

Carrageenan and pectin have a stabilizing effect on whey protein isolate, with pectin exhibiting the maximum stability. Interactions among whey proteins and polysaccharides result in mixed-gel large deformation behavior, which depends on pH, concentration, and the nature of cations added to the system. Variations in these conditions produce two types of mixed systems, one with two gelling biopolymers (whey protein/κ-carrageenan or whey protein/pectin) and another where the protein is the only gelling biopolymer. Conditions favoring incompatibility can also lead to spherical inclusions of whey protein.[59–61] Carrageenan and protein complex formation is due to the sulfate groups of the carrageenan; anion groups in the proteins produce a stable colloidal protein–carrageenan complex.[62] A typical interaction between carrageenan and casein is shown in Figure 2.5.

Whey protein concentrate (WPC) (2%) and carrageenans (1%) were shown to influence the gelation of meat exudates from massaged cured porcine muscle. The exudates were heated from 20 to 80°C and subsequently cooled after 30 min to 20°C. Analysis of the viscoelastic properties of the exudate samples showed that combinations of WPC and carrageenans increased storage modulus (G') values in comparison with samples in the absence of either carrageenan or WPC. Significant synergies were observed upon blending

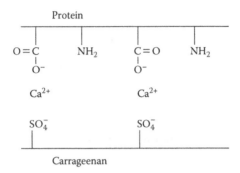

FIGURE 2.5
Examples of interactions of carrageenan with differently charged proteins.

high-gelling WPCs with carrageenan.[63] Myofibrillar protein together with whey protein concentrate form stronger gels in the presence of alginate and carrageenan, the gelation being enhanced by the presence of $CaCl_2$; electrostatic interactions seem to be the main forces involved in the formation and stability of these gels.[64] Figure 2.6 illustrates the segregation and association of polysaccharides and proteins.

The effects of specific protein–polysaccharide interactions on the foaming properties of highly viscous Newtonian types of foods have been reported. Foams were produced by either 0.1% guar or xanthan in the absence of proteins, but 0.1% pectin allowed a total incorporation of the gas phase with large bubbles. Whey protein isolate (WPI) at 2% (w/v) was able to give foams with the desired overrun and small bubbles. Overrun was reduced in WPI–xanthan mixtures, probably because the matrix exhibited viscoelastic trends. WPI–pectin mixtures provided abundant and stable foams with the best stability. Bubble diameters in foams were governed by process parameters.[65] The search for new combinations of polysaccharides and proteins is continuing with a view toward developing sheared gel products having novel rheological characteristics.[62,66]

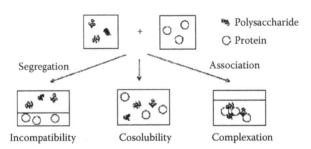

FIGURE 2.6
Segregation and association of polysaccharides and proteins. (From De Kruif, C.G. and Tuinier, R., *Food Hydrocoll.*, 15, 555, 2001. With permission from Elsevier/Rightslink.)

2.4.2 Polysaccharide–Polysaccharide Interactions

Food polysaccharides may interact among themselves to produce mixed-polymer gels with novel rheological characteristics. These interactions can be due to either association of the various hydrocolloid molecules or non-association, which results in precipitation or gelation. If either or both of the hydrocolloids can form a gel independently, phase separation and gelation will occur simultaneously. The characteristics of the resultant gel will depend on the relative rates of these two processes. If the two hydrocolloids do not associate, as is commonly the case, then at low concentrations they will appear to exist as a single homogeneous phase; at higher concentrations, they will separate into two liquid phases, each enriched in one of the hydrocolloids.

The rheological behavior of mixed systems differs noticeably from that of pure biopolymer solutions. Normally, the minimum concentration for heat-induced gelation of biopolymers varies from 0.1 to 15% (w/w). The minimum concentration for gelation usually decreases when another incompatible biopolymer is added, presumably due to an excluded volume effect. Phase separation in mixed biopolymer solutions is always accompanied by rheological changes.[6] Careful selection of hydrocolloids and their concentrations can, therefore, lead to the formation of a broad range of gel textures. Classic examples include the addition of locust bean gum to κ-carrageenan to yield softer, more transparent gels that are less brittle, more elastic, and less prone to syneresis. Similarly, addition of locust bean gum to xanthan gum induces gel formation of the latter.[8,67]

Polysaccharides can display mutual antagonism under certain dispersion conditions. Mixed dispersions of polysaccharides may evoke entirely new oral sensations compared with those containing individual biopolymers. The contribution of each component to a particular sensory parameter may be additive, complementary, or synergistic; for example, combinations of starch and carrageenan elicit a variety of textures in cream desserts. Pectin is believed to contribute flavor-releasing characteristics to yogurts superior to those of starch.[68] Methodologies to study polysaccharide–polysaccharide and polysaccharide–protein interactions include differential scanning calorimetry (DSC), rheometry, ultraviolet absorption, and circular dichroism (CD) measurements.[56]

2.4.3 Other Interactions

Non-enzymatic browning reactions (Maillard reactions) occur between reducing sugars present at the end of polysaccharide chains and amino groups in foods during processing and storage. These reactions are temperature dependent and extensive at intermediate water activities. They can influence the nutritive value of food, as they may diminish the bioavailability of amino acids, especially lysine. Complexing polysaccharides with minerals has many consequences for the taste, nutritional value, and technological quality of foods.

TABLE.2.4

Diverse Functions of Polysaccharides in Food

Function	Application
Adhesive	Icings and glazes
Binding, texture modification	Pet foods
Coating	Confectionery
Emulsification	Salad dressings
Encapsulation	Powdered flavors
Film formation	Protective coatings, sausage casings
Fining (colloid precipitation)	Wine and beer
Stabilization of foam	Beer
Gelling agent	Confectionery, milk-based desserts, jellies, pie and pastry fillings
Inhibition of ice-crystal formation	Frozen foods, pastilles, sugar syrups
Stabilization	Ice cream, salad dressings
Swelling agent	Processed meat products
Syneresis inhibition	Cheeses, frozen foods
Synergistic gel formation	Synthetic meat gels
Thickening agent	Jams, sauces, syrups, pie fillings

Source: Adapted from Sutherland, I.W., in *Biopolymers*, Vol. 5, *Polysaccharides I: Polysaccharides from Prokaryotes*, Vandamme, E., De Baets, S., and Steinbuchel, A., Eds., Wiley–VCH Publishers, Weinheim, 2002, p. 1; Sutherland, I.W., *Int. Dairy J.*, 11, 663, 2001; Sutherland, I.W., *Trends Biotechnol.*, 16, 41, 1998.

The complexing of Cu^{2+} by hydrogenated glucose syrups, for example, increases as the degree of polymerization decreases. Iron salts are also complexed by various simple carbohydrate molecules. The taste of iron salts is masked by complexing with carbohydrates at high pH, whereas the opposite occurs at low pH. Vitamin C is protected from Cu^{2+}-catalyzed oxidation when the cation is complexed with various carbohydrates or their hydrogenated derivatives. These reactions can be demonstrated by measurement of conductivity, specific rotation, redox potential, or ligand-exchange chromatography.[69] Such reactions have implications for the design, formulation, and storage of food items. Table 2.4 shows various factors determining the functions of polysaccharides in food systems.

2.5 Major Food Applications of Polysaccharides

Polysaccharides can be excellent additives to food formulations due to the various functional properties that allow them to serve as gelling and thickening agents, stabilizers, water-retention compounds, emulsifiers, ingredient

TABLE 2.5

Factors Determining the Functions of Polysaccharide
in a Food System

Interactions (additive, antagonistic, or synergistic) with other food components (e.g., proteins, other polysaccharides)
Influence of other factors (e.g., pH, temperature, cations, pressure)
Influence of the manufacturing process (e.g., cooling, shearing, dehydration)
Complex composite properties in final structure
Influence of the polysaccharide on final texture, phase separation, flavor, and overall stability of the food

binders, viscosity modifiers, and foam stabilizers (see Table 2.5). They retard crystal growth in ice cream and in confections, and products containing fluid and mixed gels improve satiety. Weak gels of biopolymers are frequently used in food applications; being of natural origin, they are quite safe, unlike many synthetic food additives.[70–72] Because of their interesting functional properties, polysaccharides, together with sugars, corn syrup, and dextrose, represent 90% of additives used. The food-related functions of polysaccharides are briefly discussed below.

2.5.1 Texture Improvement

The food industry uses large amounts of polysaccharides as thickeners and stabilizers in various foods, including sauces and dressings, due to their desirable properties at concentrations as low as 0.5 to 1%. The viscosity and thickening properties are dependent on the polymer concentration, molar mass, stiffness of the polymer, temperature, shear rate, and solvent characteristics (e.g., ion concentrations, nature of the ions, pH). These factors can be judiciously manipulated to produce the required food texture. In mixed polymer systems, electrostatic interactions and thermodynamic incompatibility determine the functional properties (rheology, surface hydrophobicity, emulsification power) of the blends. The hydration properties (solubility, viscosity), structural properties (aggregation, gelation), and surface properties (foaming, emulsifying) of these complexes determine their use in food formulations (e.g., as fat replacers or texturing agents), as well as their use in the synthesis of edible films and as encapsulation materials for nutraceuticals.

Protein–polysaccharide mixtures subject to phase separation are interesting with regard to the creation of structures having distinct rheological properties. The effect of phase separation can be suppressed, if necessary, by using various biopolymer concentrations in the stable regions. Introducing simple shear flow in food structuring processes can lead to a much broader range of structures, thereby better utilizing the full potential of food ingredients to create novel structures in food materials. A number of shearing devices are available that allow processing of biopolymer systems under simple shear flow.[73]

Polysaccharides are used to improve the stability of foods, control syneresis, enhance flavor, replace fat, and improve fiber contents in foods. Foods whose texture is influenced by polysaccharides and their interactions with water include frozen desserts, confectioneries, salad dressings, puddings, gravies, cheese, pie fillings, and a variety of diet foods.[72] Entrapment of abundant water or air in gel matrices reduces caloric density but improves satiety.[76] Polysaccharide foams and gels have attracted modern chefs wishing to develop light and exquisite textures.[74] Research on the structure of semisolid foods and the influence of structure on texture has addressed emulsion rheology, the behavior of biopolymers, key aspects of product development and enhancement, and the use of emulsions and gels in texture formulation.[71,72]

Polysaccharides have found a number of uses in bakery products. Hydrocolloids, including polysaccharides, are used in commercial baking to facilitate processing, compensate for variations in raw materials, guarantee constant quality, and preserve freshness and food properties. These additives enhance dough-handling properties and stability, viscoelastic properties, and other quality-related criteria, such as water absorption and specific loaf volume. They can also replace the wheat protein gluten[84] (see Section 2.6.5).

The development of fruit analogs is a flourishing area in the food industry. Advantages of these fruit analogs are uniformity in size and shape, decreased sugar content, improved flavor, and color stability during storage. Ideally, the products should mimic the crunchy texture of fresh fruits, but poor texture has been a significant problem. Polysaccharides influence the flow properties of fluid fruit fillings. The fluid portions of commercial fruit fillings generally have yield stress values between 39 and 51 Pa, a consistency index between 52 and 104 Pa·s, and a flow index of around 0.4. The addition of gums to fruit fillings influences their apparent viscosity, varying with the type of gum, amount added, and shear rate. Additions of guar gum, locust bean gum, and carboxymethylcellulose increase the consistency and flow indices, while xanthan gum and κ-carrageenan decrease these properties.[6,75]

The sensory perception of commercial food products that contain polysaccharide-based thickeners can be somewhat unpredictable. This can be attributed not only to structural variations in the polysaccharides themselves but also to variations in production. The composition of an individual's saliva can also affect sensory perception, as amylase in the saliva rapidly degrades polysaccharides in the oral cavity. Starches with a smaller degree of cross-linking are more susceptible to such degradation. The viscosity of starch decreases significantly upon incubation with human saliva for a period as short as 10 sec; however, a low friction coefficient is retained, which can be attributed to undigested starch granules remaining in the oral cavity.

2.5.2 Oil Emulsification

Polysaccharides contribute significantly to the development of food emulsions and foams. Gels of polysaccharides in the presence of a dispersed fat phase (emulsion gels) function as texture modifiers by thickening or gelling the continuous phase. Polysaccharide emulsifiers have certain general functional characteristics resembling those of other food emulsifying agents—namely, proteins, surfactants, and solid particles. Polysaccharides, either alone or in combination with other hydrocolloids, are useful in stabilizing foams and dispersions. In the absence of these stabilizers, aqueous dispersions of oil are inherently unstable because of the formation of aggregates. The most widely used polysaccharide emulsifiers in food applications are gum arabic (*Acacia senegal*), modified starches and celluloses, pectin, and galactomannans. Electrostatic complexation between oppositely charged proteins and polysaccharides improves the emulsifying properties of proteins by thickening the layer at the interface of the oil droplets. Examples of such mixed systems include blends of gum arabic, galactomannans, pectins and protein, caseins, whey, soya, and gelatin. Because of their enhanced emulsifying capacity, these systems improve taste and offer controlled release and targeted delivery.

Several semisolid foods are combinations of high-moisture gel matrices and dispersed microstructural elements such as fiber, globules, or air bubbles (e.g., frankfurters, cheeses, mousses). Polysaccharides improve texture by keeping solids dispersed in a medium, such as chocolate in milk, air in whipping creams, and fat in salad dressings, canned meats or fish, marshmallows, and jelled candies. Other product applications include carbonated soft drinks, beverage emulsions, ice cream, sauces, and dressings.[11,14]

2.5.3 Flavor Release

Polysaccharides may be added to foods to deliberately change the aroma and flavor (as well as texture). The release of aroma compounds is influenced by the viscosity of the food which, in turn, is controlled by the presence of hydrocolloids.[77] Polysaccharides retain volatile flavor compounds in many food systems, ranging from wine to salad dressings to dessert gels. Because flavors are usually present in foods at low levels, a relatively small amount of binding can have a significant effect on perceived flavor; however, in food products containing high levels of water, the binding of flavors to most polysaccharides is minimal.[6] The food microstructure also influences the release of flavor compounds. Emulsions encapsulated in microstructured gels suppress the release of lipophilic flavor compounds in low-fat products. The gelling biopolymer present in an emulsion affects the release of these compounds during mastication. Reducing droplet size in emulsions enhances the release of nonpolar flavors but has no effect on polar aromas. Development of extensional viscosity methods, electronic noses, and electronic tongue technology, along with traditional texture measurements such as compression and steady shear, can provide further insights into our perception of food flavor. Novel raw materials, molecular assemblies, processes, and applications for flavor emulsions have been discussed.[78,79]

2.5.4 Polysaccharides as Dietary Fiber

Dietary fiber is a material consisting of plant cell walls, structurally complex and chemically diverse non-starch polysaccharides, and other associated substances. Unlike other nutrients, fiber is not attacked by the enzymes of the stomach and small intestine and reaches the colon undegraded. In 2001, the American Association of Cereal Chemists defined dietary fiber as "that fraction of the edible part of plants or their extracts or synthetic analogues that are resistant to digestion and absorption in the human small intestine, usually with complete or partial fermentation in the large intestine." Of late, the definition has been broadened to include not only inedible parts of vegetables but also fibers of animal origin such as chitosans.[80]

The main benefits of dietary fiber are a reduction in the intestinal absorption of nutrients, a reduction in colonic luminal toxicity and systemic effects, alteration of colonic microflora, and direct action on colonic mucosa. Increased fiber intake leads to decreased food transit time and adds bulk to the stool.[81] Soluble fiber assists in plasma cholesterol reduction and large bowel fermentation. Foods containing soluble fiber include whole-grain foods such as breakfast cereals, multigrain bread, vegetables such as carrot and celery, oatmeal, nuts, legumes, whole-grain barley, and pears, ripe strawberries, and bananas. The soluble fiber forms a viscous indigestible mass in the gut and helps trap digestive enzymes, cholesterol, starch, glucose, and excess bile acids and toxins, which are then expelled through the feces. By reducing the absorption of food including high-fat items, soluble fiber can help obese and diabetics reduce the amount of starch digestion and glucose uptake from food.[81]

According to the American Dietetic Association, the current recommended fiber intake for adults ranges from 25 to 30 g per day or 10 to 13 g per 1000 kcal, and the insoluble/soluble ratio should be 3:1. In Europe, the currently recommended fiber intake is 20 g per person per day, but in developing countries the range is 60 to 120 g per day. The U.S. Food and Drug Administration (FDA) observed that food containing barley reduces the risk of coronary heart disease.[82]

Despite the recognized benefits of dietary fiber, the intake of fiber around the world appears to be far from adequate. A number of polysaccharides, including chitosan, alginate, and carrageenan, can function as dietary fiber.[69] A method for the measurement of dietary fiber, including resistant starch, nondigestible oligosaccharides, and available carbohydrates, has been reported recently that depends on their susceptibility to digestion by α-amylase and amyloglucosidase.[83] Table 2.6 lists the advantages of fiber in foods, and Table 2.7 lists the technological functional properties of dietary fibers.

TABLE 2.6

Advantages of Fiber in Foods

Fiber increases bulk quantity.
Fiber reduces calorific value of diets.
Soluble fibers such as gums form a viscous indigestible mass.
Fiber reduces absorption of cholesterol and hence coronary heart disease.
Fiber reduces glucose absorption and controls diabetes.
Fiber controls obesity.
Fiber reduces the absorption of any toxins in the food.

2.5.5 Gluten-Free Bakery Products

The main wheat component responsible for bread quality is gluten, which is an essential structure-binding protein. In pasta products, gluten forms a viscoelastic network that surrounds the starch granules, thus restricting swelling and leaching during boiling and making this product group a comparatively rich source of resistant starch. In breadmaking, interactions between the gluten protein and dough polysaccharides (starch, pentosans) are important in making the resultant batter firm or flowing, thus determining the product texture. Approximately 1% of the world's population suffers from celiac disease (gluten-sensitive enteropathy) and must avoid gluten in their diet. For this reason, demand has grown for gluten-free products, but finding alternatives to gluten that have similar properties has proven a significant challenge for food scientists.

TABLE 2.7

Technological Functionality of Dietary Fibers

Functional Property	Advantage
Water-holding/binding capacity	Soluble fibers such as algal fibers, pectin, gums, and glucans have a higher WHC than cellulosic fibers. Algae fibers, depending on type, can bind water at 20 times their own weight.
Fat-binding capacity	The porosity of the fiber rather than molecular affinity affects the fat-binding capacity. Water-soaked fibers have more fat-binding capacity.
Viscosity	Soluble fibers from algae form highly viscous solutions, a property that also makes them useful as thickeners in foods.
Gel-forming capacity	Fibers such as carrageenans, chitosan, and pectin form gel networks that absorb water and solutes in the network. Network formation depends on factors such as temperature, concentration, ions, and pH.
Chelating capacity	Many types of fibers possess the capacity of binding minerals, favoring reduced metal-induced functions such as lipid oxidation.

The sale of gluten-free products has grown at an average annual rate of 28% since 2004 and reached \$1.56 billion in 2008. Most of the gluten-free products (e.g., biscuits, cakes, pasta, pizza) available today are based on pure starches, resulting in a dry, sandy mouth feel and poor overall eating quality; however, a number of water-soluble polysaccharides such as xanthan mimic the viscoelastic properties of gluten and can be used for the development of gluten-free products.[84] A number of gluten-free products, including cookies, muffins, and cakes, have recently been introduced that have the taste, texture, and visual appeal of gluten-containing products. They contain a minimum of 25% β-glucan, which was used to replace gluten. Products containing the ingredient have qualified for the FDA barley heart health claim.[85-87]

Polysaccharides have an influence on intestinal microorganisms, which metabolize dietary components, principally complex carbohydrates, that are not hydrolyzed or absorbed in the upper gastrointestinal tract, thus releasing energy through fermentation. Some bacteria are thought to possess important health-promoting activities, especially with respect to their influence on mucosal and systemic immune responses to disease. These bioactivities can be modulated by substrates, including polysaccharides, that support and influence microbial development, growth, and survival.[93]

Carbohydrates also have an impact on the glycemic index (GI) value and satiety of a food. The glycemic index ranks carbohydrates according to their effect on blood glucose levels. Low GI carbohydrates produce only small fluctuations in blood glucose and insulin levels, thus contributing to sustainable weight loss, reducing the risk of heart disease and diabetes, and improving long-term health. High GI carbohydrates are associated with a more immediate reduction in appetite and food intake, whereas the satiating effects of lower GI carbohydrates appear to be delayed (up to 3 hours); however, there is no consistent evidence that an increase in blood glucose, either acute or sustained, is the primary determinant of food intake and satiety.[94]

Glucose has been reported to improve memory, and recent evidence supports the positive effects of other monosaccharides such as mannose and fructose on cognitive performance. When their metabolism is blocked, memory formation is inhibited and amnesia occurs. Adequate saccharide intake in middle age may help against such a decline.[95,96]

2.5.6 Control of Starch Retrogradation

Cereal starches are important polysaccharides that have numerous applications in the development of foods. Products made from the diverse types of cereal starch (e.g., wheat, rice, maize, sorghum, barley, oats) include bakery products (bread, biscuits, pancakes, cookies, muffins, naan), breakfast cereals (roasted grains, flakes, grits, waffle mixes, snack foods), spaghetti and semolina (noodles, pasta, chips, boiled porridges, extruded products), flour products (unleavened flat bread, soups, purees, baby foods, sauces, puddings, desserts), fermented products popular in Asian countries (idli, dosa, koji), boiled meals, puddings, and roasted grains, among many others. Usually, the starch granules are not completely dissolved during food processing.[76]

When heated in the presence of water, starch granules undergo gelatinization leading to irreversible loss of crystalline structure, depending on the water content. Gelatinization dramatically increases the availability of starch for digestion by amylolytic enzymes. The degree of gelatinization achieved by most commonly used food processes is sufficient to permit rapid digestion of the starch. If the viscous starch solution is cooled or left at a lower temperature for a prolonged period, the linear amylose and amylopectin chains align themselves parallel to each other, forming hydrogen bridges and resulting in a more crystalline structure. The retrogradation of the amylopectin component is a phenomenon that occurs gradually upon storage of starchy foods. The rate of freezing is also known to affect the retrogradation rate. Repeated freeze–thaw cycles that involve subjecting samples to repeated freezing and intermittent thawing to room temperature over a period of 2 to 4 hours are known to accelerate retrogradation and syneresis drastically. In bakery products, retrogradation results in staling of the bread, as retrogradation can expel water out of the polymer network by syneresis. This undesirable property of starch to retrograde can be prevented by the addition of certain compounds, such as monoglycerides of fatty acids or xanthan, that complex with starch.[84,88]

2.5.7 Control of Syneresis

Syneresis is an undesirable separation of water observed in many foods. Freeze–thaw stability is an important property of starch-based products that allows them to withstand the physical changes that occur during freezing and thawing; however, when starch pastes or gels are frozen, phase separation invariably occurs due to the formation of ice crystals. Thawing leads to syneresis because the water can be easily expressed from the dense network of starch pastes and gels.

Water separation is due to amylopectin retrogradation in the starch-rich phase. The amount of syneresis water can be used as an indicator of the tendency of starch to retrograde. This property may be evaluated by gravimetric measurement of the water of syneresis that separates from starch pastes or gels. Because of their exceptional ability to bind water, polysaccharides are able to control syneresis; for example, tapioca starch forms a clear paste during processing that has a bland taste and high viscosity, which are advantageous in many food applications. Tapioca starch, however, is prone to retrogradation during freezing and thawing.[1,8]

2.5.8 Polysaccharides as Films for Coating Food Products

Polysaccharide gels can be used to form biodegradable and edible films possessing excellent moisture and oxygen barrier properties, in addition to antimicrobial and antioxidant properties. Several polysaccharides have been successfully used for this purpose on a limited scale (see Chapter 9).

2.5.9 Stability of Polysaccharides to Processing

The stability of polysaccharides used in food product development is important, as many carbohydrates are susceptible to changes during heat treatment at 100°C or higher. Starch, for example, when heated in the absence of water, experiences thermal degradation resulting in lower paste viscosity when the product is subsequently cooked.[90]

Polysaccharides may also undergo hydrolysis due to interactions with acids in foods, including acetic acid (vinegar), citric acid (fruit juices), and potassium acid tartrate (cream of tartar). In most cases, such hydrolysis is undesirable as it can lead to a loss of the ability of polysaccharides to gel or thicken the food. Carrageenan is very susceptible to acid interactions, but alginate is not. Polysaccharides may also interact with ionic species in foods, especially cations. The effect may sometimes decrease the dispersion viscosity.[67]

2.6 Factors to Be Considered When Using Polysaccharides in Food Systems

For the successful use of polysaccharides in food products, it is important to understand their molecular properties, interactions among the food ingredients, and the influence of processing conditions, as discussed earlier (see Table 2.8). Adding and replacing various ingredients can lead to changes in the food structure; adding even a single polysaccharide requires a thorough understanding of the hydrocolloidal structure and function in real food systems. Mixed systems of biopolymers are often used in food formulations, as the incorporation of two or more hydrocolloids improves the texture of a food through the synergistic effects of their gelation properties. Factors to be considered when incorporating polysaccharides in food products include their viscosity and gel properties, their gelation kinetics, influences of other systems on the polymers, effects of the manufacturing process (e.g., cooling, shearing, dehydration), how amenable they are to processing, and other complex composite properties, as well as their effects on texture development, phase separation, flavor release, and the overall stability and quality of the final product. Proper use of additives under appropriate processing conditions can lead to a food microstructure that produces an ideal texture and stability.[6,91] Incorporating polysaccharides also requires an understanding of the relationship between food structure and its sensory properties. As pointed out earlier, commercial food products that contain polysaccharide-based thickeners can be less than desirable with regard to their sensory perception.[76]

2.7 Commercial Status of Food Polysaccharides as Additives

The food hydrocolloid industry, including polysaccharides, represents a market of over US$3 billion. The top two hydrocolloids are starch and gelatin, accounting for about 50% of the total value. Polysaccharides for industrial use come from plants (starch, guar gum, gum arabic, pectins), seaweed (alginate, carrageenan, agar), and crustaceans (chitin), in addition to microbial polysaccharides (xanthan gum, curdlan, gellan gum, pullulan, bacterial alginate). According to 2002 figures, the market value of polysaccharides includes xanthan (US$219 million), carrageenan, (US$39.4 million); alginate (US$21.5 million), gelatin (US$21.5 million), guar (US$5.8 million), and agar (US$3.7 million) (see also Chapter 7).[18] Most of these are marketed as dry

TABLE 2.8

Factors Influencing Use of a Particular Polysaccharide
in Food Product Development

Factors	Influence
Molecular properties	The molecular diversity of a simple hydrocolloid influences textural properties in various ways through its effect on gelation, oil emulsification, and foam stability.
Rheological properties	The effect of a polysaccharide on texture is greatly related to its hydrodynamic properties—namely, the volume it sweeps out in solution and its effects on shear thinning at a high shear rate.
Salt effects	Incorporation of salt to consumer taste influences the behavior of hydrocolloids; for example, cations such as Na^+, K^+, and Ca^{2+} influence the gelation of alginate and κ-carrageenan.
Sugar effects	The texture of such products as ice cream and desserts is significantly influenced by the presence of sugar. Gelation is also influenced by the sugar reactivity of some hydrocolloids.
Process conditions	Temperature and shear rate greatly influence the rheological properties and hence the texture of food products.
Mixed hydrocolloid systems	The effects could be additive (formation of mixed gels), antagonistic (phase separation), or synergistic, depending on the type of hydrocolloid; for example, the gel strength of xanthan is synergistically enhanced by locust bean gum (LBG).

powders. Recent years have witnessed a marked growth in preparations of chitosan for diverse dietetic and therapeutic uses (see Chapters 6 and 11). Polysaccharides are ubiquitous compounds. The ample availability of edible polysaccharides and their variety of functions in diverse formulation and processing systems provide a wealth of opportunities for the development of fabricated foods.

References

1. Mitchell, J. R., Water and food macromolecules, in *Functional Properties of Food Macromolecules*, Hill, S. E., Ledward, D. A., and Mitchell, J. R., Eds., Aspen Publishers, New York, 1998, pp. 50–76.
2. Sánchez, V. E., Bartholomai, G. B., and Pilosof, A. M. R., Rheological properties of food gums as related to their water binding capacity and to soy protein interaction, *Lebensm. Wiss. Technol.*, 28, 380, 1995.
3. Tharanathan, R. N., Food-derived carbohydrates: structural complexity and functional diversity, *Crit. Rev. Biotechnol.*, 22, 65, 2002.
4. Kavanagh, G. M. and Ross-Murphy, S. B., Rheological characterization of polymer gels, *Progr. Polym. Sci.*, 23, 533, 1998.
5. Rogovina, L. Z., Vasieliev, V. G., and Baudo, E. E., Definition of the concept of polymer gel, *Polym. Sci. Ser. C*, 50, 85, 2008.
6. Renard, D. et al., The gap between gel structure, texture and perception, *Food Hydrocoll.*, 20, 423, 2006.
7. Nishinari, K. and Zhang, H., Recent advances in the understanding of heat set gelling polysaccharides, *Trends Food Sci. Technol.*, 15, 305, 2004.
8. Williams, P. A. and Phillips, G. O., Introduction to food hydrocolloids, in *Handbook of Food Hydrocolloids*, 2nd ed., Phillips, G. O. and Williams, P. A., Eds., CRC Press, Boca Raton, FL, 2010, pp. 1–20.
9. Morris, V. J., Gelation of polysaccharides, in *Functional Properties of Food Macromolecules*, Hill, S. E., Ledward, D. A., and Mitchell, J. R., Eds., Aspen Publishers, New York, 1998, pp. 143–226.
10. Frith, W. J. et al., Microstructural origin of the rheology of fluid gels, in *Gums and Stabilisers for the Food Industry 11*, Williams, P. A. and Phillips, G. O., Eds., The Royal Society of Chemistry, Cambridge, 2002, pp. 95–103.
11. Aguilera, J. M., Stanley, D. W., and Baker, K. W., New dimensions in microstructure of food products, *Trends Food Sci. Technol.*, 11, 3, 2000.
12. Hart, T. D. et al., A stray field magnetic resonance study of water diffusion in bacterial exopolysaccharides, *Enz. Microbiol. Technol.*, 24, 339, 2004.
13. Krog, N. J. and Sparso, F. V., Food emulsifiers: their chemical and physical properties, in *Food Emulsions*, Vol. 1, Friberg, S., Larsson, K., and Sjoblom, J., Eds., Taylor & Francis, Boca Raton, FL, 2007, pp. 45–92.
14. Benichou, A., Asenin, A., and Garti, N., Protein–polysaccharide interactions for stabilization of food emulsions, *J. Disp. Sci. Technol.*, 23, 93, 2002.
15. McClements, D. J., Formation, stability and properties of multilayer emulsions for application in the food industry, *Adv. Colloid Interface Sci.*, 128, 227, 2006.

16. Singh, M. and Kaur, G., Nanotechnology in food science, *Beverage Food World (India)*, January, 2009, p. 39.
17. FDA, FDA Readies for More "Nanoscale" Challenges [press release], U.S. Food and Drug Administration, Washington, D.C., July 25, 2007 (http://www.fda.gov/ScienceResearch/SpecialTopics/Nanotechnology/ucm153723.htm).
18. Norton, I. T. and Foster, T. J., Hydrocolloids in real food systems, in *Gums and Stabilisers for the Food Industry 11*, Williams, P. A. and Phillips, G. O., Eds., The Royal Society of Chemistry, Cambridge, U.K., 2002, pp. 187–200.
19. McClements, D. J., Critical review of techniques and methodologies for characterization of emulsion stability, *Crit. Rev. Food Sci. Nutr.*, 47, 611, 2007.
20. Campbell, G. M. and Mongeot, L., Creation and characterization of aerated food products, *Trends Food Sci. Technol.*, 10, 283, 1999.
21. Zuniga, R. N. and Aguilera, J. M., Aerated food gels: fabrication and potential applications, *Trends Food Sci. Technol.*, 19, 176, 2008.
22. Aguilera, J. M. and Kessler, H. G., Properties of mixed and filled type gels, *J. Food Sci.*, 54, 1213, 1998.
23. Indrawati, L. et al., Effect of processing parameters on foam formation using a continuous system with a mechanical whipper, *J. Food Eng.*, 88, 65, 2008.
24. Foegeding, E. A., Luck, P. J., and Davis, J. P., Factors determining the physical properties of protein foams, *Food Hydrocoll.*, 20, 284, 2006.
25. Szezestiniak, A. S., Textural perceptions of food quality, *J. Food Qual.*, 14, 75, 2001.
26. Bourne, M. C., *Food Texture and Viscosity: Concept and Measurement*, Academic Press, New York, 1982.
27. ISO, *Sensory Analysis: Vocabulary*, ISO 5492, International Organization for Standardization, Geneva, 1992.
28. Wilkinson, C., Dijksterhuis, G. B., and Minekus M., From food structure to texture, *Trends Food Sci Technol.*, 11, 442, 2000.
29. Edmondson, P., Why is microstructure important in food systems?, *New Food*, 4, 36, 2005.
30. Foegeding, E. A., Rheology and sensory texture of biopolymer gels, *Curr. Opin. Colloid. Interface Sci.*, 12, 242, 2007.
31. Stanley, D. W., Stone, A. P., and Tung, M. A., Mechanical properties of food, in *Handbook of Food Analysis*, Nollett, L., Ed., Marcel Dekker, New York, 1996, pp. 93–137.
32. Szczesniak, A. S., Brandt, M. A., and Friedman, H. H., Development of standard rating scales for mechanical parameters of texture and correlation between the objective and the sensory methods of texture evaluation, *J. Food Sci.*, 28, 397, 1963.
33. Meullenet, J. F. and Carpenter, J. A., Perception of crisp, crunchy and crackly textures, *Trends Food Sci. Technol.*, 12, 17, 2001.
34. Hamann, D. D. et al., Analysis of compression, tension and torsion for testing food gel fracture properties, *J. Texture Stud.*, 37, 620, 2006.
35. Meullenet, J. F.-C. et al., Bicyclical instrument for assessing texture profile parameters and its relationship to sensory evaluation of texture, *J. Text. Stud.*, 28, 101, 1997.
36. Chen, J., Food oral processing: a review, *Food Hydrocoll.*, 23, 1, 2009.
37. Kokini, J. L., Rheological properties of foods, in *Handbook of Food Engineering*, Heldman, D. R. and Lund, D. B., Eds., Marcel Dekker, New York, 1992, chap. 1.

38. McKenna, B. M. and Lyng, J. G., Introduction to food rheology and its measurement, in *Texture in Food*. Vol. 1. *Semi-Solid Foods*, McKenna, B., Ed., CRC Press, Boca Raton, FL, 2003, pp. 130–160.

39. Brown, M. C., *Food Texture and Viscosity: Concept and Measurement*, Academic Press, Orlando, FL, 1982.

40. Young, N. W. G., *The Rheological Magnifying Glass*, IFIS Publishing, Reading, Berkshire, U.K., December, 2008.

41. Francis, F. J., Ed., *Wiley Encyclopedia of Food Science and Technology*, Wiley-Interscience, New York.

42. Rao, M. A., *Rheology of Fluid and Semisolid Foods: Principles and Applications*, Aspen Publishers, New York, 1999.

43. Grigorescu, G. and Kulicke, W.-M., Prediction of viscoelastic properties and shear stability of polymers in solution, *Adv. Polym. Sci.*, 152, 1, 2000.

44. Shama, F. and Sherman, P., Evaluation of some textural properties of foods with the Instron universal testing machine, *J. Text. Stud.*, 4, 344, 1973.

45. Malone, M. E., Appelqvist, I. A. M., and Norton, I. T., Oral behavior of food hydrocolloids and emulsions. Part 1. Lubrication and deposition considerations, *Food Hydrocoll.*, 17, 763, 2003.

46. de Jong, S. and van de Velde, F., Charge density of polysaccharide controls microstructure and large deformation properties of mixed gels, *Food Hydrocoll.*, 21, 1172, 2007.

47. Marcotte, M. et al., Evaluation of rheological properties of selected salt enriched food hydrocolloids, *J. Food Eng.*, 48, 157, 2001.

48. Oakenfull, D. A., A method for using measurements of shear modulus to estimate the size and thermodynamic stability of junction zones in noncovalently cross-linked gels, *J. Food Sci.* 49, 1103, 1984.

49. Nussinovitch, A., Kopelman, I. J., and Mizrahi, S., Evaluation of force deformation data as indices to hydrocolloid gel strength and perceived texture, *Int. J. Food Sci. Technol.*, 25, 692, 1990.

50. Yaseen, E. I. et al., Rheological properties of selected gum solutions, *Food Res. Int.*, 38, 111, 2005.

51. Young, N. W. G., The yield stress phenomenon and related issues: an industrial view, in *Gums and Stabilizers for the Food Industry 11*, Williams, P. A. and Phillips, G. O., Eds., The Royal Society of Chemistry, Cambridge, U.K., 2002, pp. 226–234.

52. Lefebvre, J. and Doublier, J.-L., Rheological behaviour of polysaccharide aqueous systems, in *Polysaccharides: Structural Diversity and Functional Versatility*, 2nd ed., Dumitriu, S., Ed., CRC Press, Boca Raton, FL, 2004, pp. 357–394.

53. Bruin, S. and Jongen, T. R. G., Food process engineering: the last 25 years and challenges ahead, *Comprehensive Rev. Food Sci. Food Safety*, 2, 42, 2003.

54. Mason, S., *Challenges of In-Process Food Viscosity Measurement*, IFIS Publishing, Reading, Berkshire, U.K., February, 2007.

55. Roberts, I., In-line and on-line rheology measurement of food, in *Texture in Food*. Vol. 1. *Semi-Solid Foods*, McKenna, B., Ed., CRC Press, Boca Raton, FL, 2003, pp. 161–182.

56. Doublier, J.-L. et al., Protein–polysaccharide interactions, *Curr. Opin. Coll. Interface Sci.*, 5, 202, 2000.

57. Tolstoguzov, V. B., Functional properties of protein–polysaccharide mixtures, in *Functional Properties of Food Macromolecules*, Hill, S. E., Ledward, D. A., and Mitchell, J. R., Eds., Chapman & Hall, New York, 1998, pp. 252–277.
58. Martinez, K. D. et al., Effect of limited hydrolysis of sunflower protein on the interactions with polysaccharides in foams, *Food Hydrocoll.*, 19, 361, 2005.
59. Sarkar, A. and Singh, H., Milk protein–polysaccharide interactions, in *Milk Proteins: From Expression to Food*, Thompson, A., Boland, M., and Singh, H., Eds., Elsevier, London, 2009, chap. 12.
60. Turgeon, S. L. and Beaulieu, M., Improvement and modification of whey protein gel texture using polysaccharides, *Food Hydrocoll.*, 15, 583, 2001.
61. Ibanoglu, E., Effect of hydrocolloids on the thermal denaturation of proteins, *Food Chem.*, 90, 621, 2005.
62. Schmitt, C. et al., Structure and technofunctional properties of protein–polysaccharide complexes: a review, *Crit. Rev. Food Sci. Nutr.*, 38, 689, 1998.
63. Kerry, J. F. et al., The rheological properties of exudates from cured porcine muscle: effects of added carrageenans and whey protein concentrate/carrageenan blends, *J. Sci. Food Agr.*, 79, 71, 1999.
64. De Kruif, C. G. and Tuinier, R., Protein–polysaccharide interactions, *Food Hydrocoll.*, 15, 555, 2001.
65. Narchi, I. et al., Effect of protein–polysaccharide mixtures on the continuous manufacturing of foamed food products, *Food Hydrocoll.*, 23, 188, 2009.
66. Dickinson, E., Interfacial structure and stability of food emulsions as affected by protein–polysaccharide interactions, *Soft Matter*, 4, 932, 2008.
67. Daniel, J. R. and Whistler, R. L., Chemical and physical properties of polysaccharides, in *Encyclopedia of Food Science, Food Technology, and Nutrition*, Vol. 1, Macrae, R., Robinson, R. K., and Sadler, M. J., Eds., Academic Press, London, 1993, p. 691.
68. Setser, C. S., Sensory evaluation, in *Advances in Baking Technology*, Kamel, B. S. and Stauffer, C. E., Eds., Blackie Academic, London, 1993, pp. 254–291.
69. FAO, Effects of food processing on dietary carbohydrates, *Asia Pacific Food Indust.*, 20(4), 64–69, 2008.
70. Norton, I. T., Fritz, W. J., and Ablett, S., Fluid gels, mixed gels and satiety, *Food Hydrocoll.*, 20, 229, 2006.
71. Walkenström, P., The creation of new food structures and textures by processing, in *Texture in Food*. Vol. 1. *Semi-Solid Foods*, McKenna, B. M., Ed., CRC Press, Boca Raton, FL, 2003, pp. 201–215.
72. Williams, P. A. and Phillips, G. O., The use of hydrocolloids to improve food texture, in *Texture in Food*. Vol. 1. *Semi-Solid Foods*, McKenna, B. M., Ed., CRC Press, Boca Raton, FL, 2003, pp. 251–274.
73. van der Goot, A. J. et al., Creating novel structures in food materials: the role of well-defined shear flow, *Food Biophys.*, 3, 120, 2008.
74. Windhab, E. J., Tailor-made foods: structural modification of foods, *Ann. Rev. Food Sci. Technol.*, 2, 2011 (in press).
75. Sikora, M. et al., Sauces and dressings: a review of properties and applications, *Crit. Rev. Food Sci. Nutr.*, 48, 50, 2008.
76. Khatkar, B. S., Panghal, A., and Singh, U., Applications of cereal starches in food processing, *Indian Food Indust.*, 28, 37, 2009.

77. Bylaite, E. and Meyer, A. S., Role of viscosity and hydrocolloid in flavor release from thickened food model systems, *Dev. Food Sci.*, 43, 395, 2006.

78. Koliandris, A. et al., Relationship between structure of hydrocolloid gels and solutions and flavor release, *Food Hydrocoll.*, 22, 623, 2008.

79. Clark, R., Influence of hydrocolloids on flavour release and sensory–instrumental correlations, in *Gums and Stabilizers for the Food Industry 11*, Williams, P.A. and Phillips, G. O., Eds., The Royal Society of Chemistry, Cambridge, U.K., 2002, pp. 217–225.

80. Food and Nutrition Board, *Dietary Reference Intakes for Energy, Carbohydrates, Fiber, Fat, Fatty Acids, Cholesterol, Protein, and Amino Acids*, National Academy of Sciences, Washington, D.C., 2005, chap. 7.

81. Hoebler, C. et al., Supplementation of pig diet with algal fiber changes the chemical and physicochemical characteristics of digesta, *J. Sci. Food Agric.*, 80, 1357, 2000.

82. FDA, FDA adopts soluble fiber health claim, *Fed. Reg.*, 71(98), 29248, 2006.

83. McCleary, B. V., An integrated procedure for the measurement of total dietary fibre (including resistant starch), non-digestible oligosaccharides and available carbohydrates, *Anal. Bioanal. Chem.*, 38, 291, 2007.

84. Kohajdová, Z. and Karovicová, J., Application of hydrocolloids as baking improvers, *Chem. Papers*, 63, 26, 2009.

85. Anon., *The Heart of the Batter*, IFIS Publishing, Reading, Berkshire, U.K., October, 2008.

86. BeMiller, J. N., Hydrocolloids, in *Gluten-Free Cereal Products and Beverages*, Arendt, E. and Dal Bello, F., Eds., Academic Press, Burlington, MA, 2008, pp. 203–216.

87. Anton, A. A. and Artfield, S. R., Hydrocolloids in gluten-free breads: a review, *Int. J. Food Sci. Nutr.*, 59, 11, 2008.

88. Guarda, A. et al., Different hydrocolloids as bread improvers and antistaling agents, *Food Hydrocoll.*, 18, 241, 2004.

89. Muadklay, J. and Charoenrein, S., Effects of hydrocolloids and freezing rates on freeze–thaw stability of tapioca starch gels, *Food Hydrocoll.*, 22, 1268, 2008.

90. Sundaram, J. and Durance, T. D., Influence of processing methods on mechanical and structural characteristics of vacuum microwave dried biopolymer foams, *Food Bioproducts Proc.*, 85, 264, 2007.

91. De Vuys, L. and Vaningelgem, F., Developing new polysaccharides, in *Texture in Food*. Vol. 1. *Semi-Solid Foods*, McKenna, B. M., Ed., CRC Press, Boca Raton, FL, 2003, pp. 275–320.

92. Narsimhan, G. and Wang, Z, Guidelines for processing emulsion-based foods, in *Food Emulsifiers and Their Applications*, Hasenhuettl, G. L. and Hartel, R. W., Eds., Springer, New York, 2008, pp. 349–394.

93. Rabu, B. A. and Gibson, G. R., Carbohydrates: a limit on bacterial diversity within the colon, *Biol. Rev.*, 77, 443, 2002.

94. Anderson, G. H. and Woodend, D., Effect of glycemic carbohydrates on short-term satiety and food intake, *Nutr. Rev.*, 61, S17, 2003.

95. Best, T., Kemps, E., and Bryan, J., A role for dietary saccharide intake in cognitive performance, *Nutri. Neurosci.*, 10, 113, 2007.

96. Anon., *Do Saccharides Improve Brain Power?*, IFIS Publishing, Reading, Berkshire, U.K., March, 2008.

97. Belitz, H.-D., Grosch, W., and Schieberle, P., *Food Chemistry*, 3rd ed., Springer-Verlag, Heidelberg, 2004, p. 245.

3

Crustacean Polysaccharides: Chitin and Chitosan

3.1 Introduction

The adverse environmental impact of fisheries has been a concern for regulatory agencies all over the world, prompting them to adopt measures and regulations with a view to containing these problems. These measures relate to treating effluents from aquaculture farms, dumping seafood processing discards, and fishing craft and gear hygiene, among others. Efforts to address environmental issues during the past few decades have led to the recognition that fish seafood offal can be a valuable resource, and its proper bioprocessing could yield industrially important products for a variety of applications. Some of these products include animal feed, biogas, versatile polysaccharides, pigments, collagen and gelatin, and enzymes. Success in commercial-scale recovery of at least a few of these compounds can be economically beneficial besides offering solutions to seafood-related environmental pollution.[1-6]

3.2 Crustacean Processing Wastes as Source of Chitin

Seafood processing plants in several countries process shrimp essentially for export purposes. Aquaculture of shrimp is also widespread in many developing countries. These operations generate about 35 to 45% by weight of whole shrimp as waste. Utilization of these wastes for the production of polysaccharides such as chitin and its deacetylated form, chitosan, as well as other derivatives, can be of major economic significance.[7-14] Chitin and its derivative chitosan are known for their biocompatibility, biodegradability, and nontoxicity, in addition to their antimicrobial, metal chelating, and gel-forming properties. They also offer the potential of being chemically modified to produce diverse functionally active derivatives. These polysaccharides are likely to be increasingly exploited in coming years for a variety of applications in the areas of food, medicines, cosmetics, textiles, and paper, among others.

Over the last 25 to 30 years, much research has been conducted on the biological realm of chitin and chitosan, particularly in the field of food and medicine. Some of the promising areas relate to the use of chitosan as a food additive, the development of chitosan-based edible packaging for food preservation, the use of chitan as a dietary supplement, therapeutic applications such as anticholesterolemic agents, and a variety of applications in blood coagulation, wound healing, bone regeneration, and immunoadjuvant activity. Related fields include the chemical derivatization of chitosan, combinations of these derivatives with natural and synthetic polymers, and the isolation and characterization of chitinases and chitosanases for specific uses. Developments in these areas have been summarized in several recent reviews.[8-10] This chapter discusses the isolation and characterization of chitin, chitosan, and related products from crustacean shell waste; their food uses are discussed in Chapter 6.

3.2.1 Global Availability of Crustacean Waste

The global availability of shellfish waste is enormous, generated by the catch of wild and aquacultured organisms and their centralized commercial processing; 24 to 51% of the raw material ends up as waste. On a wet weight basis, crab processing wastes can be as high as 51%, krill and shrimp processing results in 40% shell waste, and squid processing wastes are 24%. On a dry weight basis, 10% of the total processed shrimp and crab weight is shell waste, with values of 6% for squid and 15% for crab. Table 3.1 summarizes the global availability of shellfish waste.

3.2.2 Composition

The major useful components in commercial crustacean wastes are chitin, protein, minerals, and carotenoids. On a dry weight basis, about 25% of shrimp, krill, and crab wastes contain chitin, while the value is about 40% for squid pen (see Table 3.2). The shell (without the head portion) contains higher levels of chitin than the head waste. On a wet weight basis, 17% chitin, 41% protein, and 148 µg carotene per gram have been reported in shrimp head.[12-15] The most exploited sources of chitin are offal of the shrimp species *Pandalus borealis*, *Crangon crangon*, and *Penaeus monodon* and the crabs *Callinectes sapidus* and *Chionoecetes opilio*. On a dry weight basis, the chitin, protein, ash, and lipid contents of shell wastes of these shrimp species are in the range of 17 to 18%, 42 to 47%, 23 to 34%, and 1 to 5%, respectively.

The composition of shrimp shell does not vary significantly with respect to season of harvest, as shown in a recent study on the shrimp *Pandalus borealis*.[15] The annual harvest of this deepwater shrimp in Norway is estimated to average about 60000 t. They are peeled mechanically after harvest. The shell from shrimp harvested in 2000 from January to December had an average dry matter content of 22%, protein contents varying between 33 and 40%,

TABLE 3.1

Global Availability of Crustacean Waste

Resource	Total Landing (mt)[a]	Available Waste[b]	Dry Waste[c]	Chitin Content[d]
Shrimp	1,292,476	516,990	129,475	32,311
Squid	398,219	99,531	24,882	1244
Crabs	943,826	482,744	144,823	28,964
Krill	150,000[e]	60,000	15,000	3750

[a] Total landing during 2002.
[b] Assuming 40% is waste.
[c] Multiplication factor = 0.25 for calculating dry waste from wet waste.
[d] Multiplication factor = 0.25 for calculating chitin from dry waste.
[e] Average annual global krill landing.
Source: Adapted from Subasinghe, S., *Infofish Int.*, 3, 58, 1999.

and chitin levels between 17 and 20%. The shell had negligible lipid. The shell ash consisted mainly of calcium carbonate. Astaxanthin varied from 14 to 39 mg per kilogram of wet shrimp shell. This study found no clear seasonal variations for chitin, protein, and ash contents of the shell. The molecular weights and intrinsic viscosities of the extracted chitin also did not show variations.

TABLE 3.2

Chitin Contents of Selected Crustacean and Molluscan Organisms

Organism	Chitin Contents (%)
Shrimp head	11[a]
Shrimp shell	27[a]
Commercial shrimp waste	12–18[a]
Atlantic crab	26.6
Crawfish	13.2
Blue crab	14.0[a]
Crangan (shrimp)	69.0
Alaska shrimp	28.0
Nephrops (lobster)	69.8
Clam shell	6.1
Oyster shell	3.6
Squid, skeleton pen	41.0
Krill	24.0

[a] Wet body weight.
Source: Data from Arvanitoyannis and Kassaveti,[3] Synowiecki and Al-Khateeb,[16] Naczk et al.,[17] Ramachandran et al.,[21] Shahidi and Abuzaytoun,[22] Percot et al.,[23] Marquardt and Carreno,[24] Cano-Lopez et al.[28]

The chitin content in crab waste is located mostly in the legs, shoulders, and tips of the crab. On a dry weight basis, the chitin, protein, ash, and lipid contents of wastes of the crab *Callinectes sapidus* have been measured as 13.5, 25.1, 58.6, and 2.1%, respectively. Waste for the Atlantic crab *Chionoecetes opilio* was found to have a slightly higher chitin content (26.6%) and comparatively less ash (40%). The chitin, protein, ash, and lipid contents of wastes for the crawfish *Procamborus clarkia* have been measured as 13.2%, 29.8%, 46.6%, and 5.6%, respectively; for the krill *Euphausia superba*, the percentages are 24%, 41%, 23%, and 11.6%, respectively.[3,16,17,21–24]

3.3 Isolation of Chitin

The second most abundant natural polysaccharide, chitin is found in shellfish exoskeleton complexed with mineral salts. It is also present in the cell walls of fungi, insects, and marine diatoms. Chitin was first discovered in mushrooms in France by 1811 by Henri Braconnot and was later isolated from insects in the 1830s. The name "chitin" is derived from the Greek word *chiton* ("coat of mail," referring to a knight's segmented armor) and generally refers to the skeletal material of invertebrates. It has been estimated that at least 1.1×10^{13} kg of chitin are present in the biosphere. Marine shellfish, including lobster, crab, krill, cuttlefish, shrimp, and prawn, are richer in chitin compared to terrestrial organisms such as insects and fungi. In shellfish, chitin forms the outer protective coating as a covalently bound network with proteins and some metals and carotenoids. Because crustacean wastes are available in large amounts at large-scale shrimp processing facilities, it has been estimated that around 76,000 t of chitin could be available annually.[6,11]

The majority of chitin is derived from crustacean shell wastes, primarily because of their voluminous availability at a low price. In these raw materials, chitin is combined with other compounds such as proteins, pigments, and minerals. In dried and deproteinized shell waste, minerals and chitin are present in nearly equal amounts. Aggressive treatment is necessary to isolate the chitin. The isolation process consists of three steps: demineralization, deproteinization, and bleaching. Demineralization can be achieved in 1 to 3 hours using diluted (1 to 8%) hydrochloric acid at room temperature. The use of other extractants such as 90% formic acid has also been suggested. To avoid depolymerization of chitin, ethylenediaminetetraacetic acid (EDTA) is used during demineralization. Because of the significant quantity of calcium in crustacean shell, the demineralization step produces appreciable amounts of calcium chloride.[16,18,19,21,22]

The demineralized material is subjected to deproteinization by treating the shell with 4 to $5M$ sodium or potassium hydroxide at 65 to 100°C at a shell-to-alkali ratio of 1:4 for periods ranging from 1 to 6 hours. Decreasing the concentration of alkali to $1M$ requires a longer deproteinization time of 24 hours at 70°C. An increase in the shell-to-alkali ratio above 1:4 (w/v) has been found to have only a minor effect on the efficiency of deproteinization. It may be noted that prolonged alkaline digestion causes depolymerization and deacetylation of the polysaccharide (see Section 3.2.2.1). When chitin is used for chitosan production, the protein residue can be easily removed using concentrated alkaline solutions and substantial deacetylation of chitin is achieved simultaneously.

Pigment residue from chitin can be extracted at room temperature with acetone, chloroform, ethyl acetate, or a mixture of ethanol and ether. Decolonization is usually carried out using a bleaching treatment with sodium hypochlorite or hydrogen peroxide solutions. After the treatment, the chitin is washed and dried. Sun drying of the chitin can result in bleaching of the carotenoids, giving an almost colorless preparation. Alternatively, pigments can be removed by solvent extraction employing acetone or ethanol.[7,12,16,23]

The process of chitin extraction is modified to suit the raw materials being processed. To extract chitin from the shrimp *Pandalus borealis*, the wet shell was demineralized by treatment with ice-cold $0.25M$ HCl for 5 minutes at a shell-to-acid ratio of 1:6. The suspension was filtered and the same volume of fresh, cold acid was added to the residue. After 35 minutes, the suspension was filtered again and the residue was washed with water. After demineralization, the shell was subjected to deproteinization with twice the amount of hot (95°C) $1M$ NaOH for 2 hours. The suspension was then cooled to room temperature and filtered; this alkali treatment was repeated twice under the same conditions. The residue was then washed with water until the pH was neutral. The final washing of the chitin with ethanol (96%) was followed by drying at 80°C.[15,20]

The abundantly available Antarctic krill (*Euphausia superba*) is a good source of chitin; however, it is difficult to produce colorless chitin from this shellfish due to pigments from the eyes of krill which impart an intense pink color to the final product. A modified demineralization process for extracting chitin from krill involves removing a sticky substance formed by the eyes during the process that retains the pigment. The mass is usually attached to walls of the reactor and can be separated easily from the suspension of solid shell residue. Treatment of these residues with acetone followed by deproteinization yields colorless chitin.[24]

Squid pens (a waste byproduct of squid processing) are a novel source of chitin. Various parameters, such as the size of the particles, processing time, number of steps, and the concentration of sodium hydroxide, influence the molecular weight and the extent of acetylation of chitin isolated from squid (*Illex argentinus*) pens. A concentration of $1M$ NaOH has been found to be ideal for deproteinization.[25,26]

Demineralization, deproteinization, and deacetylation have been used to isolate high-quality chitin from *Squilla* shrimp caught by Indian Ocean fisheries. The chitin was found to have an ash content as low as 1% after treatment with 4% HCl for 12 hours at 50°C. A protein content of less than 1% could be achieved by treatment with 4% NaOH for 12 hours at 70°C or higher. This three-step treatment appeared to be successful in achieving a mineral content and protein content below 1% within 30 hours and at temperatures not exceeding 50°C. The chitosan obtained had a degree of deacetylation of 77 to 86%, a viscosity of 8.2 to 16.2 × 10^2 cps, solubility of 98%, and molecular weight of 1 × 10^6 Da.[27]

3.3.1 Novel Methods

The conventional harsh conditions used for extraction could adversely affect the quality of the chitin.[25] Removal of salt by demineralization with acids can result in some deacetylation of chitin. Harsh alkaline conditions for protein removal can cause depolymerization, cleavage of glycosidic linkages, and deacetylation of the chitin. Furthermore, alkali-extracted protein could be of limited use, as undesirable reactions between amino acids occur in strongly alkaline media, as well as racemization of the amino acids. Also, treating the alkali wash effluent is essential to avoid environmental pollution. To overcome these limitations, novel methods are being developed to replace conventional demineralization and deproteinization to extract chitin. Mild enzyme treatment of demineralized material with pepsin, papain, trypsin, or pronase or bacterial proteases can remove about 90% of the protein and carotenoids from shrimp waste, without affecting the quality of the chitin. The efficiency of enzymatic deproteinization depends on the source of the crustacean offal and the process conditions. For efficient enzymatic recovery of proteins, preliminary demineralization of the shell seems to be beneficial. It increases the tissue permeability for enzyme penetration and removes the minerals, which otherwise can act as enzyme inhibitors.[28,29] Deproteinization of shrimp shells by Alcalase® led to the isolation of chitin containing about 4 to 5% protein impurities and the recovery of protein hydrolyzate. The carotenoprotein produced is useful for feed supplementation.[16] The use of Alcalase® and pancreatin to extract chitin, protein, and astaxanthin from industrial shrimp waste (*Xiphopenaeus kroyeri*) resulted in 65% protein recovery in the form of hydrolysates, in addition to providing suitable conditions for the recovery of astaxanthin and chitin.[30] Figure 3.1 illustrates the process of chitin recovery from shrimp wastes by enzyme hydrolysis.

In another novel method, chitin was concurrently extracted from fungi and shrimp shell. When *Aspergillus niger* and shrimp shell powder were combined in a single reactor, the release of protease by the fungi facilitated the deproteinization of the shrimp shell powder and the release of hydrolyzed proteins. The hydrolyzed proteins in turn were utilized as a nitrogen source for fungal growth, leading to a lowering of the pH of the fermentation medium, thereby further enhancing the demineralization of the shrimp shell powder. The shrimp shell powders and fungal mycelia were separated after fermentation and used for chitin extraction. Chitin isolates from the

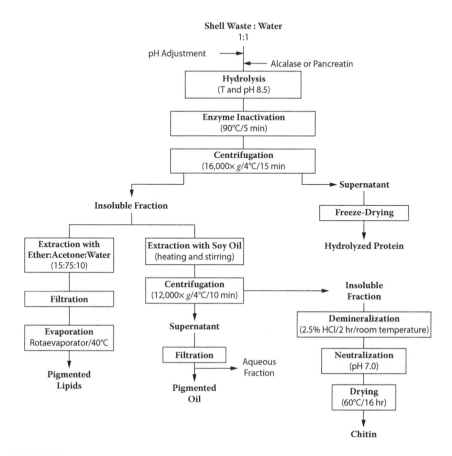

FIGURE 3.1
Chitin and protein extraction. (From Holanda, H.D. and Metto, F.M., *J. Food Sci.*, 71, C298, 2006. With permission.)

shells had protein contents as low as 5%; chitin isolates from the fungal mycelia strains had higher protein levels, in the range of 10 to 15%. The average molecular weight of the chitin samples was 10^5 Da.[31]

The production of chitin and chitosan via fermentation of shrimp shell in jaggery broth using *Bacillus subtilis* has been reported. The protease secreted by the microorganism was responsible for the deproteinization of the shell. The organism also produced sufficient quantities of acid to remove the minerals from the shell and prevent spoilage organisms. About 84% of the protein and 72% of the minerals were removed from the shrimp shell after fermentation. Mild acid and alkali treatments were used after fermentation to improve the quality and appearance of the crude chitin.[32]

Ultrasonication was used to extract chitin from the shells of the North Atlantic shrimp *Pandalus borealis*. Shells were peeled, washed, lyophilized, ground, and demineralized by treating them with 0.25*M* HCl for 4 hours at a shell-to-acid ratio of 1:40 at 40°C. The sample was then subjected to ultra-sonication at 41 W/cm^2 for 1 to 4 hours. The yield of chitin from the shrimp decreased from 16.5 to 11.4% when ultrasonication was conducted for 1 hour. Ultrasonication did not enhance the removal of minerals but did enhance the removal of proteins. The degree of chitin acetylation was unaffected by ultrasonication. The crystallinity of chitin was slightly affected by the treatment.[33]

The mycelia of various marine fungi are also suitable for chitin isolation. Fungal chitin content depends on the species, with *Aspergillus niger* having the maximum chitin content (42%) on dry weight basis. Apart from chitin, cell walls of mold mycelia also contain significant quantities of chitosan and various acidic polysaccharides. Currently, fungi are not generally used as a source of these polysaccharides, although obtaining chitin and chitosan from fungi is advantageous because they are readily available.[34]

3.3.2 Structure

Chitin is a cationic polysaccharide having units of N-acetyl-D-glucosamine, joined by β-(1,4) linkages. Its structure is β-(1,4)-N-acetyl-D-glucosamine, or β-(1,4)-N-acetyl-2-amino-2-deoxy-D-glucose. It may be also regarded as a derivative of cellulose, in which, the C-2 hydroxyl group is substituted by an acetyl amino group. Chitin occurs in three polymorphic forms, α, β, and γ, which differ in the arrangement of their molecular chains. By far the most abundant form is α-chitin, which is present in fungal and yeast cell walls, insect cuticles, egg shells of nematodes and rotifers, the radulae of mollusks, and cuticles of orthropods. This form of chitin is also present in krill, lobster, crab tendons and shells, and shrimp shells, as well as in other marine organisms such as the harpoons of cone snails and the filaments ejected by *Phacocystis* seaweed. The rarer β-chitin is found in association with proteins in squid pens and the diatom *Thalassiosira fluviatilis*.

After biosynthesis, the chitin molecules associate with one another by hydrogen bonds between >NH groups of the one molecule and >C=O groups of the adjacent chain. The extent of formation of these hydrogen bonds accounts for the structural differences of chitinous microfibrils in α- and β-chitin. Crystallographic studies of α- and β-chitin reveal that α-chitin consists of two N,N-diacetylchitobiose units in an antiparallel arrangement, whereas in β-chitin the two N,N-diacetylchitobiose units are parallel. The final structure has extensive intermolecular hydrogen bonding, with the exclusion of water, leading to great stability. Because it has a lower content of intersheet hydrogen bonds than β-chitin, α-chitin swells readily in water. Structural differences in α- and β-chitin also influence the viscosity of chitosan prepared from the chitins. Compared with that of shrimp, the crystal-line structure of squid chitin is easily destroyed by deacetylation; moreover, the resulting squid chitosan is amorphous, again unlike crystalline shrimp chitosan.[26] Transforming β-chitin into α-chitin is achieved through treatment with strong aqueous HCl (>7 M) and washing with water. The β-chitin is more reactive than α-chitin, an important property with regard to enzymatic and chemical transformations of chitin. A third form of chitin, γ-chitin, in which two of the three chains are parallel and the third antiparallel (↑↑↓) has been reported in the past; however, its existence appears to be controversial today.[9,16,35]

3.3.3 Properties

Chitin is a very light, powdery, flaky product that is white or yellowish in color. Natural chitin demonstrates a variable degree of crystallinity, depending on its structure, and varying amounts of deacetylation and cross-linking with other molecules. A low level of deacetylation (with a low content of glucosamine units) has generally been noted in natural chitin, although it varies by source. Chain lengths and the degree of acetylation differ depending on isolation conditions and sources of the chitin. Due to the high density of hydrogen bonds in its structure, α-chitin from shellfish is insoluble in water, weakly acidic aqueous media, and almost all common organic solvents. This insolubility is a major problem confronting the development of chitin applications. In cold alkaline media, chitin can swell when deacetylation occurs.[9,16,36]

Chitin becomes soluble in acetic acid and water when deacetylated (about 28% and 49%, respectively). The solubility of the partially deacetylated chitins has a close relationship with their crystal structure as well as glucosamine content. A 28% deacetylated chitin, for example, retains the crystal structure of α-chitin; chitin with 49% deacetylation has a crystal structure similar to that of β-chitin rather than α-chitin or chitosan, suggesting that the homogeneous deacetylation transformed the crystal structure of chitin from α-chitin to β-chitin. Some alterations in the hydrogen bonds also occur.[37] The solubility of chitin can be improved by gamma irradiation at doses above 10 kGy, which causes depolymerization of the material. Chitin from squid is characterized by its remarkable affinity for water and some organic solvents; this high affinity for water suggests its use as a highly hydrophilic material.[26,36]

The nitrogen and ash contents, degree of deacetylation, bulk density, and viscosity of commercial chitin and chitosan products differ. Water-binding capacities range from 281 to 673% for chitins and from 458 to 805% for chitosans. Dye-binding capacities vary, depending on the product, although the average binding capacity of chitosans (63%) was higher than that of chitins (54%). Fat-binding capacities of chitins are relatively consistent, at around 320%. Significant correlations were observed between water-binding capacity and the bulk density of chitin products. Both water- and fat-binding capacities of chitosan products were significantly correlated positively with ash and negatively with bulk density.[37]

3.4 Chitosan

Chitosan, a family of deacetylated chitins, was discovered by C. Rouget in 1859. Chitosan is obtained by partial deacetylation of chitin by chemical or enzymatic methods. It has been demonstrated that β-chitin exhibits much higher reactivity upon deacetylation than α-chitin. A minimum deacetylation of 70% is required to produce chitosan that is acceptable for various purposes. Interest in uses for chitosan peaked in the 1930s and early 1940s but grew again beginning in the 1970s due to the need to better utilize shellfish shells.[6]

3.4.1 Isolation

Chitosan is produced by deacetylation of chitin with alkali. Generally, treatment with 30 to 60% (w/v) NaOH or KOH at 80 to 140°C is employed, and NaOH is the preferred alkali. Characteristics of the final product in terms of molecular weight and extent of deacetylation depend on the treatment conditions. If chitin is extracted with NaOH powder at 180°C, highly deacetylated chitosan is obtained. Increasing the temperature enhances deacetylation but results in fragmentation of the chitosan, affecting its final uses. Deacetylation of chitin with equal amounts of aqueous 40% NaOH for 2 hours at 100°C has been found to give the product good viscosity, an important quality criterion. After deacetylation, the chitosan is washed well to remove alkali and is dried to give flakes. The crude product may be further purified by dissolving it in dilute acetic acid, followed by reprecipitation with alkali and washing and drying. It may be noted that decolorization treatment during this process can result in a significant loss of quality in terms of viscosity, probably due to changes in molecular weight. Decolorization, however, may not always be essential for an acceptable chitosan product.[5,6] Crab chitosan was recently prepared by deacetylation of crab chitin for 60 to 120 minutes with a yield in the range of 30 to 32%. With prolonged reaction time, the degree of deacetylation increased from 83 to 93%, but the average molecular weight decreased from 483 to 526 kDa.[36,39] Chitosan from squid pen has also been prepared by repeated alkaline treatments under mild conditions. It is colorless and has a low ash content.[40] As noted earlier, chitosan and chitin have been isolated from *Squilla* shrimp caught by Indian Ocean fisheries employing demineralization, deproteinization, and deacetylation. The chitosan obtained had 77 to 86% deacetylation.[27] Chitosan was prepared recently from the shells of brine shrimp (*Artemia urmiana*) that contained 29 to 35% chitin on a dry weight basis.[41]

3.4.2 Process Modifications

Harsh deacetylation using NaOH affects chitosan properties, such as molecular weight, degree of deacetylation, viscosity, and reactive terminal groups. An alternative is environmentally friendly deacetylation using chitin deacetylase enzymes from fungi such as *Mucor rouxii*, *M. mechei*, and *Aspergillus niger*. Enzyme-treated chitosan also has better functional properties. Under dry conditions, fungal chitin deacetylases are able to perform only limited heterogeneous deacetylation; however, deacetylation can be enhanced by dissolving chitin in specific solvents followed by fast precipitation to produce smaller crystals. The crystallized chitin, after pretreatment with 18% formic acid, is amenable to 90% deacetylation by the fungal deacetylase.[38]

3.4.3 Properties of Chitosan

Chitosan is a white, nontoxic, biodegradable solid that is insoluble in pure water; unlike chitin, it is soluble in weakly acidic aqueous media (<pH 6.0). This is attributed to its semicrystalline nature, derived mainly from inter- and intramolecular hydrogen bonds. In acidic conditions, its amino groups can be partially protonated, resulting in repulsion between positively charged chains and allowing diffusion of water molecules and subsequent solvation of macromolecules. Chitosan is a polycationic, long-chain biopolymer with a natural affinity for negatively charged molecules. Its cationic nature is unique relative to other neutral or negatively charged polysaccharides. In an acid environment, the amino group NH_2 in chitosan can be protonated to yield NH_4^+, which yields antifungal and antimicrobial activities, as cations can bind to anionic sites on bacterial and fungal cell wall surfaces. The binding capacities of chitosan with water, fat, and different dyes varies depending on the sources. The solubility of chitosan is usually tested in 1% acetic acid, and the viscosity of 1% chitosan in 1% acetic acid is usually taken to determine the quality of the product. Viscosity increases with increase in molecular weight and concentration—the higher the viscosity, the better the product. Significant correlations have been observed between molecular weight and viscosity and between nitrogen and degree of deacetylation. Fat-binding capacity shows a significant correlation with viscosity.[54]

Properties of chitosan vary with respect to source. Squid chitosan, because of its specific structural properties, has a superior water-retention capacity compared to shrimp chitosan; hence, it could be useful as a highly hydrophilic material. It has also significant thickening and suspending properties. The flocculation capacity of squid chitosan could be further improved by decreasing the degree of acetylation. These properties make squid chitosan highly suitable for medical and analytical applications.[36,40]

The flocculation capacity of squid chitosan, which is low in comparison with its crustacean counterparts, could be enhanced by decreasing the degree of acetylation.[40] Chitosan obtained from *Squilla* has been found to have a degree of deacetylation of 77 to 86%, a viscosity of 820 to 1620 cps, solubility of 98%, and molecular weight of 1×10^6 Da.[27] *Squilla* species can give better chitosan as compared to that from shrimp shell. A 1% solution of *Squilla* chitosan in 1% acetic acid has a viscosity of 340 cps vs. a value of 180 to 200

TABLE 3.3

Physicochemical Properties of Chitosan Prepared from Crawfish Shell

Chitosan Type	Molecular Weight (kDa)	Deacetylation (%)	Viscosity (cps)
DPMCA	454	86.7	35
DPMA	1462	86.1	1164
DMCA	950	84.6	259
DMA	1054	84.2	1054

Abbreviations: DPMCA, deproteinized, demineralized, decolorized, and deacetylated; DPMA, deproteinized, demineralized, and deacetylated; DMCA, deproteinized, decolorized, and deacetylated; DMA, demineralized and deacetylated.

Source: Adapted from: Natarajah, N. et al., *J. Food Sci.*, 71, E33, 2006. With permission from Wiley InterScience.

for the shrimp product.[21] Deproteinized, demineralized, and deacetylated (DMCA) and deproteinized, decolorized, and deacetylated (DPCA) crawfish chitosans with molecular weights of 1462 and 859 kDa have viscosities as high as 1000 cps, as shown in Table 3.3.

Recent studies on chitosans isolated from different crab species have revealed some of their unique features using element analysis, differential scanning calorimetry, scanning electron microscopy, and x-ray diffraction patterns. Crab chitosan has a higher molecular weight and a melting temperature of 152 to 159°C.[39] Crab chitosan harvested from 2004 to 2007 had protein in the range of 22 to 27%; chitin, 17 to 20%; and ash, 49 to 51%. In one study,[40] the physicochemical characteristics of chitosans were found to differ somewhat depending on the year of harvest. The degree of deacetylation ranged from 82 to 89%. The highest water-binding capacity (555%), dye-binding capacity (65.5%), and 2,2-diphenyl-1-picrylhydrazyl (DPPH) radical scavenging activity (18.1%) were observed with chitosan from the 2004 sample, but this chitosan had the lowest viscosity (200 mPa·s).

Brine shrimp (*Artemia*) chitosan has a molecular weight varying from 4.5 to 5.7×10^5 Da, deacetylation ranging from 67 to 74%, and a viscosity range of 29 to 91 cps. The physicochemical characteristics (e.g., ash, nitrogen, and molecular weight) and functional properties (e.g., water-binding capacity and antibacterial activity) of *Artemia* chitosan depend on the extraction procedure.[41]

3.4.4 Structure

Chitosan (deacetylated chitin) is a polysaccharide composed of β-(1,4)-linked D-glucosamine and N-acetyl-D-glucosamine, or β-(1,4)-linked 2-amino-2-deoxy-D-glucose. It is very similar to cellulose, with the only difference

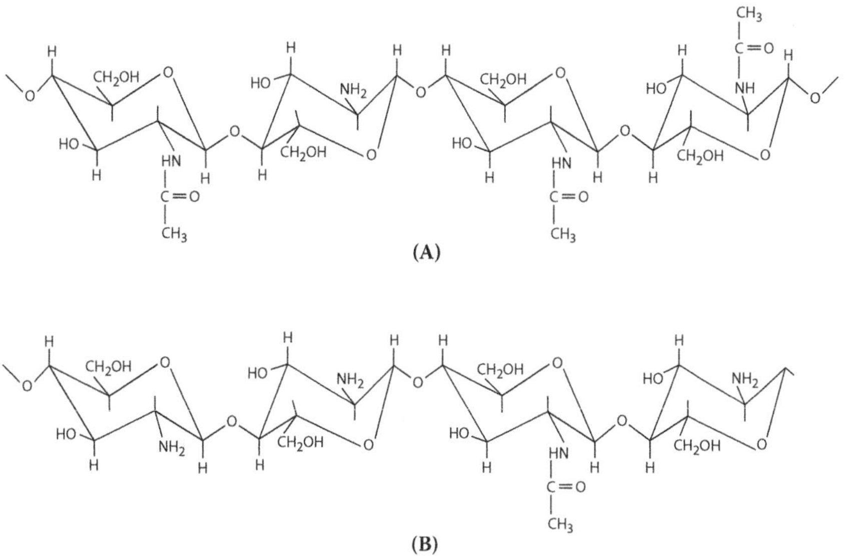

FIGURE 3.2
Structures of (A) chitin and (B) chitosan.

between chitosan and cellulose being that the amine ($-NH_2$) group is in the C-2 position of chitosan and the hydroxyl ($-OH$) group is in that position in cellulose. Chitosan has a free amino group and two free hydroxyl groups for each glucose ring. Chitosan is a primary aliphatic amine that can be protonated by selected acids.[8] Figure 3.2 compares the structures of chitosan and chitin.

3.4.4.1 Ionic Properties

The amino groups of chitosan have a pK_a value of approximately 6.5, giving the molecule an overall positive charge and solubility in acidic conditions. This also gives chitosan the ability to chemically bind with negatively charged fats, lipids, cholesterol, metal ions, proteins, and macromolecules. Because of its cationic nature, chitosan is compatible in solution with most anionic water-soluble gums such as alginates, pectate, sulfated carrageenan, and carboxymethylcellulose. Chitosan acid solutions are also compatible with nonionic water-soluble gums such as starch, dextrin, glucose, polyhydric alcohols, oils, fats, and nonionic emulsifiers. Due to these properties it interacts with various biomolecules. These ionic properties offer the potential for commercial applications of chitosan in the areas of waste treatment, food preservation (see Chapter 6), nutraceutical and drug delivery (see Chapter 10), and cosmetics.[9,22,43–45]

3.4.4.2 Degree of Deacetylation

Commercial chitosans vary in their degree of deacetylation, usually having a minimum deacetylation of 70%, and molecular weights ranging from 100 to 1000 kDa. The electrostatic and solution properties of chitosan are influenced by the degree of acetylation, average molecular weight, and distribution of acetyl groups along the main chain.[46] The degree of acetylation of chitosan can be determined by: (1) spectroscopy (infrared, ultraviolet, or ^1H, ^{13}C, ^{15}N nuclear magnetic resonance); (2) conventional methods (various types of titration, conductometry, potentiometry, ninhydrin assay, adsorption of free amino groups of chitosan by picric acid); and (3) destructive methods (elemental analysis or acid or enzymatic hydrolysis of chitin or chitosan) followed by colorimetric methods or high-performance liquid chromatography, pyrolysis gas chromatography, and thermal analysis using differential scanning calorimetry.[47] A picric acid-based colorimetric method is also available to determine the degree of acetylation in chitin and chitosan.[48]

3.4.4.3 Stability

Chitosan is biodegradable by the enzymes chitosanase and lysozyme. The nonspecific activity of some digestive enzymes such as amylases and lipases can also lead to digestion of the chitosan molecule, giving rise to soluble oligosaccharides.[49] Chitosan, depending on its structural features, may lose functionality during the course of storage. In addition to the degree of acetylation, temperature (5 to 60°C) and acid concentration were found to also affect the stability of chitosan with a degree of deacetylation of 81 or 88% suspended in acid solution. At 28°C, the rates of hydrolysis for both chitosan samples were four to five times lower than those at 60°C. At 5°C, chain degradation was not significant. The first-order rate constant of chain hydrolysis of 88% deacetylated chitosan at 60°C was about 1.4 times higher than that of the 81% sample. Acetic acid caused significantly higher chain scission than formic acid.[50]

Exposure of chitosan to steam greatly darkens the color of chitosan powder. Similar to heat, gamma irradiation strongly depolymerizes chitosan, especially at irradiation doses above 10 kGy. Exposure to an electron beam produces minor changes in the crystalline and network structures.[51]

Physicochemical and functional properties of two chitosan products having low (16.8 mPa·s) and high (369.4 mPa·s) viscosities were evaluated during 9 months of storage at room temperature. As storage time increased, increased moisture content and DPPH radical scavenging activity and decreased viscosity and water-binding capacity were observed. The effect was more pronounced with high-viscosity chitosan. Significant correlations were observed between water-binding capacity and viscosity and between DPPH radical scavenging activity and viscosity. Although significant differences in $L^*a^*b^*$ values were observed during storage, color differences were not easily discerned visually.[52]

3.4.4.4 Emulsification Capacity

Chitosan is able to enhance the emulsification capacities of certain proteins; for example, a study on the influence of chitosan content (0 to 0.5%) on particle size distribution, creaming stability, apparent viscosity, and microstructure of oil-in-water emulsions (40% rapeseed oil) containing 4% whey protein isolate (WPI) at pH 3 revealed that the WPI–chitosan mixture had a slightly higher emulsifying activity than whey protein had alone. An increase in chitosan content resulted in a decreased average particle size, higher viscosity, and increased creaming stability of emulsions. The microstructure analysis indicated that increasing the concentration of chitosan resulted in the formation of a flocculated droplet network.[53,54] For most purposes, chitosan is usually dissolved in aqueous solutions of acetic acid; however, the presence of the acid alone is insufficient to allow the use of chitosan solutions as food additives, especially as emulsifiers. It has been suggested that chitosan could be emulsified in an acid-free aqueous medium using nonionic surfactants (see Chapter 6).[55,56]

3.4.4.5 Derivatives of Chitin and Chitosan

In view of the interesting properties of chitin and chitosan, considerable attention has been focused on developing diverse chemical derivatives of these polysaccharides for a variety of potential applications. The advantage of chitosan over other polysaccharides (e.g., cellulose, starch, galactomannans) is that its chemical structure allows specific modifications at the C-2 position without too much difficulty. These reactions included hydrolysis of the main polysaccharide chain, deacetylation, acylation, tosylation, alkylation, Schiff base formation, reductive alkylation, O-carboxymethylation, N-carboxyalkylation, silylation, and graft copolymerization, among others. The functional advantages of such materials are their improved biodegradable nature, antibacterial activities, and hydrophilic character introduced by the addition of polar groups able to form secondary interactions (i.e., –OH and $-NH_2$ groups involved in hydrogen bonds with other polymers).

Because chitin, unlike chitosan, is inert and water insoluble, many of the approaches have aimed at modifications of chitin structure. The β-chitin is used for most of these modifications, as it has a higher chemical reactivity than the α-polymorphic form due to the presence of weak intramolecular bonds. These efforts have led the preparation of various chitin derivatives, such as carboxymethyl chitin, hydroxyethyl chitin, ethyl chitin, chitin sulfate, glycol chitin, and glucosylated chitin. Other products have been developed through modification of hydroxyl groups employing alkyl and acyl halides or isocyanates to yield ethers, esters, or carbamate derivatives. These compounds are generally more polar and consequently more soluble than native chitin. Long residues introduced into chitin molecules through acylation have yielded soluble products.

Carboxymethyl chitosan (CM-chitosan) is the most well-studied derivative of chitosan. It is an amphoteric polymer whose water solubility depends on pH. It is obtained by treating chitin initially soaked in 40% aqueous sodium hydroxide with monochloroacetic acid at 0 to 15°C; the low temperature controls deacetylation and degradation of the polysaccharide. CM-chitosan and other chitosan derivatives in the form of acetate, ascorbate, lactate, and malate are also water soluble.[44,57–59] Water-soluble chitosan produced using the Maillard reaction may be a promising commercial substitute for acid-soluble chitosan.[59] A novel fiber-reactive chitosan derivative was synthesized from chitosan with a low degree of acetylation. The process consisted of the initial preparation of a water-soluble chitosan derivative, N-[(2-hydroxy-3-trimethylammonium) propyl] chitosan chloride (HTCC), by introducing quaternary ammonium salt groups on the amino groups of chitosan. This derivative was then modified by introducing functional (acrylamidomethyl) groups on the primary alcohol groups (C-6) of the chitosan backbone. The chitosan derivative showed significant inhibition of *Staphylococcus aureus* and *Escherichia coli*.[60]

Hydrophobic derivatives of chitosan have been prepared by esterification reactions with acyl chlorides through reactions on hydroxyl and amine groups of polysaccharides by incorporation of long alkyl chains.[18] Three different acyl thiourea derivatives of chitosan were synthesized that had antimicrobial activities against the bacteria *Escherichia coli, Pseudomonas aeruginosa, Staphylococcus aureus*, and *Sarcina* and four crop-threatening pathogenic fungi, including *Alternaria* and *Fusarium* species. All of the acyl thiourea derivatives had a significant inhibitory effect on the fungi at concentrations as low as 50 µg per mL.[61] Phosphorylated derivatives were produced by the reaction of chitin and chitosan with phosphorus pentoxide in methanesulfonic acid. The degree of substitution of the products increased with increases in the amount of phosphorus pentoxide (P_2O_5) employed. Chitin phosphates are easily soluble in water and behave as typical polyelectrolytes in terms of viscosity, and they have good metal-binding ability.[62]

When N-sulfated chitosans with varying degrees of acetylation (0.04, 0.10, 0.22) were prepared under a variety of reaction conditions, the prepared derivatives differed in the degree of sulfation. All of the compounds were soluble in water, but their rheological properties varied markedly based on sulfation. Both ionic strength and pH had an effect on their solubility properties and on interactions they exhibited with carboxymethylcellulose, xanthan gum, and heparin.[44,57,63–66] Figure 3.3 shows various chitosan derivatives.

3.4.5 Chitin and Chitosan Oligosaccharides

Controlled acid or enzymatic hydrolysis of chitin or chitosan produces chitin oligosaccharides. Degradation of chitin can be accomplished enzymatically using chitinase. Low-molecular-weight chitosan (LMC) can be prepared

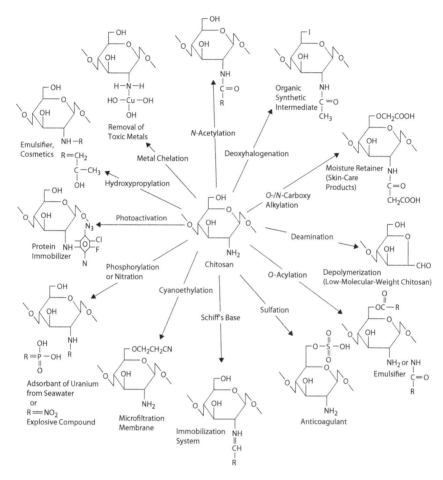

FIGURE 3.3
Derivatives of chitosan. (From Tharanathan, R.N. and Kuttur, F.S., *Crit. Rev. Food Sci. Nutr.*, 43, 61, 2003. With permission from Taylor & Francis, Ltd.)

enzymatically by the depolymerization of chitosan with *Aspergillus niger* pectinase (polygalacturonase) at pH 3.0 and 37°C. The compounds have molecular weights in the range of 5000 to 20,000 Da. Drying of the product at elevated temperatures results in a loss of water solubility, presumably due to changes in chain conformation.[67] Crude enzyme from *Bacillus cereus* NTU-FC-4 was used to hydrolyze chitosan with 66% deacetylation in a membrane reactor operated at 45°C and pH 5 to produce chitooligosaccharides. Major oligomers created in the reactor were chitobiose, chitotriose, chitotetraose, chitopentaose, and chitohexaose. When the membrane reactor was operated at an enzyme-to-substrate ratio of 0.2 (unit/mg) and residence time of 100 minutes, it reached steady state in 2.5 hours. The system could be operated for 15 hours to provide a stable product composition.[68]

A novel method using hydrogen peroxide at low concentrations has been used to partially degrade deacetylated chitin and chitosan from shrimp and squid. The treatment resulted in breaking the β-(1,4) glycosidic linkages. The formation of glucosamine and chitooligosaccharides depended on the concentration of H_2O_2, temperature, and physicochemical properties of chitin and chitosan substrates and was in accordance with first-order kinetics. Degradation rates were faster than those from ultrasonic degradation and were comparable to enzymic hydrolysis of chitosan.[69]

An alternative to chemical and enzymatic treatment is ozonation. Ozone has been shown to be able to degrade macromolecules and remove pigments due to its high oxidation potential. Crawfish chitosan was treated with 12% ozone in water and acetic acid solution for durations extending from 0 to 20 minutes at room temperature. Ozonation for 20 minutes reduced the molecular weight of the chitosan by 92% (104 kDa) compared to the untreated chitosan (1333 kDa), with a decrease in viscosity of the chitosan solution. Ozonation for 5 minutes markedly increased the whiteness of chitosan with a molecular weight of 432 kDa; however, further ozonation resulted in some development of yellowness. In the case of ozonation in water, no significant differences were found among ozone-treated chitosans with regard to molecular weight and color.[70]

Not only are oligosaccharides water soluble but their solutions also have low viscosity. The oligomers with high degrees of polymerization, especially those with six residues or more, exhibit strong physiological activities. Chitooligosaccharides, therefore, are of interest for nutritional and food preservative applications; however, unlike chitin and chitosan, the oligosaccharides may be absorbed in the human intestine so there is a need to evaluate their safety.

3.4.6 Glucosamine

Glucosamine is the end product of the hydrolysis of chitosan. If it is produced from chitin, it has to be first deacetylated to give chitosan; however, treatment of chitin with 10M HCl under vacuum for 10 minutes followed by heat treatment at 140°C for 60 minutes can directly give glucosamine. The preparation and properties of glucosamine produced from shrimp shell have been reported.[58,71] It may be noted that glucosamine is a natural amino sugar found in large concentrations in certain foods such as milk, eggs, liver, yeast, and molasses. Glucosamine is absorbed easily by the human intestine. In the human body, it is synthesized from L-glutamine and glucose.

3.4.7 Chitosan-Based Materials

Chitosan is used to prepare hydrogels, films, fibers, or sponges. Most of these materials are used in the biomedical domain, where biocompatibility is essential. Chitosan is much easier to process than chitin, but the stability of chitosan materials is generally lower, due to their being more hydrophilic and pH sensitive. Various techniques are used to control both their mechanical and chemical properties. A commonly used technique employs other polymers such as starches and specific cross-linking agents such as epichlorohydrin and diisocyanate to cross-link chitosan. Some important products are briefly mentioned below.

3.4.7.1 Composite Gels

Chitosan is a non-gellable polysaccharide; however, composite hydrogels of chitosan together with other gel-forming materials can be prepared. Many chitosan hydrogels are obtained by treatment with multivalent anions such as glycerol phosphate and tripolyphosphate, as discussed below. Hydrogels are three-dimensional and hydrophilic polymer networks capable of swelling in water or biological fluids and retaining a large amount of fluids in the swollen state. The water content in the swelling equilibrium affects different properties of the hydrogels—namely, permeability, mechanical properties, surface properties, and biocompatibility. The utility of hydrogels as biomaterials lies in the similarity of their physical properties to those of living tissues. Their unique properties of biocompatibility, biodegradability, bioadhesion, and tissue-promoting ability make them suitable for a wide variety of food and pharmaceutical applications.

The rheological behavior of these gels depends on the concentrations of the polymers and conditions of polyelectrolyte complex formation. The blending of these polymers in a solid state results in dispersion of polymer particles, amorphization, and depolymerization, as well as chemical interactions resulting in the formation of branched and cross-linked structures.[72] Composite hydrogels of chitosan in combination with many natural polymers such as alginate and gelatin have been prepared. Depending on the concentration, the chitosan–alginate solutions demonstrated unexpected rheological properties, including a dynamic storage modulus (G') and loss modulus (G'').[73] A novel injectable *in situ*, gelling, thermosensitive hydrogel based on chitosan and gelatin blends has recently been designed that could be very effective for several applications, including industrial wastewater treatment and antibacterial cosmetic preparations. The gels are heterogeneous and porous. Gelation provided buffering and other physicochemical conditions, including control of hydrophobic interactions and hydrogen bonding, which are necessary to retain chitosan in solution at neutral pH near 4°C. Because the gelation occurs at body temperature, the process holds promise for the sustained delivery of protein drugs.[74]

Transglutaminase catalyzes the formation of covalent cross-links among proteins. A microbial transglutaminase was observed to catalyze the formation of strong and permanent gels from gelatin solutions. In the presence of chitosan, gel formation was faster, and the resulting gels were stronger. The transglutaminase-catalyzed gelatin–chitosan gels do not have the ability to undergo thermally reversible transitions (i.e., sol–gel transitions) characteristic of gelatin. Mushroom tyrosinase was also observed to catalyze gel formation for gelatin–chitosan blends. The strength of both transglutaminase-catalyzed and tyrosinase-catalyzed gels could be adjusted by altering the gelatin and chitosan compositions.[75]

Chitosan and also other non-gellable polysaccharides, including hyaluronate and cyclodextrin, can also be gelled in the presence of a novel silica precursor that is completely water soluble and compatible with biopolymers. The gelation was produced by the mineralization of carbohydrate macromolecules which strengthens them and provides the cross-linking.[76]

Polyvinyl alcohol (PVA)–water-soluble chitosan hydrogels were prepared by a combination of gamma irradiation and freeze–thawing. Irradiation reduced the crystallinity of PVA, whereas freeze–thawing increased it. Hydrogels made by freeze–thawing followed by irradiation had higher degrees of crystallinity and higher melting temperatures than those made by irradiation followed by freeze–thawing. All of the hydrogels showed shear-thinning behavior in the frequency range of 0.2 to 100 rad/sec. Hydrogels made by freeze–thawing dissolved into sol solutions at about 80°C, whereas those made by irradiation showed no temperature dependence up to 100°C. The chemical cross-linking density of the hydrogels made by irradiation followed by freeze–thawing was much greater than that of hydrogels made by freeze–thawing followed by irradiation.[51]

Chitosan hydrogels were prepared from a water-soluble methacrylic acid (MA) derivative of chitosan by photo-initiating polymerization. The chitosan derivative was obtained by amidation of the amine groups of chitosan with lactic acid and methacrylic acid. The gelation time of the hydrogel was adjusted within a range of 5 to 50 minutes and controlled by factors such as the degree of MA substitution, initiator concentration, existence of oxygen, and salt. The dry hydrogel adsorbed tens to hundred times quantities of water, forming a highly hydrated gel. Rheological studies showed that the hydrogel is viscoelastic, with a storage modulus and loss modulus of 0.8 to 7 kPa and 10 to 100 Pa, respectively. The chitosan hydrogels have better biocompatibility and are more suitable for biomedical applications. Blending under the joint action of high pressure and shear deformation can be advantageous compared to the conventional techniques of polysaccharide mixture production.[72,77]

A water-soluble form of chitosan at neutral pH and ambient temperature was obtained in the presence of glycerol-2-phosphate. Heating at 37°C (body temperature) resulted in the formation of a gel. When 0.5 to 2% (w/v) chitosan was mixed with 5 to 20% (w/v) β-glycerol phosphate (GP) solutions, the product gelled at 37°C. High GP and chitosan concentrations resulted in faster gelation. The chitosan–GP hydrogel extracts can stimulate mesenchymal stem cell proliferation at certain concentrations. The sol–gel transition was partially reversible, and the gelation temperature depended slightly on experimental conditions. This material is a promising vehicle for cell encapsulation and injectable tissue-engineering applications (see Chapter 9).[78,79]

3.4.7.2 Microcrystalline Chitosan

Chitosan exists in amorphous, semicrystalline, or crystalline forms, depending on the degree of deacetylation. It crystallizes in a number of different forms depending on the experimental conditions. In most of these cases, chitosan adopts a twofold-screw, cellulose-like, helical conformation with a fiber repeat ranging from 10.1 to 10.5 A. During crystallization, chitosan molecules imbibe small molecules of water and inorganic acids. Microcrystalline chitosan is prepared from low-molecular-weight chitosan with a degree of polymerization of 35. It is a special multifunctional polymeric material that offers several advantages when compared to standard chitosan, including its ability to form hydrogen bonds, high intrinsic surface and water retention value, dispersion stability, crystallinity up to 95%, biodegradability, bioactivity, film-forming behavior, and high reactivity and chelating properties. The product is manufactured by the aggregation of glucosamine macromolecules from solution. It has a molecular weight of 10 to 300 kDa. It may be in a gelatinous aqueous dispersion or powder form. The former is characterized by an average particle dimension of 0.1 to 100 μm and water retention values ranging from 500 to 5000%. The powder form, on the other hand, has a polymer content of 85 to 95, particle dimension of 0.1 to 50 μm, and lower water retention values ranging from 200 to 800. Microcrystalline chitosan is a unique multifunctional polymeric material with numerous potential applications. Microcrystalline chitosan films, for example, stick on every type of surface, including glass.[80]

3.4.7.3 Beads

Chitosan beads have been prepared from crab and squid chitosan. Chitosan from crab should preferably have a degree of acetylation of 8% and molecular weight of 7×10^5 Da, whereas squid chitosan should ideally have a degree of acetylation of 5% and molecular weight of 2×10^5 Da. The aqueous solutions of these chitosans are obtained by dissolving 2.5 g of the powder in 100 mL of a solution of 5.5M acetic acid. The solution is added dropwise into 4M NaOH using a 0.8-mm syringe. The chitosan beads formed are stored in the alkaline solution for 2 hours and then filtered off and washed with water.[81]

Composite spherical chitosan beads (1 to 2 mm in diameter) have been developed that offer biocatalytic properties, pH stability, and biocompatibility. They are prepared by ionic gelation using sodium tripolyphosphate (TPP). The mechanical strength of the chitosan–TPP beads is improved by the addition of clay or cassava starch granules. The chitosan–starch hydrogel beads are significantly firmer compared to chitosan–clay beads. Swelling studies show that the particles expand at pH 1.2 and contract at pH 7.4. When immobilization of fungal (*Aspergillus*) protease was accomplished using glutaraldehyde cross-linking, beads immobilized with the enzyme retained activity as high as 70% even after eight cycles of repeat use. Above 3% TPP, the activity of the enzyme decreased. Chitosan–starch hydrogel beads exhibited degradation peaks at about 90 to 110°C in thermogravimetric analysis. The freeze-dried beads has good storage stability and can be used either as artificial bioreactor systems in detergent or in therapeutic formulations.[82]

3.4.7.4 Films

Chitin and chitosan can form films that are tough, flexible, and transparent. Films can be extruded from acidic solutions of chitosan into a 70°C coagulating bath containing caustic soda and sulfonic acid esters of high-molecular-weight alcohols. Films made with squid chitosan were more elastic than crustacean chitosan with improved functional properties.[40] Films of N-sulfofurfuryl chitosan with negative charges on their surface exhibited absorption of both negatively charged proteins (albumin and fibrinogen) and positively charged ones (ribonuclease, lysozyme). The quantity of the adsorbed protein tended to increase as a function of the swelling ratio of the positively charged film.[83]

3.4.7.5 Sponges

Preparation of chitosan sponge involves dissolution of chitosan in an aqueous solution of an acid, addition of a softening agent, and removal of the acid by vacuum freeze drying, neutralizing the solution with an alkali and then subjecting the chitosan to vacuum freeze drying, or general drying. The chitosan sponge is useful as food packaging paper, wound-covering material, face pack material, etc.[84]

3.4.7.6 Fibers

Electrospinning technology opens up enormous possibilities for the implementation of bio-based materials and food hydrocolloids, including chitosan, in numerous applications. The electrospinning of chitin was performed using 1,1,1,3,3,3-hexafluoro-2-propanol (HFIP) as the spinning solvent. The spun chitin nanofibers had diameters ranging from 40 to 640 nm, most of them less than 100 nm. The fibers could be treated with a 40% aqueous NaOH solution at 60°C or 100°C, which converted chitin matrix into chitosan matrix with a degree of deacetylation of 85% without any shrinkage of the fiber. The prepared chitosan fibers offer antimicrobial properties and can be used for wound dressings.[85,86]

3.4.7.7 Nanoparticles

Nanotechnology can be used for the development of versatile chitin-based materials. The formation of biocompatible nanoparticles via the self-assembly of chitosan and modified lecithin has been reported. Stable nanoparticles in the size range of 123 to 350 nm were formed over a wide range of molar mixing ratios of chitosan and modified lecithin solutions (amino group/phosphate group; NH_3^+/PO_3^-) and total polyelectrolyte (PE) concentrations (0.1 to 1 wt%) except at intermediate molar ratios, when the surface charge was close to neutrality. Zeta-potentials of the nanoparticles were found to be independent of the total PE concentrations. The nanoparticles exhibited excellent stability at pH 3 to 6 and high ionic strength. The particle size and zeta-potential of the nanoparticles increased with the molecular weight of the chitosan derivative. Transmission electron microscopy indicated that the nanoparticles were generally spherical in shape. The nanoparticles could encapsulate both positively and negatively charged molecules to various capacities. The nanoparticle suspensions could be converted to lyophilized powder or concentrated suspensions.[87]

An aqueous emulsion of chitosan nanoparticles encapsulating silver oxide was prepared from silver nitrate and chitosan. The nanoparticles were positively charged with an average diameter of 300 nm. The dried particles had a spherical shape and a 100-nm diameter. The emulsion applied on cotton delivers a durable antibacterial activity against *Staphylococcus aureus* and *Escherichia coli*, effective even after 20 washings.[88] Chitosan film reinforced with carbon nanotubes exhibits a large increase in the tensile modulus with the incorporation of only 0.8% multiwalled carbon nanotubes. The most promising developments at present are in pharmaceutical and biological areas, followed by cosmetics.[9] Some of the future research needs in commercial applications of chitosan are pointed out in Table 3.4.

TABLE 3.4

Future Research Needs in Commercial Applications of Chitosan

Area	Description
Process standardization	Traditional methods with respect to deproteinization, demineralization, decolorization, and deacetylation influence molecular weight, degree of deacetylation, viscosity, fat and water absorption, and hydrophilicity.
Simpler and novel processes for chitosan and chitosan oligomers	These cost-effective processes would encourage further applications.
Improvements in film casting techniques; incorporate plasticizers and antimicrobial additives	Such improvements would provide antimicrobial properties and better stability against humidity.
Removal of astringent and bitter taste by such techniques as ozone technology	Better taste would result in wider food applications.
Quality standards	Various applications reported so far have used chitosan with diverse properties; a need exists for common standards for universal applications.

Source: Adapted from No, H.K. and Meyers, S.P., *J. Aquat. Food Prod. Technol.*, 4, 27, 1999; No, H.K. et al., *J. Food Sci.*, 65, 1134, 2000.

TABLE 3.5

Specifications for Chitin and Chitosan

Properties	Food-Grade Chitin	Pharmaceutical-Grade Chitosan	Technical-Grade Liquid Chitosan
Appearance	White/yellow flake	White/yellow powder or flake	Clear yellow liquid
Odor and taste	Odorless, tasteless	Odorless, tasteless	Odorless, tasteless
Moisture	<10%	<10%	—
Ash	<2.5%	<0.2%	<0.5%
Protein	<1.0%	<0.3%	<0.5%
Deacetylation	None	70–100%	>90%
Viscosity (0.5% solution)	600 cps	<5 cps	50 cps
pH	7.9	7.9	<5.5
Heavy metals (arsenic and lead)	<10 ppm	<10 ppm	<10 ppm

Source: Adapted from Subasinghe, S., *Infofish Int.*, 3, 58, 1999. With permission from Infofish International, Malaysia.

3.5 Summary

The huge amounts of crustacean waste available globally could be processed to isolate chitin and chitosan, which could be further converted into its various derivatives. These products could find wide application in the food and biomedical areas; however, because the types of raw materials differ, the properties of the isolate will vary. It is important to ensure that the isolated material meets specifications required for particular applications. In the case of chitosan, the degree of deacetylation has yet to be standardized. Currently, chitosan is available in a wide variety of commercial products with various deacetylation grades, molecular weights, and viscosities. Table 3.5 presents some specifications for chitin and chitosan.

References

1. Venugopal, V., *Marine Products for Healthcare: Functional and Bioactive Nutraceutical Compounds from the Ocean*, CRC Press, Boca Raton, FL, 2008, chap. 6.
2. Blanco, M. et al., Towards sustainable and efficient use of fishery resources: present and future trends, *Trends Food Sci. Technol.*, 18, 29, 2007.
3. Arvanitoyannis, I. S. and Kassaveti, A., Fish industry waste treatments, environmental impacts, current and potential uses, *Int. J. Food Sci. Technol.*, 43, 726, 2008.

4. Peter M. G., Chitin and chitosan from animal sources, in *Polysaccharides and Polyamides in the Food Industry: Properties, Production, and Patents*, Steinbüchel, A. and Rhee, S. K., Eds., John Wiley & Sons, New York, 2005, pp. 115–208.
5. Rasmussen, R. S. and Morrissey, M. T., Marine biotechnology for production of food ingredients, *Adv. Food Nutr. Res.*, 52, 237, 2007.
6. Subasinghe, S., Chitin from shellfish waste: health benefits overshadowing industrial uses, *Infofish Int.*, 3, 58, 1999.
7. Hayes, M. et al., Mining marine shellfish wastes for bioactive molecules: chitin and chitosan. Part A. Extraction methods, *Biotechnol. J.*, 3, 871, 2008.
8. Muzzarelli, R. A. A. and Muzzarelli, C., Chitosan chemistry: relevance to the biomedical sciences, *Adv. Polym. Sci.*, 186, 151, 2005.
9. Rinaudo, M., Chitin and chitosan: properties and applications, *Prog. Polym. Sci.*, 31, 603, 2006.
10. Domard, A. and Guibal, E., The international conference on chitin and chitosan, *Int. J. Biol. Macromol.*, 43, 1, 2008.
11. Tharanathan, R. N., Kittur, T., and Kittur, F. S., Chitin: the undisputed biomolecule of great potential, *Crit. Rev. Food Sci. Nutr.*, 43, 61, 2002.
12. No, H. K. and Meyers, S. P., Preparation and characterization of chitin and chitosan: a review, *J. Aquat. Food Prod. Technol.*, 4, 27, 1999.
13. Knorr, D., Chitin: a biomaterial in waiting, *Curr. Opin. Solid State Mater. Sci.*, 6, 313, 2002.
14. Sen, D. P., *Advances in Fish Processing Technology*, Allied Publishers, New Delhi, 2005, chap. 18.
15. Rødde, R. H., Einbu, A., and Varum, K. M., A seasonal study of the chemical composition and chitin quality of shrimp shells obtained from northern shrimp (*Pandalus borealis*), *Carb. Polym.*, 71, 388, 2008.
16. Synowiecki, J. and Al-Khateeb, N. A., Production, properties, and some new applications of chitin and its derivatives, *Crit. Rev. Food Sci. Nutr.*, 43, 145, 2003.
17. Naczk, M., Synowiecki, J., and Sikorski, Z. E., The gross chemical composition of Antarctic krill shell waste, *Food Chem.*, 7, 175, 1981.
18. Archer, M., *Fish Waste Production in the United Kingdom: The Quantities Produced and Opportunities for Better Utilization*, Report No. SR 537, Sea Fish Industry Authority, Edinburgh, Scotland, 2001.
19. Mathur, N. K. and Narang, C. K., Chitin and chitosan, versatile polysaccharides from marine animals, *J. Chem. Edu.*, 67, 938, 1990.
20. Muffler, K. and Ulber, R., Downstream processing in marine biotechnology, *Adv. Biochem. Eng. Biotechnol.*, 97, 63, 2005.
21. Ramachandran Nair, K. G. et al., Chitin preparation from prawn shell, *Indian J. Poultry Sci.*, 22, 40, 1987.
22. Shahidi, F. and Abuzaytoun, R., Chitin, chitosan, and co-products: chemistry, production, applications, and health effects, *Adv. Food Nutr. Res.*, 49, 93, 2005.
23. Percot, A., Viton, C., and Domard, A., Optimization of chitin extraction from shrimp shells, *Biomacromolecules*, 4, 12, 2003.
24. Marquardt, F. H. and Carreno, R. R., The production of colorless chitin from Antarctic krill (*Euphausia superba*) shell waste, *Arch. Fisch. Wiss.*, 41, 159, 1992.
25. Chaussard, G. and Domard, A., New aspects of the extraction of chitin from squid pens, *Biomacromolecules*, 5, 559, 2004.
26. Susano Cortizo, M. et al., Characterization of chitin from *Illex argentinus* squid pen, *Carb. Polym.*, 74, 10, 2008.

27. Rao, M. S. et al., Optimum parameters for production of chitin and chitosan from *Squilla* (*S. empusa*), *J. Appl. Polym. Sci.*, 103, 3694, 2007.
28. Cano-Lopez, A., Simpson, B. K., and Haard, N. F., Extraction of carotenoprotein from shrimp process wastes with the aid of trypsin from Atlantic cod, *J. Food Sci.*, 52, 503, 1987.
29. Wang, S. L. and Chio, S. H., Deproteinisation of shrimp and crab shell with the protease of *Pseudomonas aeruginosa* K-187, *Enzyme Microbial Technol.*, 22, 629, 1988.
30. Holanda, H. D. and Metto, F. M., Recovery of components from shrimp (*Xiphopenaeus kroyeri*) processing waste by enzymatic hydrolysis, *J. Food Sci.*, 71, C298, 2006.
31. Teng, W. L. et al., Concurrent production of chitin from shrimp shells and fungi, *Carb. Res.*, 332, 305, 2001.
32. Sini, T. K., Santhosh, S., and Mathew, P. T., Study on the production of chitin and chitosan from shrimp shell by using *Bacillus subtilis* fermentation, *Carb. Res.*, 342, 2423, 2007.
33. Kjartansson, K. J. et al., Ultrasonication-assisted extraction of chitin from North Atlantic shrimps (*Pandalus borealis*), *J. Agric. Food Chem.*, 54, 5894, 2006.
34. Synowiecki, J. and Al-Khateeb, N., Mycelia of *Mucor rouxii* as a source of chitin and chitosan, *Food Chem.*, 60, 605, 1997.
35. Minke, R. and Blackwell, J., The structure of α-chitin, *J. Mol. Biol.*, 120, 167, 1978.
36. Kunita, K. et al., Deacetylation behavior and characteristic properties, *J. Polym. Sci.*, 31, 485, 1993.
37. Cho, Y. I. et al., Physicochemical characteristics and functional properties of various commercial chitin and chitosan products, *J. Agric. Food Chem.*, 46, 3839, 1998.
38. Win, N. N. and Stevens, W. F., Shrimp chitin as substrate for fungal chitin deacetylase, *Appl. Microbiol. Biotechnol.*, 57, 334, 2001.
39. Yen, M-T. et al., Physicochemical characterization of chitin and chitosan from crab shells, *Carb. Polym.*, 75, 15, 2009.
40. Youn, D. K. et al., Physicochemical and functional properties of chitosans prepared from shells of crabs harvested in three different years, *Carb. Polym.*, 78, 41, 2009.
41. Tajik, H. et al., Preparation of chitosan from brine shrimp (*Artemia urmiana*) cyst shells and effects of different chemical processing sequences on the physicochemical and functional properties of the product, *Molecules*, 13, 1263, 2008.
42. Harding, S. E., Analysis of polysaccharides by ultracentrifugation, size, conformation and interactions in solution, *Adv. Polym. Sci.*, 186, 211, 2005.
43. Agullo, E. et al., Present and future role of chitin and chitosan in food, *Macromol. Biosci.*, 3, 521, 2003.
44. Kurita, K., Chitin and chitosan: functional biopolymers from crustaceans, *Mar. Biotechnol.*, 8, 203, 2006.
45. D'Ayata, G. G. et al., Marine derived polysaccharides for biomedical applications: chemical modification approaches, *Molecules*, 13, 2069, 2008.
46. Sorlier, P. et al., Relation between the degree of acetylation and the electrostatic properties of chitin and chitosan, *Biomacromolecules*, 2, 765, 2001.
47. Kasaai, M. R., Various methods for determination of the degree of *N*-acetylation of chitin and chitosan: a review, *J. Agric. Food Chem.*, 57, 1667, 2009.

48. Neugebauer, W. A., Neugebauer, E., and Brezinski, R., Determination of the degree of acetylation of chitin–chitosan with picric acid, *Carbohydr. Res.*, 189, 363, 1989.
49. Jeon, Y. J., Shahidi, F., and Kim, S. K., Preparation of chitin and chitosan oligomers and their applications in physiological functional foods, *Food Rev. Int.*, 16, 159, 2000.
50. Nguyen, T. T. B. et al., Molecular stability of chitosan in acid solutions stored at various conditions, *J. Appl. Polym. Sci.*, 107, 2588, 2008.
51. Yang, Y. M. et al., The effect of different sterilization procedures on chitosan dried powder, *J. Appl. Polym. Sci.*, 104, 1968, 2007.
52. No, H. K. and Prinyawiwatkul, W., Stability of chitosan powder during long-term storage at room temperature, *J. Agric. Food Chem.*, 57, 8434, 2009.
53. Speiciene, V. et al., The effect of chitosan on the properties of emulsions stabilized by whey proteins, *Food Chem.*, 102, 1048, 2007.
54. No, H. K., Lee, K. S., and Meyers, S. P., Correlation between physicochemical characteristics and binding capacities of chitosan products, *J. Food Sci.*, 65, 1134, 2000.
55. Kakiuchi, R. et al., Preparation and application of a colloidal solution of chitosan, *Polym. Preprints Jpn.*, 55, 2247, 2006.
56. Yun, H. G., Park, P. J., and Kim, S. K., Preparation and characterization of emulsions using carboxymethyl chitin, *J. Chitin Chitosan*, 6, 95, 2001.
57. Kurita, K., Controlled functionalization of the polysaccharide chitin, *Prog. Polym. Sci.*, 26, 1921, 2001.
58. Santosh, S. and Mathew, P. T., Preparation and properties of glucosamine and carboxymethyl chitin from shrimp shell, *J. Appl. Polym. Sci.*, 107, 280, 2007.
59. Chung, Y- C. et al., Preparation and important functional properties of water-soluble chitosan produced through Maillard reaction, *Bioresource Technol.*, 96, 1473, 2005.
60. Lim, S.-H. and Hudson, S. M., Synthesis and antimicrobial activity of a water-soluble chitosan derivative with a fiber-reactive group, *Carb. Res.*, 339, 313, 2004.
61. Zhong, Z. et al., Synthesis and antimicrobial activity of a water-soluble chitosan derivative with a fiber-reactive group, *Carb. Res.*, 343, 566, 2008.
62. Nishi, N. et al., Highly phosphorylated derivatives of chitin, partially deacetylated chitin and chitosan as new functional polymers: preparation and characterization, *Int. J. Biol. Macromol.*, 8, 311, 1986.
63. Holme, K. R. and Perlin, A. S., Chitosan N-sulfate, a water-soluble polyelectrolyte, *Carb. Res.*, 302, 7, 1997.
64. Yin, Q. et al., Synthesis and rheological behavior of a novel N-sulfonate ampholyte chitosan, *J. Appl. Polym. Sci.*, 113, 3382, 2009.
65. Kim, S. J. et al., Synthesis and characteristics of polyelectrolyte complexes composed of chitosan and hyaluronic acid, *J. Appl. Polym. Sci.*, 91, 2908, 2004.
66. Prashanth, K. V. H. and Tharanathan, R. N., Chitin/chitosan: modifications and their unlimited application potential—an overview, *Trends Food Sci. Technol.*, 18, 117, 2007.
67. Kittur, F. S., Vishnu Kumar, A. B., and Tharanathan, R. N., Low molecular weight chitosans: preparation by depolymerization with *Aspergillus niger* pectinase and characterization, *Carbohydr. Res.*, 338, 1283, 2003.

68. Kuo, C. H., Chen, C. C., and Chiang, B. H., Process characteristics of hydrolysis of chitosan in a continuous enzymatic membrane reactor, *J. Food Sci.*, 69, 332, 2003.
69. Chang, Ke-L. B., Tai, M. C., and Cheng, H., Kinetics and products of the degradation of chitosan by hydrogen peroxide, *J. Agric. Food Chem.*, 49, 4845, 2001.
70. Deo, S. et al., Simultaneous depolymerization and decolorization of chitosan by ozone treatment, *J. Food Sci.*, 72, C522, 2007.
71. Shu, C.-K., Degradation products formed from glucosamine in water, *J. Agric. Food Chem.*, 46, 1129, 1998.
72. Rogovina, S. Z. and Galina, A., and Vikhoreva, G. A., Polysaccharide-based polymer blends: methods of their production, *Geoconjugate J.*, 23, 611, 2006.
73. Shon, S. O. et al., Viscoelastic sol–gel state of the chitosan and alginate solution mixture, *J. Appl. Polym. Sci.*, 103, 1408, 2007.
74. Chang, Y., Xiou, L., and Tang, Q., Preparation and characterization of a novel thermosensitive hydrogel based on chitosan and gelatin blends, *J. Appl. Polym. Sci.*, 113, 400, 2009.
75. Chen, T. et al., Enzyme-catalyzed gel formation of gelatin and chitosan: potential for *in situ* applications, *Biomaterials*, 24, 2831, 2003.
76. Yurii, A. et al., Gelling of otherwise nongellable polysaccharides, *J. Coll. Interface Sci.*, 287, 373, 2005.
77. Hu, X. and Gao, C., Photoinitiating polymerization to prepare biocompatible chitosan hydrogels, *J. Appl. Polym. Sci.*, 110, 1059, 2008.
78. Chenite, A. et al., Novel injectable neutral solutions of chitosan form biodegradable gels *in situ*, *Biomaterials*, 21, 2155, 2000.
79. Ahmadi, R. and de Bruijn, J. D., Biocompatibility and gelation of chitosan-glycerol phosphate hydrogels, *J. Biomed. Mater. Res.*, 86A, 824, 2008.
80. Struszczyk, H. and Kivekas, O., Recent developments in microcrystalline chitosan applications: biological materials for wound healing, in *Advances in Chitin and Chitosan*, Brine, C. J. and Sandford, P. A., Eds., Elsevier, London, 1992, pp. 549–555.
81. Quignard, F. et al., Aerogel materials from marine polysaccharides, *New J. Chem.*, 32, 1300, 2008.
82. Sangeetha, K. and Abraham, T. E., Investigation on the development of sturdy bioactive hydrogel beads, *J. Appl. Polym. Sci.*, 107, 2899, 2008.
83. Hoven, V. P. et al., Surface-charged chitosan preparation and protein adsorption, *Carbohydr. Poly.*, 68, 44, 2007.
84. Yoshimori, T. et al., Chitosan Sponge and Method for Producing the Same, European Patent No. JP2003292501 (A), 2003.
85. Min, B. M. et al., Chitin and chitosan nanofibers: electrospinning of chitin and deacetylation of chitin nanofibers, *Polymer*, 45, 7137, 2004.
86. Tores-Giner, S. et al., Development of active antimicrobial fiber based chitosan polysaccharide nanostructures using electrospinning, *Eng. Life Sci.*, 8, 303, 2008.
87. Chuah, A. H. et al., Formation of biocompatible nanoparticles via the self-assembly of chitosan and modified lecithin, *J. Food Sci.*, 74, H1, 2009.
88. Hu, Z. et al., Nanocomposite of chitosan and silver oxide and its antibacterial property, *J. Appl. Polym. Sci.*, 108, 52, 2008.

4

Polysaccharides from Seaweed and Microalgae

4.1 Introduction

Marine algae, including both macroalgae (seaweed) and microalgae, contain important polysaccharides. Seaweed, the multicellular algae rich in minerals and vitamins found in marine waters, has been an important component of food, feed, and medicine in the Orient for several centuries; however, very few of the world's available seaweed species are used globally as food sources. In the Western world, seaweed is almost exclusively used for the extraction of important food hydrocolloids, including carrageenan, alginic acid, and agar, which are traditionally used for food product development. The major food polysaccharides are agar, alginates, and carrageenans; the minor polysaccharides include sulfated fucose (brown seaweed), xylans (certain red and green seaweed), and cellulose (which occurs in all genera but at lower levels than found in higher plants), laminarin (brown species), and Floridean starch (amylopectin such as glucan), the storage polysaccharides most notably found in red seaweed.[1] At least a few species of microalgae have been recognized as functional foods because of the presence of significant amounts of nutrients such as vitamins, proteins, and carotenoids in them. They also contain significant amounts of polysaccharides. Nevertheless, only a few selected species are currently being used in foods.[2] This chapter discusses the isolation and characterization of polysaccharides from both seaweed and microalgae.

4.2 Seaweed Species Important as Food

Some of the important seaweed species used for food purposes belong to Rhodophyceae, including *Chondrus*, *Gelidium*, *Gigartina*, *Gracilaria*, *Eucheuma*, and *Kappaphycus*. *Chondrus crispus*, popularly known as Irish moss, is an abundant red seaweed of the North Atlantic. The seaweed has a flattened, dark violet thalus, and it reaches a size of up to 25 cm. The basal disc of *Chondrus* attaches to rocks down to a depth of 12 to 15 cm. *Gelidium* is a small, agar-containing seaweed with a length of up to 30 cm and thin, rigid branches; it grows in wave-exposed coasts and can be found in many parts of the world. *Gigartina* species are large plants up to 5 m in length that grow in the cold, deep coastal waters off Chile and Peru. *Gigartina radula*, found abundantly in Chile, is flat with dark red leaves; it grows from a basal disc attached to rocks. It can reach up to 3 m in length but is normally 25 to 100 cm long. The genus *Gracilaria* (Gracilariales) is another important alga that is a major source of agar. *Eucheuma cottonii* (new name, *Kappaphycus alvarezii*) has a bushy thallus with a bright green to dark brown color. It can grow to a size of more than 1 m in diameter but is normally 20 to 30 cm. *Eucheuma denticulatum* has the same general morphology as *E. cottonii* except that the branches have 3- to 4-mm spines all over the surface of the thallus. It can reach a size of 75 cm in diameter.

Undaria pinnatifida, a brown seaweed, occurs on rocky shores and bays in the temperate zones of Japan, the Republic of Korea, and China. It grows on rocks and reefs in the sublittoral zone, down to about 7 m. It grows best between 5 and 15°C and stops growing if the water temperature rises above 25°C. It has spread, probably via ship ballast water, to France, New Zealand, and Australia. *Kombu* is the Japanese name for the dried seaweed that is derived from a mixture of *Laminaria* species.[3,4]

4.2.1 Proximate Composition

The proximate compositions of several seaweed species have been examined with respect to their uses as food.[5] Generally, their proximate compositions depend on species, growth environments, and harvesting season. Red algae contain 80 to 85% water, 15 to 30% minerals, 15 to 20% carbohydrates (agar and carrageenan), 8 to 25% proteins, and 2 to 4% lipids.[3,6,7] All essential amino acids are present in both red and brown seaweed. Red algae species have uniquely high concentrations of taurine when compared to brown algae varieties.[7,8] The amino acid content per 100 g proteins in the edible seaweed *Durvillaea antarctica* (frond and stem) and dried *Ulva lactuca* varied significantly, ranging from 508 to 2020 mg. Seaweed is rich in polysaccharides, which function as dietary fiber; the dietary fiber content is comparable among red and brown algae classes. The lowest and highest concentrations of dietary fiber are found in *Laminaria* spp. and *Hizikia fusiforme*, respectively. The edible seaweed *Durvillaea antarctica* (frond and stem) and dried *Ulva lactuca* have both soluble and insoluble dietary fiber. *U. lactuca* was found to contain 60.5% total dietary fiber (TDF), while *D. antarctica* frond and stem contained 71.4% and 56.4% TDF, respectively.[9,10] The crude fiber contents of the red seaweed *Hypnea charoides* and *H. japonica* and the green seaweed *Ulva lactuca* are in the range of 46 to 55%, on a dry weight basis.[8]

The proportion of crude lipids in various species varies from 1.2 to 4.9% and that of phospholipids from 2.9 to 19.7%. Major phospholipids are phosphatidyl-choline, phosphatidylglycerol, and phosphatidylethanolamine.[11] Analysis of the total lipid, protein, ash, and individual fatty acid contents of canned edible seaweed (*Saccorhiza polyschides* and *Himanthalia elongate*) or dried seaweed (*H. elongate, Laminaria ochroleuca, Undaria pinnatifida, Palmaria, Porphyra*) demon-strated a total lipid content ranging from 0.70 to 1.80% on a dry weight basis. Unsaturated fatty acids predominated in all of the brown seaweed studied and saturated fatty acids in the red seaweed, but both groups were found to be balanced sources of omega-3 and omega-6 acids. Ash contents ranged from 19.07 to 34%, and protein contents from 5.46 to 24.11% on a dry weight basis.[8,12] It is interesting to note that polyunsaturated fatty acids (PUFAs) from seaweed have been found to have cytotoxic effects against human cancer cell lines.[23]

When eleven species of macroalgae (including four species from commer-cially important genera) were analyzed for proximate compositions, *Corallina officinalis*, on a dry weight basis, had low protein (6.9%) and high ash (77.8%) and calcium (182 ppm) contents, with a low caloric value of 2.7 MJ/kg. In contrast, the *Porphyra* species had low ash (9.3%), high protein (44.0%), and low calcium (19.9 ppm) content, with a high caloric value (18.3 MJ/kg). Other species examined had intermediate values but tended to be more similar to *Porphyra* than to *Corallina* species.[13] *Utricularia rigida* is rich in protein, carbo-hydrates, fiber, vitamins, and minerals and has a low lipid content. When the compositions of the red seaweed species *Hypnea charoides* and *H. japonica* and the green seaweed species *Ulva lactuca* were compared, the total ash contents ranged between 21.3 and 22.8% on a dry weight basis, and crude lipids were in the range of 1.4 to 1.6%.[12,13]

The Mexican seaweed *Enteromorpha* has 9 to 14% protein, 2 to 3.6% ether extract, and 32 to 36% ash. It also contains 10.4% and 10.9% omega-3 and omega-6 fatty acids, respectively, per 100 g of total fatty acids. The protein has a high digestibility of 98%. This seaweed could be a potential food in view of its composition. *Hizikia fusiforme* is a brown seaweed with a finer frond (leaf) structure than *Undaria pinnatifida* (wakame) and kelp (kombu). It is collected from the wild in Japan and cultivated in the Republic of Korea. The protein, fat, carbohydrate, and vitamin contents are similar to those found in kombu, although most of the vitamins are destroyed in the processing of the raw seaweed. The iron, copper, and manganese contents are relatively high, cer-tainly higher than in kombu. Like most brown seaweed, the fat content of *U. pinnatifida* is low (1.5%), but 20 to 25% of the fatty acid is eicosapentaenoic acid (EPA).[1,2] Seaweed is also a source of interesting bioactive components, such as vitamins (e.g., A, B_1, B_2, C), carotenoids (e.g., fucoxanthin), and omega-3 fatty acids, many of which have nutritional importance.[1,14–16]

Seaweed, in general, has a characteristic flavor. The major flavor com-pound common in all seaweed groups (red, brown, and green) is dimethyl-sulfide (DMS). Green seaweed flavor is primarily due to DMS and a group of unsaturated fatty aldehydes, (8Z, 11Z, 14Z)-heptadecatrienal. Brown seaweed flavor, on the other hand, is due to β-ionone and cubenol. Flavor of the popular seaweed product nori is mainly due to DMS, carotenoid derivatives, and aldehydes.

4.2.2 Nutritional Value

The proximate compositions of seaweed generally indicate their nutritional value. Most species contain significant amounts of ash, consisting of diverse minerals and high fiber. They have moderate amounts of fatty acids, but protein contents may be somewhat low. Seaweed proteins contain all of the essential amino acids at levels that are sufficient to meet normal nutritional requirements.[7,8] Edible seaweed contains 33 to 50% total fiber, which is higher than the levels found in higher plants. The rich fiber in seaweed contributes to the protein bioavailability. Seaweed can function as major dietary fiber when incorporated in animal feed.[17] The alginate and carrageenan in seaweed function as soluble fiber. In addition to its fiber and mineral content, seaweed can serve as a source of such biologically active compounds as carotenoids, phycobilins, vitamins, sterols, tocopherol, and phycocyanins, among others.[1,2,18,19] The carotenoids include β-carotene, lutein, and violaxanthin in red and green seaweed and fucoxanthin, tocopherols, and sterols (e.g., fucosterol) in brown seaweed.

A number of studies on the nutritional values of individual seaweed species have been reported. The brown alga *Undaria pinnatifida* and the red alga *Chondrus crispus* could be used as food supplements to meet the recommended daily intake of minerals (Na, K, Ca, and Mg) and trace elements (Fe, Zn, Mn, Cu), the contents of which range from 8.1 to 17.9 mg/100 g and from 5.1 to 15.2 mg/100 g, respectively.[6] *Ulva rigida* is well accepted by experimental animals and does not significantly change nutritional parameters, although it does reduce low-density lipoprotein (LDL) cholesterol.[20] Animal feeding experiments were employed for a nutritional evaluation of three types of subtropical, fiber-rich brown seaweed (*Sargassum* spp.) in terms of net protein ratio, true protein digestibility, nitrogen balance, biological value, net protein utilization, and fecal and urinary nitrogen loss. No significant differences in net protein ratio and urinary nitrogen loss were found.[21,22] In view of the presence of several nutrients, seaweed species such as *Gracilaria changii* are recommended for human consumption.[24]

4.2.3 Effects of Processing on Nutritive Value

It is important to determine yield and identify those seaweed species that maximize their commercial use. Yield is dependent on the species and the environmental conditions during their growth. Spectral characteristics and levels of 3,6-anhydro-D-galactose and sulfate are some of the factors used to determine the quality of commercial seaweed. Processing has a significant influence on the composition and quality of seaweed products. High-temperature drying and cooking may cause a significant loss of vitamin C in brown seaweed. Canning or drying does not significantly affect the nutritional value of seaweed. Several species of canned edible seaweed, including *Porphyra* spp., were found to have total lipid contents ranging from 0.7 to 1.8%, ash contents from 19 to 34%, and protein contents from 5.4 to 24.1%, on a dry weight basis.[12,26] Air-dried wakame has a vitamin content similar to that of wet seaweed and is relatively rich in B vitamins, especially niacin. Processed wakame products lose most of their vitamins, but raw wakame contains appreciable amounts of essential trace elements such as manganese, copper, cobalt, iron, nickel, and zinc, similar to kombu and hiziki.[27]

4.2.4 Quality Evaluation

An electrophoretic method has been developed to identify seaweed based on profiles of the algal proteins; for example, a band with a molecular weight above 70 kDa appears to be specific to *Porphyra* species. *Palmaria palmata* is composed of six protein bands with apparent molecular weights between 59.6 and 15.2 kDa; *Gracilaria verrucosa* has eight permanent bands.[28] Much effort has been directed toward determining relationships between the chemical structure and gelling characteristics of polysaccharides. Developments in multiple- and low-angle laser-light diffusion detectors coupled with high-performance size-exclusion chromatography have made it relatively simple to determine the molecular weights and distributions of polysaccharides. The rapid enzyme-linked lectin assay (ELLA) and enzyme-linked immunosorbent assay (ELISA) have been applied to study food-grade polysaccharides for use as thickeners in fruit jelly desserts, coating films, and pet foods.[29] A molecular method for the rapid discrimination of red seaweed involves the use of internal transcribed spacer (ITS) sequences from 5.8S rDNA.[30]

4.3 Seaweed Polysaccharides

Seaweed contains large amounts of polysaccharides in its cell wall structures. Most of these polysaccharides are not digested by humans and therefore can function as fiber in foods, as mentioned earlier. Seaweed fiber content is higher than what is found in higher plants. These fibers are rich in soluble fractions. Other minor polysaccharides are also found in the cell wall, including polysaccharides containing sulfated fucose (brown seaweed), xylans (certain red and green seaweed), and cellulose (which occur in all genera but at lower levels than in higher plants). Seaweed also contains storage polysaccharides, most notably laminarin in brown seaweed and Floridean starch

TABLE 4.1

Some Seaweed as Sources of Marine Polysaccharides

Polysaccharide	Seaweed
Agar, agarose	*Gracilaria, Gelidium, Pterocladia*
Alginic acid (alginate)	*Macrocystis, Laminaria, Ascophyllum, Sargassum*
Carrageenans	*Gigartina, Chondrus, Eucheuma*
Fucoidan	*Fucus serratus*
Laminarin	*Laminaria japonica* (brown seaweed)
Furcellaran	*Furcellaria lumbricialis, F. fastigiata*
Ulvan	*Ulva rigida, Enteromorpha compressa*

(amylopectin-like glucan) in red seaweed. Table 4.1 lists some important seaweed sources of polysaccharides, and Table 4.2 lists the major functional properties of seaweed polysaccharides.

Seaweed polysaccharides have three functions: (1) biological, (2) physiological, and (3) technological. Their biological functions include providing structure to the seaweed cells as components of the cell wall architecture, cell–cell recognition, stimulating host defenses, and hydration of intracellular fluid. Their physiological functions are closely related to their physicochemical properties and include solubility, viscosity, hydration, and ion exchange capacities

TABLE 4.2

Major Functional Properties
of Seaweed Polysaccharides

Seaweed structure
    Cell wall architecture
    Cell–cell recognition
    Hydration of intracellular fluids
    Stimulation of host defense
Food applications
    Gelling agent
    Stabilizer
    Texture modification
    Water-holding capacity modifier
    Film formation
    Inhibits syneresis
    Increases yield
Nutraceuticals
    Antioxidants
    Antithrombin activity
    Antitumor activity
    As sources of functional oligosaccharides
    Dietary fiber

in the digestive tract. The polysaccharides bind several times their weight of water, as high as 20 times their own volume, depending on the length and thickness of the polysaccharide moiety. The water-holding capacity of seaweed polysaccharides is much higher than that of cellulosic fibers. The high affinity for water qualifies the polysaccharides to be referred as *hydrocolloids*. From a technological point of view, their exceptional water-binding capacity allows them to function as texturizers, stabilizers, emulsifiers, fat reducers, film formers, stabilizers, shelf-life extenders, and viscosity modifiers. Further, they can function as additives for the inhibition of syneresis to reduce dryness and toughness and improve the yield of food products. They may also offer novel properties in combination with other gum additives such as cellulose gums, gum arabic, gum acacia, guar gum, pectin, and carboxymethylcellulose.[25]

4.4 Agar

Minoya Tarazaemon discovered agar in Japan in the year 1658. The term *agar* is synonymous with *agar–agar*, Japanese gelatin, Japanese isinglass, vegetable gelatin, and angel's hair. Agar is a polysaccharide that accumulates in the cell walls of agarophyte algae. Its content in a particular seaweed varies depending on the season.[36]

4.4.1 Extraction

Commercial agar is primarily extracted from red algae such as *Gracilaria* and *Gelidium* (Rhodophyceae). The genus *Gracilaria* is the major source of agar in Japan, the United States, Mexico, Africa, and India. In Japan, agar is primarily extracted from *Gelidium* species, usually collected from rocks exposed at low tide along the coast, but the best material is collected by trained divers. The algae is partially bleached before being sold to commercial processors for agar extraction. In the United States, agar is extracted from *Gelidium cartilagineum* and *Gracilaria confervoides*.[31]

The traditional Japanese process to extract native agar begins with careful blending of up to six or seven different types of red seaweed, which are selected according to the desired flexibility, density, smoothness, solidity, and resilience of the end product. The seaweed is extracted in boiling water in open iron cauldrons; tough seaweed (e.g., *Gelidium* spp.) is introduced first, the softest (e.g., *Gracilaria* spp.) last. The pH of the extracting water is adjusted to 5 to 6 with sulfuric acid. The mixture is kept boiling for 4 to 10 hours. A bleaching agent (e.g., hypochloride, hydrosulfite, bisulfite) is introduced during boiling. The slurry is transferred to a sedimentation tank where the impurities settle to the bottom. The supernatant is strained through wire mesh or cloths of various degrees of fineness, usually under pressure. A second boiling of the residue for about 10 hours follows, after which the combined liquor is poured into wooden trays to cool and gellify. The gel is cooled outdoors on straw mats and cut into suitable shapes and sizes. The gel pieces are subjected to repeated freezing and thawing for 3 to 6 days. Each day, the liquid that forms is drained off, as it contains salts and other impurities. The moisture of the crude agar is partially restored by sprinkling water, and the final extract is dried in the sun for 15 to 30 days. The gels are packed in either strips or threads or in a shredded or powdered form, and they are graded according to color, luster, gel strength, etc. Agar from the extract can also be precipitated by alcohol. The commercial product is white, shiny, semitransparent, tasteless, and odorless.[27,32,33] Figure 4.1 illustrates the typical process for agar extraction.

A number of modifications of the conventional method have been made to obtain agar having better qualities, particularly gel strength. A limited alkali treatment (usually $4M$ KOH) results in hydrolysis of sulfate groups by conversion of (1,4)-linked galactose-6-sulfate to anhydrogalactose form.[34] A novel photobleaching method has been developed to extract agar from *Gracilaria* that has excellent gel strength and other desirable qualities, including gelling temperature, sulfate content, and 3,6-anhydro-L-galactose content. The extracted agar has a maximum gel strength of 1913 g/cm.[2,35] Because seaweed is usually dried and stored for some time before extraction, post-harvest storage conditions influence the quality of extracted agar. Storage of dried seaweed for a maximum period of 31 months has been found to have no significant influence on yield, chemical composition, or physical and textural properties of alkali-treated agar. The gel strength of agar extracts from *Gracilaria eucheumatoides* averaged 318 g/cm^2 until the third month of storage but decreased considerably thereafter. The yield of agar was between 22.9 to 29.0%, and its relative viscosity and molecular weight varied inversely with storage time. Both physical and textural parameters of agar generally decrease with storage time; therefore, a maximum storage of up to 3 months is suggested to avoid a loss of agar quality.[36]

A number of seaweed species belonging to *Gracilaria* (e.g., *G. gracilis, G. dura, G. bursa-pastoris*) displayed yields ranging between 30 and 35% and varying physical and chemical properties. Yield and gel strength of agar from *Gracilaria vermiculophylla* decreased with an increase in the duration of alkali treatment from 0.5 to 3 hours. The highest yield (15.3%) and highest gel strength (1064 g/cm^2) were obtained after alkali treatment for 0.5 hours. The alkali-treated agar showed higher melting (92 to 100°C) and gelling (36 to 40°C) temperatures compared to non-alkali-treated agar, which had a melting temperature of 60 to 64°C and gelling temperature of 20 to 23°C. The 3,6-anhydrogalactose content decreased with increasing alkali treatment time.[37] *G. gracilis* and *G. bursa-pastoris* gave agars having the highest (630 ± 15 g/cm^2) and lowest (26 ± 3.6 g/cm^2) gel strengths, respectively. A positive correlation was found between 3,6-anhydrogalactose content and gel strength. The contents of 3,6-anhydrogalactose were comparable, indicating superior agar quality and their potential as industrial raw material.[38] The agar of *G.*

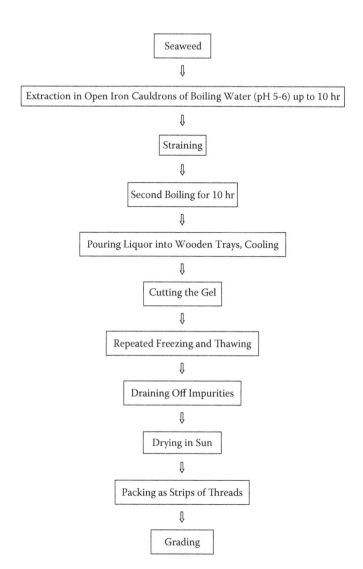

FIGURE 4.1
Process for the extraction of agar.

dura exhibited gel strengths ranging from 263 to 600 g/cm^2; the sample harvested in the month of October demonstrated the maximum gel strength. The agar was characterized by different degrees of methoxylation and sulfation, which influenced its gelling ability.[38]

Gracilaria gracilis from Argentina was sequentially extracted with water either at ambient temperature or 90°C. Both the extracts consisted of polysaccharides having a molecular weight of 5.4 kDa. Structural analysis revealed that the ambient temperature extract was mainly agarose. Alkaline treatment removed most of the sulfate group of the sample. It was observed that a high yield of good-quality agarose could be obtained after extraction with water at 70°C without alkaline pretreatment.[39] In France, the agar yield from *Gracilaria cervicornis* varied from 11 to 20%, with generally higher values being recorded during the dry season. Agar yield from *Hydropuntia cornea* ranged from 29 to 41%, with a peak recorded in June. Maximum agar biomass yield from *G. cervicornis* was observed during the summer, with a maximum value of 390 g/m^2; in the rainy season, it was only 129 g/m^2, suggesting a seasonal variation.[40] *Gracilaria edulis* collected during the onset of the rainy season was found to exhibit the highest gel strength, deformation, cohesiveness, and melting temperature—properties useful in the food industry.[36]

Agar and other polysaccharides have been isolated from Indian seaweed. Employing sequential extractions with various solvents such as water and alkali (4M KOH), agar from *Enteromorpha compressa* was isolated with a yield of 25% and a molecular weight of 55 kDa. Detailed studies on agar samples from *Gelidiella acerosa* specimens collected from 2001 to 2002 at various sites on the Indian coast showed that the best quality agar was produced by *G. acerosa* occurring in the Gulf of Mannar region in the southeast. The gel strengths and the viscosities of the agar ranged from 450 to 845 g/cm^2 and 33 to 67 cP, respectively. Agar samples extracted from the west coast had lower gel strengths (225 to 400 g/cm^2) and viscosity values (15 to 30 cP).[41–43] Table 4.3 provides data on the yields and properties of agar isolated from various types of seaweed, and Table 4.4 summarizes the properties of two commercial agar samples.

4.4.2 Structure

Agar is a hydrophilic colloid composed of two polysaccharides, agarose and agaropectin. Once deposited in the cell wall, most of the compound is enzymatically polymerized and desulfated, giving agarose. The rest remains in the form of agaropectin. Agarose consists of alternating (1,4)-linked 3,6-anhydro-α-L-galactopyranose and (1,3)-linked β-D-galactopyranose. It has a double-helical structure that aggregates into a three-dimensional framework, holding water molecules within the interstices of the framework and creating thermoreversible gels. Agarose is difficult to extract from agar. The sulfate contents of the agarose and agaropectin fractions differ. In addition to sulfate, agaropectin also contains pyruvic acid, D-glucuronic acid, and agarobiose. The type and quantity of the substituent groups in the polysaccharide chain depend on species, environmental conditions, and physiological factors.[38,44] Structural analysis of agar includes determining the monosaccharide constituents and partial depolymerization by reductive hydrolysis, identifying disaccharide repeating units by NMR spectroscopy, and sequence analysis by enzymatic degradation. Examples of the elucidation of primary structures for several complex sulfated galactans and xylogalactans are given by Usov.[45] Figure 4.2 illustrates the structures of agarose and agaropectin.

TABLE 4.3

Yields and Properties of Agar Isolated from Seaweed

Species	Yield (%)	Gel Strength and Other Properties	Refs.
Gracilaria gracilis, *G. bursa-pastoris*	30–35	Highest, 630 ± 15 g/cm^2 (*G. gracilis*); lowest, 26 ± 3.6 g/cm^2 (*G. bursa-pastoris*); contents of 3,6-anhydrogalactose similar	Marinho-Soriano and Bourret[38]
Gracilaria gracilis	—	Optimum yield when extracted in water at 70°C	Rodriguez et al.[39]
Gracilaria dura	—	263 to 600 g/cm^2; *g. dura* harvested in October had maximum anhydrogalactose content and gel strength	Marinho-Soriano et al.[40]
Gracilaria eucheumatoides	22–29	Maximum gel strength, 318 g/cm^2; pre-extraction storage for more than 4 months adversely affected yield and gel strength	Romero[36]
Gracilaria vermiculophylla	15–30	Gel strength, 1064 g/cm^2; gelling temperatures, 35.7–39.6°C and 20.4–23.4°C for alkali- and water-extracted agar, respectively	Higuera[37]
Gracilaria cervicornis	11–19	Peak yield in summer in France	Marinho-Soriano et al.[40]
Gracilaria lemaneiformis (aquacultured red seaweed)	—	Maximum yield of 1913 g/cm^2	Li et al.[35]
Hydropuntia cornea	29–41	Peak yield in June, in France	Marinho-Soriano et al.[40]
Enteromorpha compressa	25	Sequential extraction in hot alkaline water; molecular weight, 55 kDa	Chattopadhyay,[41] Ray[42]
Gelidiella acerosa	—	Gel strength, 450 to 845 g/cm^2	Prasad et al.[43]

4.4.3 Gelation

Agar forms one of the strongest gels. Gelation of agar is exothermic. The gel is formed when a hot aqueous solution of agar is cooled. During gelation, the molecules undergo a coil–helix transition followed by aggregation of the helices. The polysaccharide differs from other hydrocolloids in two properties: its high gel strength and its wide range of hysteresis (defined as the difference between the transition temperatures measured upon heating and cooling). This hysteresis property makes agar useful for applications in food, microbiological, and pharmaceutical products.[34] Gel strength, normally measured as resistance against a plunger of a 1.5% soluble agar gel at 20°C, is in the range of 200 to 1000 g/cm^2 for different agars. Typical temperatures for gelation of agar extracted from several agarophytes vary between 38 and 45°C. Agar sols are thermostable and generally lose about

TABLE 4.4

Properties of Two Commercial Agar Samples

Criteria	Grand Agar (Hispanagar, Spain)	Speed Agar-80 (Taito, Japan)
Moisture	7.16%	7.08%
Ash	1.53%	1.47%
Solution clarity at 1.5%	26 Nephelos	510 Nephelos
pH of 1.5% solution	6.8	6.2
Viscosity of 1.5% solution	6.5 cps	4 cps
Gel temperature of 1% solution	31.9°C	75.6°C
Melting temperature of 1% solution	87.5°C	75.6°C
Gel strength at 1.5%	1270 g/cm^2	590 g/cm^2
Gel strength at 1.5% (dissolved 5 minutes at 85°C)	680 g/cm^2	430 g/cm^2
Gel strength at 1.5% (dissolved 5 minutes at 90°C)	820 g/cm^2	440 g/cm^2

Source: Armisen, R. and Galatas, F., in *Handbook of Hydrocolloids*, Phillips, G.O. and Williams, P.A., Eds., CRC Press, Boca Raton, FL, 2000, p. 39. With permission from Woodhead Publishing, Ltd.; www.woodheadpublishing.com.)

5% of their strength if autoclaved for 1 hour at 120°C at pH 6.5 to 7.5. They rapidly degrade at either a very high or a very low pH.

A hot 1.5% aqueous solution of agar is clear; it congeals at 32 to 39°C to a firm, resilient gel that becomes liquid only above 80°C, indicating its remarkable stability at high temperatures. For melted aqueous agar from *Gracilaria confervoides* from North Carolina that is cooled slowly, gelation begins at 63°C with the formation of soft gel, followed by a sudden increase in firmness as temperature falls to 43°C or lower.[31] Agarose is the component responsible for gelation of agar. Agarose will gel at concentrations as low as 0.1% (w/w); 0.5% solutions will give a strong gel. Agarose gels are regarded as physical gels, and their gelation is governed by hydrogen bonding. The gels are composed

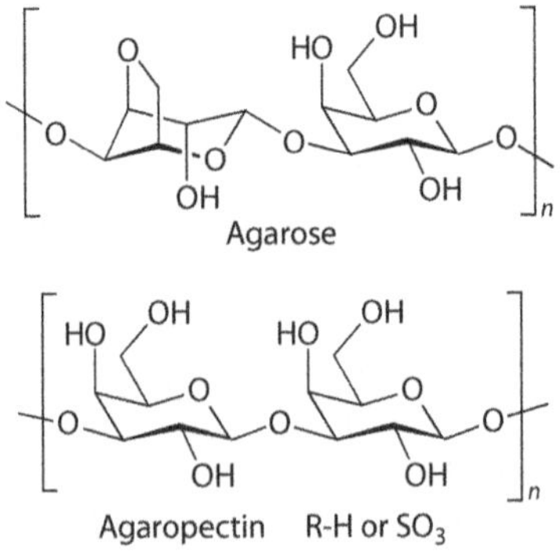

Agarose

Agaropectin   R-H or SO$_3$

FIGURE 4.2
Structure of agarose and agaropectin.

of thick bundles of agarose chains and large pores of water and exhibit high turbidity and strong elasticity. The sulfate and methoxy content of agar adversely influences its gelation; the gelling temperature required increases with an increase in either of these. Conversion of these precursor units to the corresponding 3,6-anhydrides by treatment with hot alkali at about 4*M* concentration has been found to increase gel strength. Agaropectin has a higher sulfate content and therefore a poorer gelling ability than agarose.[46]

Although insoluble in cold water, dry agar will absorb water in large quantities accompanied by swelling and evolution of heat. The presence of solutes, particularly KCl or NaCl, in the water affects swelling.[31] Binding of water by agar is more efficient at low water activity (a_w) levels.[47] The binding of water was not as tight in agar compared with carrageenan at the same water content, suggesting that most of the water molecules absorbed by agar were free water and could not be plasticized.[48] The viscosity of agar solution is influenced by the seaweed species and processing conditions. The viscosity at 45°C is relatively constant at a pH ranging from 4.5 to 9.0. Gels of agars from *Gelidium cartilagineum* and *G. amanasii* are noted for their low viscosity, low syneresis, and uniformity at low temperatures of gelation (34 to 40°C). *Hypnea* agar is similar to that of *Gelidium* agar in that it possesses a relatively low viscosity, but it resembles *Gracilaria* agar in its high degree of syneresis.[49,50]

Rheological properties of agarose gel in both the linear viscoelastic region (small-strain) and the nonlinear region and up to fracture (large-strain) are useful for studying gel network properties of agarose. Small-strain measurements describe the network, whereas large-strain (fracture) properties are more associated with sensory texture. Small-strain behavior of agarose gels is primarily elastic, with only slight frequency dependence. The large-strain viscoelastic behavior of 1 to 2.5% (w/v) agarose gels included fracture properties and nonlinear behaviors that were dependent on agarose concentration and strain rate. The addition of glycerol causes an increase in fracture stress but also increases fracture strain. Large-strain behaviors and fracture properties are dependent on strain rate and agarose and glycerol concentrations. Increasing concentrations of agarose produce an increasingly stronger, more brittle network. This information is useful in the application of agar as a texturizer in food products.[50,51]

Functional properties of agar (and other polysaccharides) can be modified as desired through biotechnological means, encompassing both traditional biotechnology (fermentation and enzymology) and molecular biology (genetic manipulation and protein engineering).[52] The manufacture and characterization of agarose microparticles have recently been reported.[53] High-speed shearing of bulk agarose gels yielded microparticle sizes of about 100 μm. The microparticles formed a solidlike suspension at high volume fractions, becoming fluidlike above a well-defined yield stress. The mechanical properties of the microparticles were characterized employing rheological analyses (Young's modulus, stress and strain at failure). Table 4.5 shows the influence of temperature on the solubility and gelation properties of agar.

TABLE 4.5

Influence of Temperature and Salts on the Solubility and Gelation Properties of
Seaweed Polysaccharides

	Solubility		
Polysaccharide	Room Temperature	Hot Water	Gel Formation
Agar	—	Soluble	Gel formation is observed; KCl or NaCl affects swelling in water.
Algin Na-alginate	Dissolves	Dissolves	Ca^{2+} enhances gelation and viscosity.
κ-Carrageenan	Dissolves in presence of Na^+	Dissolves	Monovalent cations promote the aggregation of κ-carrageenan double helices. Combined sodium and potassium salts of κ-carrageenan give a relatively weak gel. Calcium and potassium ions distinctly raise the gelling temperature.
ι-Carrageenan	Dissolves in presence of Na^+	Dissolves	Calcium and potassium ions distinctly raise the gelling temperature.
λ-Carrageenan	—	—	No gel formation is observed.

Source: Adapted from Yuguchi, Y. et al., *Food Hydrocoll.*, 16, 515, 2002; MacArtain, P. et al., *Carbohydr. Polym.*, 53, 395, 2003; Hermansson, A.M. et al., *Carbohydr. Polym.*, 3, 297, 1991.

4.4.4 Interactions of Agar with Other Food Components

Food components can influence the functional properties of agar when incorporated in food products. Interactions of agar with other food components also influence the sensory attributes of food products. Some of the important interactions and functional characteristics of agar are discussed below. Table 4.6 summarizes the influence of various food components on the behavior of agar in food systems.

4.4.4.1 Sugar Reactivity

High concentrations of sugars are known to modify the gel characteristics of agar, a phenomenon known as *sugar reactivity*, which can influence the gel strength of gum in products containing high levels of sugar. Sugar reactivity is observed within sugar–agar complexes in the presence of sucrose or glucose, with the sugar reactivity being more pronounced in the presence of sucrose than glucose. Maximum sugar reactivity has been reported at the 50% level of sucrose and glucose in gels containing agar (1.12%) from *Gelidiella acerosa, Gracilaria edulis, Gracilaria crassa,* and *Gelidium pusillum.* The sugar reactivity of these agars was characterized by an increase in gel strength (25 to 45%) and increase in gelling and melting temperatures (2 to 3°C). In contrast, the addition of glucose resulted in only a 19 to 34% increase

TABLE 4.6

Influence of Some Food Components on the Behavior of Seaweed Polysaccharides in Food Systems

Polysaccharide	Food Components	Influence on Gelation
Agar	Sugars (sucrose and glycerol)	Gelation modified by high concentrations (sugar reactivity); glycerol influences gelation
	Starch	Gel strength decreased
	Locust bean gum	Synergistic effect on the gel strength, elasticity, and rigidity
	Tannic acid	Gelation inhibited; glycerol counteracts effect
	κ-Carrageenan	Gel strength decreased
	Various proteins	Gelation not affected; being neutrally charged, agar does not react strongly with proteins or other charged molecules
Alginate	β-Lactoglobulin	pH-dependent formation of complexes
	Gelatin	Formation of mixed gels in presence of calcium
	Pectin	Synergistic effect on gelation
	Polyphosphates	Retard gelation
Carrageenan	Proteins (lysozyme, bovine serum albumin, whey protein isolate)	Depends on type of carrageenan; favorable interactions with milk proteins and casein micelles (milk reactivity); may modify surface hyrophobicity of protein–hydrocolloid mixtures, aiding emulsion and foam formation.
	Starch	Control of starch gelatinization and retrogradation
	Locust bean gum	Synergism in gelation properties of κ-carrageenan

in the gel strength, but the increase in gelling and melting temperatures of the agar gels was the same (2 to 3°C).[54] Increasing concentrations of glycerol produced an increasingly stronger, more deformable network.[50]

4.4.4.2 Interactions with Other Hydrocolloids

Interactions with other hydrocolloids may modify the gel characteristics of agar. Sodium alginate and starch decrease the strength of agar gels, while dextrins and sucrose increase the gel strength. Locust bean gum (LBG) is a natural hydrocolloid extracted from the seeds of the carob tree (*Ceratonia siliqua* L.) after removal of the seed coat. LBG has a marked synergistic effect on the strength of agar gels. Incorporation of LBG at 0.15% can increase the strength of an agar gel by 50 to 200%. Although 1.5% agar alone has a gel strength of 900 g/cm², a mixture of locust bean gum (0.2%) and *Gelidium* agar (1.3%) gives gel strengths of about 975 g/cm². This phenomenon has practical applications; however, not all *Gracilaria* agar shows this same synergy.[34]

Studies on the rheological properties of agar and κ-carrageenan mixtures having a total polysaccharide concentration of 1.5% (w/w) suggest a temperature dependence of the storage modulus (G') of these mixtures, as they exhibit a one-step change during cooling but two-step changes upon heating. Significant thermal hysteresis was observed in all mixtures. Moreover, the observed hysteresis was influenced by those characteristics of κ-carrageenan. The gel point as determined by the storage and loss moduli (G' and G'') of these mixtures showed a decrease in temperature as the proportion of κ-carrageenan increased. Incorporation of κ-carrageenan caused a large reduction in gel rigidity. In comparison with agar gel, the mixed gels were much more deformable, with a higher failure strain, but they had lower strength, as indicated by a marked decrease in Young's modulus and failure stress. The significance of these interactions could influence the functionality of agar in food products.[54,55] Tannic acid, which is present in large amounts in certain vegetables and fruits such as squash, apple, and prune, inhibits agar gelation if the quantity is high. Adding glycerol in small amounts can counteract the tannin effect. Being neutrally charged, agar does not react strongly with proteins or other charged molecules.[56]

4.5 Alginate

The terms *algin* and *alginate* are generic names for salts of alginic acid, such as sodium, potassium, ammonium, calcium, and propylene glycol alginates. Alginate was discovered by Edward Stanford in 1883, and its commercial production began in 1929 in California. Algin occurs in all brown seaweed in the form of insoluble mixed salts of mainly calcium, with lesser amounts of magnesium, sodium, and potassium, and is concentrated in the intracellular space. The biological function of alginates in algae is primarily to provide the strength and flexibility necessary to withstand the force of water in which the seaweed grows. The most important algal sources of alginate are *Macrocystis pyrifera*, *Ascophyllum nodosum*, and *Laminaria* spp. Other important sources are *Ecklonia maxima*, *Ecklonia cava*, *Eisenia bicyclis*, *Lessonia nigrecans*, and *Sargassum* spp.[57]

4.5.1 Extraction

Much of the seaweed for alginate extraction comes from *Macrocystis, Laminaria,* and *Ascophyllum* species from the coasts of the United States, Canada, South America, Europe, Africa, and Japan. *Macrocystis pyrifera* (giant kelp), which grows along the west coast of the North American continent, is an important alga for the commercial extraction of algin. In Canada, algin is extracted from *Ascophyllum nodosum* (rockweed), and in Europe sources include

TABLE 4.7

Contents of Algin and Its Mannuronic and Guluronic Acid Residues
in Some Seaweeds

Seaweed	Algin Content (%)	PolyM (%)	PolyMG (%)	PolyG (%)
Macrocystis pyrifera (stem)	18.2	38	46	16
Laminaria andersonii (entire plant)	22.8	—	—	—
Laminaria digitata (stem)	33.3	—	—	—
Laminaria digitata (leaves)	31.3	43	32	25
Ulva stenophylla (leaves)	40.1	—	—	—
Laminaria hyperborea (stem)	—	17	26	57
Laminaria japonica	—	48	36	16

Note: The values of polyM/polyG usually vary from 1.2 to 1.8, but they can be as low as 0.45 for *L. hyperborea* (stem) and as high as 3.0 for *L. japonica*.

Source: Adapted from Humm, H.J., in *Marine Products of Commerce*, Tressler, D.K. and Lemon, J.M.W., Eds., Reinhold, New York, 1951, chap. 5; Owusu-Apenten, R.K., *Introduction to Food Chemistry*, CRC Press, Boca Raton, FL, 2004, p. 55; Clementi, F. et al., *J. Sci. Food Agric.*, 79, 602, 1999.

Laminaria hyperborea and *L. digitata*. In India, alginate is extracted primarily from *Sargassum* brown seaweed, whose alginic acid contents vary from 5.3 to 16.6% on dry weight basis. Alginic acid content is highest in the rachid (the thickest part of a plant), although other parts of the plant (e.g., vesicles, leaves) also contain fair amounts of the phycocolloid. Table 4.7 provides the algin contents and mannuronic and guluronic acid residues of some algae.

Two widely practiced methods used to produce alginate are the Green and Le Gloahec–Herter processes.[27,32] In Green's process, fresh algae are first demineralized with 0.3% aqueous HCl and then pulverized and treated with aqueous soda ash (8 to 2.0%; pH 10 to 11). The treatment is repeated a second time, followed by grinding of the solids in a hammer mill. The product is then diluted with water and allowed to settle. The supernatant is mixed with a suitable filter aid, heated to 50°C, and passed through a plate-and-frame filter press. The filtrate is mixed with 10 to 12% aqueous $CaCl_2$ when the insoluble calcium alginate that forms rises to the surface. The lower liquid layer containing soluble salts, organic matter, and other materials is discarded. Calcium alginate is bleached with aqueous sodium hypochlorite (10%), drained, and mixed with 5% HCl. The precipitated alginic acid is thoroughly washed with water to remove the calcium completely. The purified alginic acid is generally converted to the desired salt (e.g., sodium alginate) by treatment with the appropriate carbonate, oxide, or hydroxide and then dried, ground, and packed.[27,32]

In the Le Gloahec–Herter process, initial leeching is done with 0.8 to 1.0% aqueous $CaCl_2$ to eliminate salts and other impurities without damaging the algin. After washing with water, the material is soaked in 5% HCl and

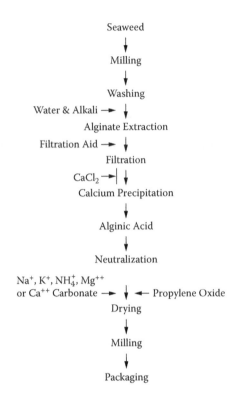

FIGURE 4.3
Extraction of alginic acid. (From Owusu-Apenten, R.K., *Introduction to Food Chemistry*, CRC Press, Boca Raton, FL, 2004, p. 55. With permission from Taylor & Francis, Ltd.)

again washed. It is then digested with a 4% solution of soda ash at 40°C and ground well for 2 to 3 hours. The paste obtained is diluted with water, bleached with H_2O_2 or ozone, and then centrifuged. The bleached liquor is treated with adsorbent materials (hydrated alumina or gelatinous silica) followed by precipitation with HCl. The separated alginic acid is washed first in water and then in ethyl alcohol to remove impurities, followed by drying. The alginic acid may be converted into appropriate salts by treating with carbonates, oxides, or hydroxides. Alginate has also been extracted using selective solvents from three species: *Saccharina longicruris*, *Ascophyllum nodosum*, and *Fucus vesiculosus*. Structural analysis (total sugars, uronic acids, sulfates, molecular weight) and rheological characterization showed important variations among species. Alginate exhibited Newtonian behavior.[57] A simple method has been devised for the separation of water-soluble polysaccharides, including alginates, laminarins, and fucoidans, based on hydrophobic chromatography.[58] One process for the extraction of alginic acid is illustrated in Figure 4.3.

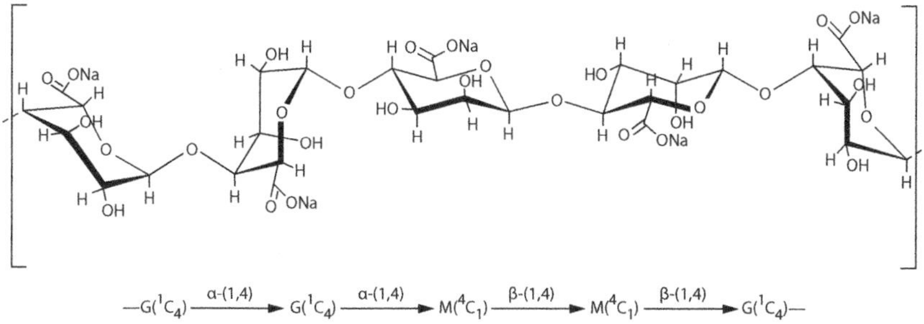

$$-G(^1C_4) \xrightarrow{\alpha\text{-}(1,4)} G(^1C_4) \xrightarrow{\alpha\text{-}(1,4)} M(^4C_1) \xrightarrow{\beta\text{-}(1,4)} M(^4C_1) \xrightarrow{\beta\text{-}(1,4)} G(^1C_4)-$$

FIGURE 4.4
Bonding of alginate repeating units mannuronate (M) and guluronate (G).

4.5.2 Composition and Structure

Alginates are linear unbranched polymers containing β-(1,4)-linked D-mannuronic acid (M) and β-(1,4)-linked L-guluronic acid (G) residues, which are negatively charged polymers like DNA. Sequences of these units may be designated as $(M)_m$, $(G)_n$, and $(M,G)_x$. Newly synthesized alginate contains entirely poly-M sequences, which are subsequently converted to guluronic acid by a mannuronic acid epimerase The ratio of D-mannuronic and L-glucuronic acids in alginic acids in a seaweed varies with its age, type of species, portions of plant used, and distance from shore. The G/M ratios are usually in the range of 1.45 to 1.85. Figure 4.4 illustrates the bonding of G and M residues in alginic acid. Mature seaweed fronds or plants harvested closer to shore have a higher G/M ratio as a result of their greater maturity and adaptation to a strong surf. Biochemical and biophysical properties of alginate are dependent on molecular weight and G/M ratios. G blocks are believed to be important to alginate structure as a function of their interaction with Ca^{2+} and H^+ binding capability. Molecular weights of alginates range between 32 and 200 kDa.[59,60]

4.5.3 Gelation and Other Properties

Alginic acid is essentially insoluble in water. Like DNA, alginate is a negatively charged polymer. The pK values of the carboxyl groups range from 3.4 to 4.4. Monovalent ions such as sodium and ammonium interact with the carboxyl groups of alginic acid to form water-soluble salts. The solutions of soluble alginates are transparent, colorless, and noncoagulable on heating, and they have a wide range of viscosity. Alginate molecules adopt an essentially random conformation in solution. The resulting viscosity increases as a function of concentration, molecular weight, characteristic G/M ratio, and polymer–polymer interactions in the semidilute regime. Addition of alkali metal ions (Ba^{2+}, Ca^{2+}, Mg^{2+}, Sr^{2+}) induces alginate gelation. The calcium

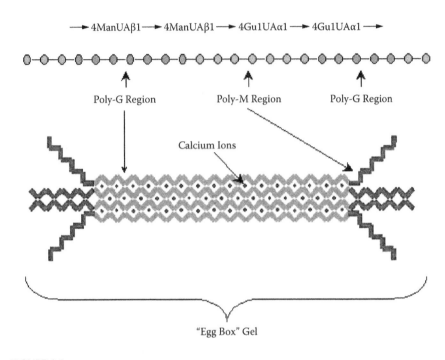

$\longrightarrow$ 4ManUAβ1 $\longrightarrow$ 4ManUAβ1 $\longrightarrow$ 4Gu1UAα1 $\longrightarrow$ 4Gu1UAα1 $\longrightarrow$

Poly-G Region   Poly-M Region   Poly-G Region

Calcium Ions

"Egg Box" Gel

FIGURE 4.5

Egg-box binding of Ca^{2+} in the gelation of alginate. (From Rastall, R., *Tailor-Made Food Ingredients: Enzymatic Modulation of Nutritional and Functional Properties*, IFIS Publishing, Reading, Berkshire, U.K., 2001. With permission from IFIS Publishing.)

salt is insoluble in water, but the magnesium salt is water soluble. Cations (except Mg^{2+}) show different affinities for alginate, and selective ion binding allows alginate to form ionotropic hydrogels. Gelation in the presence of divalent cations such as Ca^{2+} occurs without any heating or cooling. Gelation is favored by a uniform distribution of divalent cations and a polymer concentration above 0.1%, with 1% (w/w) typically being preferred.

Junction zones form due to interchain cross-linking by metal ions; for example, Ca^{2+} cross-links four guluronic acid residues from adjacent poly-G chains. The structure of the alginate gels has been described by the so-called "egg-box model," in which each divalent cation (e.g., Ca^{2+}) is coordinated to the carboxyl and hydroxyl groups of four guluronate monomers from two adjacent chains of the polymer (Figure 4.5). This structure confers a high rigidity to the parallel aggregates of polymer chains.[61]

Age and habitats of the seaweed are important in determining its composition and gelation properties. High-M alginates form turbid gels with low elastic moduli, and high-G alginates yield transparent, stiffer, and more brittle gels. Alginates with high guluronic content give gels with a higher strength than alginates with high mannuronic content. Increased numbers of polyguluronate junction zones in alginate gels result in brittle, rigid gels with syneresis, whereas fewer junction zones produce an elastic gel with a low tendency to syneresis. This has been attributed to the stronger affinity of the guluronic residues for divalent cations.[59]

Alginate undergoes gelation in the presence of cations, which is of commercial significance. The gelation can be accomplished in two ways. First, a divalent cation (usually Ca^{2+}) is diffused into a solution of sodium alginate solution. Such gels exhibit nonhomogeneous pore structures. Alternatively, Ca^{2+} ions can be released homogeneously within a sodium alginate solution. This may be achieved under neutral conditions by the dispersion of a sparingly soluble salt, such as calcium citrate.

Alginate gel beads can be prepared by dissolving sodium alginate in distilled water at a concentration of 2% (w/w). The polymer solution is added dropwise at room temperature to a stirred aqueous $0.24M$ $CaCl_2$ solution using a 0.8-mm needle syringe. The microspheres are cured in the gelation solution for 15 hours. Another method involves slowly lowering the pH of a sodium alginate solution with glucono-δ-lactone. These methods could be modified to produce many forms of alginate gel, such as fibers or films varying in molecular weights, calcium contents, particle size, and particle form (i.e., granular or fibrous).[62] Propylene glycol alginate (PGA) is the only commercially available, chemically modified alginate (coded as E405). PGA is made by bringing a partially neutralized alginic acid in contact with propylene oxide gas under pressure. The propylene oxide reacts exothermically with the alginic acid to form a mixed primary/secondary ester that is soluble in water and stable at pH 2 to 3; in the presence of calcium ions soft, elastic, nonbrittle gels are produced. PGA functions as a good foam stabilizer. Partial or total substitution of acid groups with hydroxyester reduces the capacity for gelling, facilitating the use of alginate as a densifier in acidic solutions and its potential use for biomedical applications.[59,60]

Ca^{2+}-induced gelation of alginate helps to modify the viscosity of alginate solutions. The viscosities of these solutions can range from 20 to 200 cps. Freezing and thawing of a sodium–alginate solution containing Ca^{2+} ions can result in a rise in viscosity; the viscosity decreases with increasing temperature. Viscosity is not affected by pH in the range of 5 to 10, but below pH 4.5 viscosity increases until the pH reaches 3, when insoluble alginic acid precipitates.[63] Additives such as dextran and glycerol dramatically change the viscosity of alginate solutions; however, the gelation kinetics are not affected.[64] Table 4.8 gives the characteristics of calcium alginate gel in terms of the storage and loss moduli.

Alginates are susceptible to the effects of temperature, salts, and gamma radiation. Alginate degrades when exposed to cobalt-60 radiation in a dose range of 20 to 500 kGy, both in aqueous solution or solids; the extent of degradation in solution is about 25 times higher than that of solid alginate powder. The molecular weight of alginate in 1% (w/v) solution decreased from 6×10^5 to 8×10^3 Da when exposed to 20 kGy; irradiation at 500 kGy was required for an equivalent change in molecular weight in the solid state.

TABLE 4.8

Storage and Loss Moduli of Calcium Alginate Gels After 24 hr Gelation at 23°C

Algal Source	Concentration (%)	NaCl (M)	Storage Modulus (G')	Loss Modulus (G'')
Ascophylum nodosum	1	0	20.5 ± 14.7	1.4 ± 0.1
Ascophylum nodosum	1	0.1	10.0 ± 7.6	0.8 ± 0.3
Fucus vesiculosus	2	0	60.1 ± 10.0	4.0 ± 1.3
Fucus vesiculosus	2	0.1	06.6 ± 22.0	6.0 ± 0.7

Source: Adapted from Rioux, L.-E. et al., *J. Sci. Food Agric.*, 87, 1630, 2007.

The free radicals from irradiated water were found to be responsible for the degradation in solution. Highly degraded alginate changes color to a deep brown. Ultraviolet spectra show a distinct absorption peak at 265 nm, with the absorption increasing with dose.[65]

4.5.4 Interactions with Other Food Components

Alginic acid and alginate give various material properties through their interactions with various compounds. In foods, alginate forms a gel in the presence of divalent cations such as Ca^{2+} without heating or cooling and independent of sugar contents. Interactions of alginate with major food components are discussed below.

4.5.4.1 Water

Alginate shows the highest level of water absorption among hydrocolloids, such as κ-carrageenan and xanthan, due to the presence of extensive hydroxyl groups in its structure favoring significant formation of hydrogen bonding with water. (Carrageenans and chitosan interact with water due to sulfonic groups and amino groups, respectively, as will be discussed later.) Because of this high affinity for water, a product having algin has a lesser tendency to weep. The ability of alginate to entrap water and form gels and to form and stabilize emulsions has led to many food and industrial applications. Alginate levels in food applications generally fall in the range of 0.5 to 1.5%. Sodium alginate is used as the primary source of alginate, which is a white to yellowish powder that is odorless and tasteless. It can become a thick solution when it easily dissolves in water. Dry alginate hydrogels retain water similar to the dispersion of the polymer. Applications of these hydrogels as catalysts, catalyst supports, or adsorbents allow the seaweed biomass to be used for a variety of applications, including food processing.[62] In commercial products, the molecular weights of alginate generally range between 30,000 and 200,000, primarily because of varying hydration and polymerization characteristics.[32]

4.5.4.2 Proteins

Alginate interacts with food proteins. When β-lactoglobulin and sodium alginate were mixed together at pH 3 and 4, alginate formed large complexes with protein with diameters of 1000 nm and larger due to electrostatic attractions. At pH 5, β-lactoglobulin and sodium alginate formed fairly soluble complexes, but at pH 6 and 7 the protein and sodium alginate did not form complexes due to electrostatic repulsion between the similarly charged molecules. This property could be used in the development of various food products.[66] Alginate also forms a mixed gel with gelatin in the presence of calcium. A slow release of calcium ions leads first to an irreversible alginate gel, and cooling results in a reversible gelatin gel. Between 35 and 45°C, gelation is favored by a high total biopolymer concentration or a high calcium concentration and ionic strength.[67]

4.5.4.3 Polysaccharides

The synergistic interaction between alginate and pectin has been reported. The strongest synergism was found between alginate with a high G/M ratio and pectin with a high degree of esterification (amidation). These gels showed the highest storage modulus (G′) and the fastest kinetics of gel formation. Alginate with a low G/M ratio and pectin had a lower G′ and slower rate of gelation. A relation close to 1:1 for low-G alginate and pectin resulted in gels with the highest G′.[68] Polyphosphates such as di- or trisodium phosphate and tetrasodium pyrophosphate, which are sometimes added as cryoprotectants in muscle foods, retard the gelation of alginate (see Table 4.6).[69]

4.6 Carrageenans

Carrageenan is a generic term for a complex family of anionic polysaccharides isolated from red seaweed. These water-soluble, linear biopolymers are increasingly being used as natural thickeners, formulation stabilizers, or gelling agents in applications ranging from food products to pharmaceuticals. Carrageenan is classified in three industrially relevant types: kappa (κ), iota (ι), and lambda (λ). A hybrid form consisting of κ- and ι-carrageenans is also found. Their presence in seaweed depends on the algal source, season of harvest, and extraction procedure used.[70]

4.6.1 Extraction and Characterization

Carrageenan is extracted from a wide variety of red seaweed such as *Gigartina, Chondrus, Eucheuma,* and *Furcellaria* (class Rhodophyceae) growing off the coasts of countries all around the world, including the Philippines,

Indonesia, Canada, the United States, Denmark, Chile, Spain, Japan, and France. The principal species used in the commercial production of carrageenan include *E. cottonii*, *E. spinosum*, *C. crispus* (known as Irish moss), and *G. stellata*. *Chondrus* is abundant along the Atlantic coast of North America, particularly Canada. *C. crispus* is a small bushy plant that is about 10 cm in height. *Eucheuma* occurs in the Philippines, Indonesia, and East Africa. In 2009, 169,000 t (dry weight basis) of *E. cottonii* were harvested globally, and *E. spinosum*, *Garatina*, and *C. chondrus* had a total combined production of 202,500 t.[27,31]

The seaweed is washed well to remove sand and stones and then dried quickly to prevent microbial degradation of the carrageenan. The seaweed is then shipped to processing plants. Manufacturing plants located near the harvesting site are able to utilize wet seaweed and avoid the costly drying and rehydration processes. The seaweed is subjected to extraction with dilute hot alkali (calcium or sodium hydroxide). The duration of extraction depends on the quality and condition of the raw material and other processing variables. The alkali promotes an internal rearrangement that modifies the polysaccharide backbone and gives carrageenan its gel-forming properties. The residue remaining in the extract is removed by settling. The viscous slurry is filtered employing a filter aid. Filtered liquor is concentrated by single- or multiple-stage evaporation. Carrageenan is then precipitated from the aqueous extract using isopropyl alcohol. Separated carrageenan is dried under vacuum, ground, and packed.[71]

Eucheuma cottonii contains predominantly κ-carrageenan with low levels of ι-carrageenan, methylated carrageenan, and precursor residues. The fresh seaweed is washed thoroughly with clean seawater, sun-dried, and stored under refrigeration. The dried seaweed is then soaked in water for about 24 hours to remove sand, salt, and other impurities. This process is repeated twice. The seaweed is then chopped into about 1-cm in lengths. Carrageenan has been extracted by treating 15 g dried algae in 750 mL water at temperatures ranging from 50 to 90°C for 1 to 5 hours. After extraction, the suspension was centrifuged at 12,000 rpm at 50°C for 30 minutes. One volume of supernatant was poured into two volumes of 2-propanol when the polysaccharide precipitated as long fibers. The liquor was removed by centrifugation at 12,000 rpm at 4°C for 30 minutes and the residue subjected to freeze-drying.

Generally, the molecular weight of carrageenan decreases with increasing extraction temperature. Extraction at a temperature of 50°C for 5 hours gave carrageenan with a molecular weight of 2.3×10^6 Da, and approximately 75% of the isolate contained κ-carrageenan.[72] Rather than being freeze-dried, carrageenan extracts from red algae can be dried by microwave (preferably at a frequency of 2450 MHz) at temperatures not exceeding 100°C. Optionally, the carrageenan solution can be preconcentrated to about 70% before drying. This process gives better dispersibility of the dried κ-carrageenan compared to that obtained in conventional process.[73]

Carrageenans have been isolated from other seaweed, also. *Mastocarpus stellatus* is unexploited seaweed from Portugal that could potentially be a source of κ-carrageenan. The seaweed also contained ι-carrageenan and required an alkaline pretreatment (200 g of wet algae in 8 L of $0.1M$ Na_2CO_3 at room temperature for 20 to 70 hours), followed by thorough washing and drying at 60°C for 48 hours. Carrageenan was extracted from the dried seaweed at optimum conditions—namely, treatment at 96°C for 4 hours at a pH of 8.0. The extract was filtered with metallic screens followed by cotton cloths prior to water evaporation performed at 60°C. The concentrate of carrageenan was precipitated in 95% ethanol, washed with fresh ethanol, dried at 60°C under vacuum, and milled. The resulting powder was purified by mixing in hot distilled water for 1 hour and subsequent centrifugation at 38°C. The supernatant was finally recovered and dried at 60°C under vacuum.[74]

Another source of κ-carrageenan is ibaranori, the red seaweed *Hypnea charoides* Lamoroux, which was purified by gelation with KOH. The polysaccharide gelled at 0.2% concentration and was composed of D-galactose, 3,6-anhydro-D-galactose, and ester sulfate in a molar ratio of 1.2:0.9:1.2.[75] An alkali process has been reported for the extraction of carrageenan from *Eucheuma cottonii* and *Gigartina*. The treatment did not significantly affect the gelling behavior of the polysaccharide. Optimal gel firmness (157 g/cm^2) was obtained when *E. cottonii* was subjected to extraction for 120 minutes at 100°C and pH 7.[76] The yield of carrageenan from *E. isiforme* was 57% of dry weight and decreased to 43.5% when the alga was alkali treated. The treatment also decreased viscosity from 144.6 cPs to 113.9 cPs. Alkali treatment also reduced the sulfate content by 19.3% and increased 3,6-anhydro-D-galactose content by 13%. Alkali-treated carrageenan formed very weak gels in 1.5% solutions.[33] A λ-like carrageenan was produced from *Halymenia durvillaei*, a red seaweed that grows widely in the Philippines. Maximum extraction was achieved employing a ratio of seaweed to hot water of 1:40 (w/v). An average yield of 29% was obtained using two extractions followed by precipitation of the carrageenan with isopropyl alcohol. In India, ι- and κ-carrageenans were extracted from two marine algal species, *Sarconema filiforme* and *Hypnea valentiae*, collected from Tamil Nadu. The carrageenans were extracted with water after an initial short pretreatment with cold, dilute HCl, followed by alcohol precipitation.[77,78]

4.6.2 Composition and Structure

Commercial carrageenans consist of the three types—namely, κ-, ι-, and λ-carrageenan. They have molecular weights in the range of 10^5 to 10^6 Da. The three carrageenans differ in their chemical compositions and structures. They differ prominently in their contents of sulfate groups; for example, κ-carrageenan has a 3,6-anhydrogalactose and only one sulfate ester group, making it less hydrophilic and less soluble in water. The polysaccharide is composed of D-galactose, 3,6-anhydro-D-galactose, and ester-bound sulfate in a molar ratio of 6:5:7. Also, λ-carrageenan has no

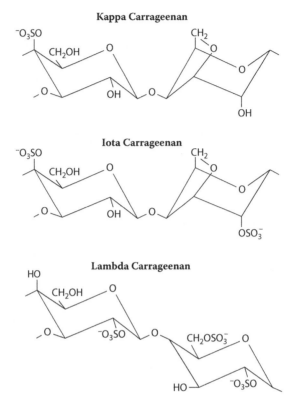

FIGURE 4.6
Chemical structures of carrageenans.

3,6-anhydrogalactose but has three sulfate groups and hence is readily soluble in water due to its strong hydrophilic character. With a 3,6-anhydrogalactose and two sulfate ester groups, ɪ-carrageenan is intermediate. The structure of ɪ-carrageenan consists of an alternating disaccharide repeating unit of (1,3)-linked β-D-galactopyranosyl-(1,4)-sulfate and (1,4)-linked 3,6-anhydro-α-D-galactopyranosyl-(1,2)-sulfate residues. In ɪ-carrageenan, the anhydro-galactose residue carries a sulfate group, but it is absent in κ-carrageenan. Both κ- and ɪ-carrageenan exist as right-handed, threefold helices that reversibly form double helices. The double-helix segments can then interact to form a three-dimensional network.[3] Figure 4.6 illustrates the chemical structures of the different carrageenans.

4.6.3 Solubility Properties and Stability

Carrageenans differ in their solubilities depending on their chemical compositions and hydrophilic character. The potassium salt of κ-carrageenan is practically insoluble in cold water, whereas the sodium salt readily dissolves in water. Similarly, although sodium salts of ι-carrageenan are soluble in cold water, K$^+$ and Ca^{2+} salts of κ- and ι-carrageenans are not soluble. However, the potassium salt of ι-carrageenan swells markedly in water. Solutions of ι-carrageenan alone will tolerate high concentrations of electrolytes such as NaCl up to 20 to 25%, whereas κ-carrageenan will be salted out. Both κ- and λ-carrageenans are soluble in hot (70°C) sucrose solutions up to 65%, but ι-carrageenan is not easily soluble in sucrose solution at any temperature; λ-carrageenan is also water soluble in all of its salt forms. Commercial preparations of λ-carrageenan mostly contain some κ-carrageenan, which makes it less salt compatible. Because of differences in their solubility properties, κ-, ι-, and λ-carrageenans vary in their thickening ability and gel strength. Acid and oxidizing agents may hydrolyze carrageenans in solution, leading to loss of properties through cleavage of glycosidic bonds. Very stable emulsions of carrageenans can be made by mechanical methods. Carrageenans at 2.5% can emulsify an equal volume of oil such as cod liver oil. Various aspects of carrageenans isolated from red seaweed have been discussed with respect to their functionality.[3]

4.6.4 Gelation

Thermal gelation is a valuable property of carrageenans that determines their diverse applications including foods and pharmaceuticals. Carrageenans differ in their ability to undergo gelation. While κ- and ι-carrageenans form gel, λ-carrageenan does not gel and behaves as a normal polyelectrolyte in solution. There is general agreement on the mechanism of gelation of the polysaccharide, which involves transition from a disordered (random coil) at higher temperatures to an ordered (helical) state upon cooling, as shown in Figure 4.7. In general, carrageenan gives an ideal gel at 40°C when dissolved in water at 1% (w/v) concentration; it exhibits strong elasticity, with a storage modulus (G') of 4485 Pa.

At high temperatures, carrageenans exist in solution in a disordered chain conformation, but on cooling a rigid ordered double helical structure is adopted which upon reheating again melts, thus giving reversible gels, like agar. Of the three types of carrageenans, κ-carrageenan gel offers the best properties. Helix formation and gelation are cation specific. κ-Carrageenan has higher gel strength than ι-carrageenan at the same concentration. In general terms, κ-carrageenan gels are hard, strong, brittle, and freeze–thaw unstable, whereas ι-carrageenan forms soft and weak gels that are freeze–thaw stable. Both κ- and ι-carrageenan helices are right-handed, double helices that interact to form a three-dimensional network. The gelling temperature of κ-carrageenan ranges from 35 to 65°C, and the melting temperature varies between 55 and 85°C. The extent of hysteresis is dependent on the type of carrageenan. For κ-carrageenan, it is 10 to 15°C; for ι-carrageenan, about 5°C. Gels of κ-carrageenan show thermoreversible setting and melting behavior.[79,80]

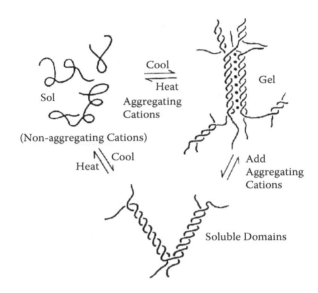

FIGURE 4.7
A model for the gelation of carrageenan. (From Rastall, R., *Tailor-Made Food Ingredients: Enzymatic Modulation of Nutritional and Functional Properties*, IFIS Publishing, Reading, Berkshire, U.K., 2001. With permission from IFIS Publishing.)

Beads of κ-carrageenan gel can be prepared when hot (80°C) droplets of a 2.5% (w/w) solution of the polysaccharide (e.g., from *Eucheuma cottonii*) are added dropwise, under stirring, into a 0.6*M* KCl solution using a syringe with a 0.8-mm-diameter needle at 5°C. The gel beads are aged for 12 hours in the solution at 5°C without stirring and finally washed with cold water. If κ-carrageenan solution is extruded into a commercial solution of potassium chloride, fibers of the polysaccharide are precipitated. The precipitated mass exudes free water and is dewatered under pressure to make "gel press" carrageenan. The fibers may be frozen and thawed to assist the dewatering step. The pressed fibers are then dried and ground to the appropriate particle size.[62]

The viscosities of carrageenans depend on concentration, temperature, the presence of other solutes, and molecular weights. The viscosity increases almost exponentially with concentration. Salts lower the viscosity of carrageenan solutions by reducing the electrostatic repulsion among the sulfate groups For the gelling types of carrageenan (κ- and ι-carrageenan), the viscosity measurement is carried out at high temperature (e.g., 75°C) to avoid the effects of gelation, usually at a concentration of 1.5% (w/v). For the coldwater-soluble, nongelling λ-carrageenan, viscosity is measured at 25°C at 1.0% concentration. Viscosity is usually measured with easily operated rotational viscometers (e.g., Brookfield). Commercial carrageenans are generally available in viscosities ranging from 5 to 800 cps. The solutions of carrageenans having viscosities less than 100 cps display Newtonian flow, whereas varying degrees of pseudoplasticity are exhibited by the sodium salt of ι-carrageenan. The calcium salt of ι-carrageenan exhibits a thixotropic character, typified by a decrease in viscosity with increasing shear or agitation and returning to normal viscosity with a decrease in shear.

Monovalent cations, such as potassium, rubidium, and cesium, strongly promote the gelation of κ-carrageenan. The sodium form has a network structure with flexible superstrands of uniform thickness. Cations such as Li^+, Na^+, K^+, or Cs^{2+} reduce electrostatic repulsion between chains, promoting formation of well-defined, double-stranded helices. The effectiveness of salts followed the sequence of $K^+ > Ca^{2+} > Na^+$. Strong synergistic effects were found between Ca^{2+} and K^+ with regard to the gel strength of κ-carrageenan. Synergistic effects were also observed when Na^+ was added to potassium κ-carrageenan. In addition to salts, molecular weights of carrageenans also influence their gelation, as shown in the case of *Eucheuma cottonii* κ-carrageenan; its aggregation rate decreases with decreasing size.[81] Enzymes can be used to modify carrageenan structure and its functionality.[79]

Rheological measurements have been helpful in understanding the behavior of carrageenan gels.[82] Typically, 1% (w/v) aqueous solutions of the polysaccharide are used. The solutions, prepared in water in closed tubes, are left overnight at 4°C and are then heated at 90°C for 30 minutes while stirring. Dynamical rheology measurement is carried out using a HAAKE Rheo Stress rheometer. The temperature dependence of the storage modulus (G') and loss modulus (G'') can be observed by using a cooling system to reduce the temperature from 85°C to 10°C. Cooling scans are performed at 1°C/min, with measurement of G' and G'' at 1 Hz and the gap set to 1 mm. (During measurement, the edge of the sample is covered with a moistened sponge to minimize water evaporation.) Dynamic torque sweeps are conducted to select a common linear viscoelastic region for all system combinations. The storage and loss moduli of the gels differ depending on the extraction conditions and molecular weights of carrageenans; for example, extraction under nonalkaline conditions results in carrageenans with higher molecular weights and hence greater gel strength.[72]

The influence of temperature on the flow behavior of carrageenan solutions was demonstrated when the shear rate of samples increased from 0 to 300 s^{-1} in 3 minutes, held at the highest rate for 10 minutes, and then decreased linearly back to 0 over 3 minutes. Shear thinning behavior was observed in all samples for the upward and downward curves of rheograms. Yield stresses were observed in carrageenan at 20°C and 40°C. The consistency coefficient and flow behavior index were both sensitive to changes in temperature and concentration.[79] Figure 4.8 illustrates the temperature dependence of the dynamic storage modulus and dynamic loss modulus of carrageenan at 50°C.[72]

Rheological studies have revealed the influence of salts on carrageenan gelation, with gel stiffness increasing with polysaccharide and salt concentrations. Flow curves of a solution of calcium salt of ι-carrageenan from the

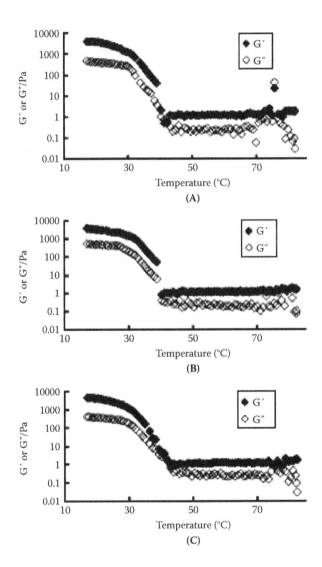

FIGURE 4.8
Temperature dependence of dynamic storage modulus G' (◆) and dynamic loss modulus G'' (◇) of carrageenan at 50°C: (A) 1 hour, (B) 2 hours, and (C) 3 hours. (From Montolalu, R.I. et al., *J. Appl. Phycol.*, 20, 521, 2008. With permission from Elsevier/Rightslink.)

red seaweed *Eucheuma serra* showed plastic behavior, and yield values were 0.4, 1.7, and 7.7 Pa at 0.1, 0.2, and 0.3% (w/v) concentrations, respectively.[83,84] Rheology of carrageenan isolated from *Mastocarpus stellatus* correlates with the degree of sulfate groups, the molecular weight distribution, and ι-carrageenan monomer content.[85]

4.6.5 Antimicrobial Activities

Carrageenans offer antimicrobial activities against foodborne pathogenic bacteria, such as *Salmonella enteritidis*, *S. typhimurium*, *Vibrio mimicus*, *Aeromonas hydrophila*, enterotoxigenic *Escherichia coli*, and *Staphylococcus aureus*. The growth of all the bacterial strains was significantly inhibited by the carrageenans, particularly by ι-carrageenan. A growth inhibition experiment using *S. enteritidis* showed that the inhibitory effect of the carrageenans was not bactericidal but bacteriostatic. The sulfate residues in carrageenan played an essential role in this effect (see Chapter 7).[86]

4.6.6 Determination and Characterization of Carrageenan in Food Products

The demand for carrageenans is steadily rising as new applications are identified. Quantitative determination of carrageenan, therefore, is of great importance and a significant challenge for manufacturers wishing to deliver constant quality and for food technologists exploring potential new applications. Various methods are available for this purpose, including colorimetric methods, light microscopy, immunological detection, electrophoresis, and nuclear magnetic resonance (NMR), as well as chromatographic methods, coupled with chemical or enzymatic depolymerization procedures or high-performance anion exchange chromatography.[87] A sensitive resorcinol reagent has been used for several decades for the colorimetric determination of fructose and of 3,6-anhydrogalactose in agar, carrageenan, and other algal polysaccharides.[88] Another method is a methylene blue binding assay that produces a color change from blue to purple when the dye binds with polysaccharides (including carrageenan) at concentrations as low as 0.02%.[89] Binding of carrageenan with another dye, Alcian blue, is a rapid method for its determination in foods such as jellies and salad dressings.[90]

A methodology for estimating carrageenan in dairy products involves the digestion of the dairy product with papain at 70°C in the presence of $1.0M$ NaCl at pH 8.0 to 8.5. The digest is filtered over glass wool using celite as a filter aid, followed by precipitation of the carrageenan with cetylpyridinium (CP) chloride in the presence of $1.0M$ KCl. The carrageenan precipitate is thoroughly washed with 0.1% CP and $0.05M$ KCl followed by its dissolution in 30% H_2SO_4. The carbohydrate content is then determined by the phenol-H_2SO_4 method.[91] Characterization of carrageenan by chemical means is long and complex. H^1 and C^{13} high-resolution NMR spectroscopy, antibody- and lectin-based assays, and Fourier transform infrared spectroscopy are some of the newer methods used to differentiate κ-, ι-, and λ-carrageenans. To determine κ- and ι-carrageenans in foods, samples are homogenized and freeze-dried prior to release of 3,6-anhydrogalactose dimethylacetal, characteristic of gelling carrageenans, by mild methanolysis; the results are subsequently quantified by reverse-phase high-performance liquid chromatography analysis. This method is not applicable for the analysis of λ-carrageenans, as they are devoid of 3,6-anhydrogalactose. Various measurements of carrageenans in foods have been compiled using the rapid enzyme-linked lectin assay (ELLA) and various other techniques.[92]

4.6.7 Interactions with Food Components

Understanding the interactions of carrageenans with food components is important to the food industry because of their possible applications as texturizers and stabilizers. These food components include water, salts, proteins, starches, and other polysaccharides. Carrageenans, being hydrocolloids, interact with water to give hydrogels, particularly in the case of κ- and ι-carrageenans; their gelation properties are influenced by salts, as discussed earlier. These properties have a profound influence on utilizing carrageenans in the development of various food products. Interactions of carrageenans with some major food components are briefly discussed below.

4.6.7.1 Proteins

Electrostatic interactions of carrageenans with proteins in food systems play an important role in determining their functional behavior. Such interactions lead to either complex formation due to attractive forces or segregation due to repulsion. These interactions are influenced by pH and ionic strength, which determine the stability and water-holding capacity of the products. Mixing carrageenan with proteins variously increased gelation temperatures and storage moduli of the polysaccharide gels. Protein addition also resulted in higher melting temperature, hardness, cohesiveness, gumminess, and springiness of carrageenan gels and reduced syneresis. Proteins such as lysozyme, bovine serum albumin, and whey protein isolate are protected by ι-carrageenan, which enhances their denaturation temperatures, presumably due to the formation of hydrogen bonds and blockage of hydrophobic binding sites of proteins, preventing aggregation. The type of protein affects the outcome of such interactions; for example, denatured soy protein has a profound effect on melting temperature but less of an effect on thermal hysteresis than native soy protein. Similarly, native soy protein increases aggregation rate and maximum viscosity compared to those of carrageenan and β-lactoglobulin gels. Similarly, the ability of carrageenan to increase the surface hydrophobicity of protein–hydrocolloid mixtures can improve their ability to adsorb at an oil–water or air–water interface.[93] λ-Carrageenan can enhance the stability of β-lactoglobulin and thereby inhibit heat denaturation and aggregation.

The foaming capacity and thermal stability of complexes of κ-carrageenan with β-lactoglobulin, native soy protein, and denatured soy protein have been investigated. Gelling and melting temperatures of the complexes, as determined by dynamic rheology, were related to the thermal stability of foams. The presence of κ-carrageenan reduced foam expansion due to the higher viscosity of the systems, and κ-carrageenan induced a greater synergistic effect with denatured soy protein, reflected in a faster gelling rate and increased foam stability.[94] Hydrolysis of sunflower protein improved its foam overrun and stability against liquid drainage and collapse. Carrageenan at 0.25% (w/w) stabilized the foam of the hydrolyzed proteins.[95]

At 0.2%, κ-carrageenan prevented calcium-induced precipitation of proteins of coconut, glandless cottonseed, peanut, and soy protein isolates in aqueous solutions at neutral pH ranges.[96] Mixtures of whey protein isolate and ι-carrageenan did not exhibit visual phase separation, which was attributed to the high gelling ability of ι-carrageenan.[97] Addition of 0.2% carrageenan led to an increase in water-holding capacity, gel strength, and hardness of salt-soluble protein isolates of ham in the presence of NaCl and KCl at pH 6.2. Although carrageenan did not interact with the meat proteins in network formation, the hydrocolloid was present in the interstitial spaces of the protein gel.[98] Recently, fish gelatin has been examined as a substitute for animal gelatin.[15] Fish gelatin, however, has relatively low gelling and melting temperatures and gel strength. The addition of 1% κ-carrageenan and KCl to 2% fish gelatin resulted in enhanced gel strength. Storage and loss moduli for the gelatin–carrageenan mixtures were higher than those for either gelatin or κ-carrageenan, suggesting that the two components reinforce each other. The Young's modulus of the gel decreased considerably at pH < 6.0, but increases in pH above the isoelectric point (approximately 8.7) did not have a major effect on gel strength.[99] Carrageenan along with starch enhanced the gel strength and rigidity of cooked giant squid muscle proteins.[100]

4.6.7.2 Milk Reactivity

Carrageenans are used extensively in dairy products because of their favorable interactions with milk proteins and casein micelles. Bulk phase separation in dairy products systems can be prevented by using any of the carrageenan types at concentrations as low as 0.03%. The effect is due to absorption of the carrageenans onto casein micelles, with λ-carrageenan adsorbing at temperatures between 20° and 60°C and κ- and ι-carrageenans adsorbing at 40°C or below. Similarly, at 60°C, κ- and ι-carrageenans induce flocculation of casein micelles above a critical concentration. On cooling, systems containing ι-carrageenan form a network at the helix–coil transition temperature. The network formation involves electrostatic interactions between negatively charged sulfate groups of carrageenan and a positively charged region of casein. In the presence of excess carrageenan, this network is further reinforced by cross-links among carrageenan moieties. As a consequence of this property, the concentration of carrageenan required for gelling in milk is much lower, a phenomenon known as *milk reactivity*.

Dilution of milk up to 75% does not cause much loss of gel strength; however, the presence of extra casein micelles does not increase gel strength much. The synergistic effect of casein micelles on carrageenan gelation is higher compared to any other proteins. In acidified milk, though, carrageenan does not show milk reactivity, probably due to destruction of the carrageenan. Semifinished carrageenan obtained from *Sarconema filiforme* in India exhibited good milk reactivity, suggesting its potential for the preparation of milk products.[77]

4.6.7.3 Starch

Retrogradation is a progressive reassociation of gelatinized starch molecules upon aging which may reduce the digestibility of the starch (see Chapter 2). A number of hydrocolloids, including carrageenan, are capable of influencing gelatinization and retrogradation of such starches as those of waxy maize, waxy rice, tapioca, regular rice, potato, and wheat, as reflected in changes in their viscosities. When suspensions of normal and waxy rice starches were heated in the presence of carrageenans, the following were observed: (1) increase in apparent pasting temperatures, while peak and final viscosities decreased in the order for the carrageenan types of $\lambda > \iota > \kappa$; (2) less solid appearance for the rice starch–hydrocolloid pastes than the control; (3) increase in apparent viscosity and consistency coefficient values of both normal and waxy starches, with a decrease in flow behavior index values; (4) increase in the hardness and adhesiveness of normal wax starch pastes by the addition of κ- and ι-carrageenans but not by λ-carrageenans; and (5) appearance of a phase-separated microstructure of starch–hydrocolloid pastes.[101]

During gelatinization of corn starch, the addition of ι-carrageenan lowered the swelling temperature with decreased peak viscosity, whereas the temperature increased in the presence of divalent cations. Similarly, the pasting peak viscosity of potato starch was greatly decreased by carrageenans and by alginate and xanthan. Most dairy dessert gels contain starch, which is added as a thickening or gelling agent mainly to improve the mouth feel. Generally, liquid dairy desserts use a combination of carrageenan and frequently nongelling waxy maize starch as the gelling system. κ-Carrageenan gelation is negatively affected by starch addition, but ι-carrageenan does not show this characteristic. Incorporation of carrageenan offers a low processing viscosity, reducing negative effects of high starch levels on flavor and mouth feel, in addition to giving dairy desserts better texture. These interactions are therefore of great consequence when carrageenans are used in food products.[101–103]

4.6.7.4 Other Polysaccharides

The behavior of mixed gels of carrageenan and other polysaccharides differs from that of individual components. Blends of κ-carrageenan and locust bean gum show synergism in gelation properties. The two gums together produce a much more elastic gel with markedly greater gel strength and less syneresis. Although κ-carrageenan at 0.5% cannot form a strong gel, it readily forms a gel in the presence of LBG or taragum at a ratio of 1:1. Because of their higher strength, carrageenan–LBG gels find use in foods such as canned pet foods, fruit gels, and processed meat and seafood. The synergistic effect of carrageenan and LBG also enhances the texture of low-fat yogurt, cheese spreads, cottage cheese, cream cheese, dips, ice creams, and whipped toppings.[3,103,104] Mixed gels of carrageenan and agar, with a total polysaccharide concentration of 1.5% (w/w), are formed when gelation occurs at temperatures required by the agar. κ-Carrageenan causes a large reduction in gel rigidity. In comparison with agar gel, the mixed gels are more deformable, with a higher failure strain, but have less strength, as indicated by a marked decrease in Young's modulus and failure stress. A decrease in pH below 4.5 was found to cause a sudden drop in the gelling temperature and G', causing a weaker and more brittle gel.[54]

Sweeteners such as sucrose influence κ-carrageenan gelation. The addition of 10% sucrose increased G' and G'' values of soft κ-carrageenan gels slightly; however, no marked changes in the viscoelastic properties were observed upon the addition of aspartame.[105] Rheological studies indicate that the dynamic Young's modulus and melting point of κ-carrageenan gels increase with increase in sugar content; however, the excessive addition of sugar immobilizes the free water necessary for junction zone formation. Furthermore, synergistic interactions with other hydrocolloids such as LBG and agar, as discussed above, could also be beneficially applied in food product developments.[64] These studies have clearly established that the effect of carrageenans as food additives on food texture and stability is strongly influenced by its interactions with other food components.

4.7 Other Seaweed Hydrocolloids

In addition to the hydrocolloids discussed above, certain other polysaccharides derived from seaweed have important food and allied applications. These are briefly discussed below.

4.7.1 Ulvan

Marine green algae (Ulvales, Chlorophyta), common on all seashores, are the sources of ulvan, a soluble sulfated polysaccharide that represents about 8 to 29% of the algae dry weight. The physicochemical, rheological, and biological properties of ulvan offer novel potential applications.[106] When ulvan from the green seaweed *Utricularia rigida* was extracted sequentially with oxalate, 1*M* and 4*M* KOH, sodium chlorite, and 4*M* KOH again, it proved to be the main component. The polysaccharide contained 23 to 35% sulfate ester groups, 10 to 15% uronic acid, and 3.8 to 4.5% protein.[42] Ulvan was recently extracted from Indian samples of *Enteromorpha compressa*, another important green alga. The polysaccharide was treated for depigmentation by treatment of the dry algal flour with acetone and ethanol for a period of 48 hours each. The depigmented algal powder was then extracted with 100-mM ammonium oxalate in 50-mM sodium acetate buffer at pH 5 and 80°C at a solvent-to-powder ratio of 150:1; 23% of the weight of the dried powder was ulvan.[41] Ulvan mainly consists of rhamnose with variable contents of glucose and fucose and trace amounts of xylose, galactose, and mannose. It is a gelling polysaccharide of potential economical value which forms a weak gel at 1.6% (w/v) in deionized water. It also plays a nutritional role as dietary fiber.[10]

4.7.2 Fucoidan

Fucoidan is a sulfated polysaccharide having an average molecular weight of 20,000 Da; it is found mainly in various species of brown seaweed dominating the North Sea. Until recently, it was considered to be a byproduct of the alginate industry. Fucoidan may constitute 25 to 30% of the algae dry weight depending on the seaweed species and season. Fucoidan can be extracted from algae cell walls by treatment with hot mild acid. The crude fucoidan extracted with hot water (60 to 70°C) is further purified by hydrophobic chromatography, followed by fractional precipitation with cetyltrimethyl-ammonium hydroxide or cetylpyridinium chloride, further solubilization with acetic acid to release the polysaccharide from its precipitated salt, and subsequent precipitation with ethanol or KCl. A protease treatment can accelerate the release of the polysaccharide from bound proteins. The purified fucoidan can be subjected to enzymatic or acid hydrolysis to obtain low-molecular-weight fucoidan oligomers for biological applications.[107] The yield of fucoidan extracted from *Fucus serratus* L. and *F. vesiculosus* by 0.1-N HCl was 0.5% based on wet weight of the seaweed. The total carbohydrate, sulfuric acid, ash, and moisture contents of fucoidan having a molecular weight of 2.4×10^5 Da were 0.5, 30.8, 22.3, and 3.8%, respectively.[108]

Fucoidan exists in two distinct forms. F-Fucoidan represents more than 95% of the fucoidan in seaweed and is composed of sulfated esters of L-fucose. In F-fucoidan, depending on the source of the algae, the sulfated polysaccharide consists mainly of L-fucose units, but it can also contain minor amounts of sugars such as galactose, mannose, xylose, or uronic acid and sometimes proteins (comprising less than 10%). Fucoidan has an α-(1,3)-backbone of repeating disaccharide units of α-(1,3)- and α-(1,4)-linked fucose residues. Depending on the structure of the main chain, fucoidan may be sulfated at C_4, C_2, or both C_2 and C_4 positions of fucose units. Some may also be both sulfated and acetylated. The other fucoidan, U-fucoidan, contains glucoronic acid, approximately 20% of its weight.[109,110] The commercially available fucoidan from *Fucus vesiculosus* is a heterogeneous mixture of more than 15 different fucans with varied properties of industrial monosaccharide moieties. Forms of fucoidan have also been found in the sea cucumber *Nemacystus decipiens*, which is cultured in Japan at the annual rate of 2000 t.[111] Fucoidan exhibits appreciable viscosity in water. The *Fucus vesiculosus* fucoidan exhibited Newtonian behavior and had the highest viscosity. Gelation of fucoidan was not observed up to 25% concentration. A sol–gel transition is induced by the addition of glycerol in aqueous solutions containing high concentrations of fucoidan. Mixed gels of fucoidan and gelatin having a soft texture can be prepared. Similarly, bovine serum albumin, when mixed with fucoidan, gives a viscoelastic solution. Fucoidan has been consumed for a long time in Japan, China, and Korea as part of whole seaweed, and it is used as a nutraceutical in Australia and the United States.[57]

4.7.3 Furcellaran

Furcellaran is also called *Danish agar*. In contrast with agar, however, which has a very low sulfate content (below 4.5%), furcellaran contains significant sulfate content (12 to 16%). *Furcellaria* species that produce the polysaccharide are found in the cold waters around Northern Europe and Asia. Furcellaran is generally produced by the red seaweed *Furcellaria fastigiata*. After alkali treatment of the algae, the polysaccharide is isolated using hot water. The extract is then concentrated under vacuum and seeded with 1 to 15% KCl solution. The separated gel threads are concentrated further by freezing, and the excess of water is removed by centrifugation or pressing, following by drying. The product is a potassium salt of the polysaccharide containing 8 to 15% KCl. Furcellaran is composed of D-galactose (46 to 53%), 3,6-anhydro-D-galactose (30 to 35%), and sulfated portions of both of these sugars (16 to 20%).

The structure of furcellaran is similar to κ-carrageenan. The essential difference is that κ-carrageenan has one sulfate ester per two sugar residues, and furcellaran has one sulfate ester residue for every three to four sugar residues. Under the appropriate conditions, aqueous preparations of furcellaran can be induced to form thermoreversible gels. At a given rigidity, the turbidity, thermal hysteresis, and tendency for syneresis follow the order of furcellaran > κ-carrageenan > ι-carrageenan, implying that these characteristics are favored by a low level of sulfation. Similarly, the concentration required to produce a gel of defined properties follows the order ι-carrageenan > κ-carrageenan > furcellaran. Gelation can be induced by either mono- or divalent cations. Gelation of furcellaran and the carrageenan is accompanied by a reversible change in optical rotation at visible wavelengths.[112]

4.7.4 Floridean Starch from Red Algae

Isolation and physicochemical characterization of a carbohydrate known as Floridean starch (amylopectin-like glucan) from three species of agarophytic red macroalgae have been reported. The starch is glycogen in fungal and

TABLE 4.9

Future Priorities in Seaweed Research

Taxonomic studies, both classical and molecular, for cataloging as well as improvement

Diversified uses of seaweed as feed, fodder, fertilizer, etc.

Creation of a biodiversity database

Refinement and quality control of processes required to prepare agar, alginate, and other products

Ecological and enzyme immunoassay studies pertaining to the introduction of exotic species

Improvement of strains through biotechnological intervention such as tissue culture, genetic transformation by protoplast fusion and hybrid strain production, gene and molecular mapping

---

Source: Adapted from NAAS, *Seaweed Cultivation and Utilization*, NAAS Documents on Policy Issues, National Academy of Agricultural Sciences, New Delhi, India, 2003. With permission from the National Academy of Agricultural Sciences.

animal cells and is similar to plant starches. Unlike plant starches, Floridean starch does not have amylase activity. Floridean starch has a low level of covalently linked phosphate.[113] Future priorities in seaweed research with a view to enhance their varied applications, including isolation of diverse polysaccharides, are provided in Table 4.9.

4.8 Polysaccharides from Microalgae

As discussed in Chapter 1, microalgae is the largest primary biomass, covering almost three quarters of the Earth's surface and serving as the very foundation of marine food. Several species of microalgae, including cyanobacteria (blue–green algae), synthesize copious amounts of polysaccharides, which are present in their slimes or are released into the growth media. These polysaccharides can have interesting biological activities and potential commercial applications; therefore, attempts have been made to isolate and cultivate these organisms to isolate the polysaccharides and other compounds.[114]

The type and amount of polysaccharides produced depend on the species and cultivation conditions. Some of the microalgae so far commercially cultivated include *Chlorella* and *Spirulina*, among other species of cyanobacteria. *Phormidium* is the best producer of extracellular polysaccharide.[115,116] The microalga *Chlorella pyrenoidosa* has been shown to yield polysaccharides at a rate of 44.8 g/kg under ideal cultivation conditions.[117] Cyanobacterial strains of *Cyanothece*, *Oscillatoria*, and *Nostoc* were studied for the physicochemical composition of the extracellular polysaccharides (EPSs) secreted during controlled growth conditions. The EPSs, upon hydrolysis, gave mannose, glucose, xylose, and ribose, in varying proportions. The polysaccharides had high thermal stability. In the presence of a 0.1M NaCl aqueous solution, the intrinsic viscosity of polysaccharides from *Oscillatoria* and *Nostoc* decreased 1.6-fold, in comparison with a 3- to 5-fold reduction in intrinsic viscosity of commercially available guar and xanthan gum.[118,119]

Microbial mats present in two shallow atolls of French Polynesia were characterized by high amounts of extracellular polysaccharide associated with cyanobacteria as the predominating species. Cyanobacterial isolates recovered from Polynesian microbial mats were cultured using closed photobioreactors.[123] The cyanobacterium *Anabaena* produced a heteropolysaccharide, with xylose, glucose, galactose, and mannose as the main neutral sugars, during the stationary growth phase in batch culture.[119] Similarly, some marine diatoms isolated from Portugal waters have been cultured for extracellular polysaccharides. Nutrient status and light intensity influence polysaccharide biosynthesis by these organisms.

The exopolysaccharides (mainly polymers of uronic acids) produced by three species of diatoms were comparable.[120] Cells of the diatom *Amphora coffeaeformis* were grown in batch cultures under continuous illumination at 18°C for 10 days. The microalgal cells were removed by centrifugation and lyophilized, and the exopolysaccharides were isolated.[121] EPSs from the benthic marine epipelic diatoms *Navicula salinarum* and *Cylindrotheca closterium* have also been isolated. The EPSs consisted predominantly of polysaccharide, but small quantities of protein were also present. The EPSs of both species contained glucose and xylose as the main constituents together with other monosaccharides in smaller quantities. The exopolysaccharides also contained uronic acids and SO_4 groups. The rate of EPS production in batch culture was highest during the transition from exponential growth to stationary growth.

Drying polysaccharides during their isolation, especially at temperatures above 90°C, can cause a significant decline in their biological activities and rheological properties, which need to be maintained for optimal utilization. High-temperature drying of the polysaccharide from the red microalga *Porphyridium* damaged its structure and was associated with significant conformational alterations in the polymer chains and changes in the interaction between the polysaccharide and the glycoprotein to which it is noncovalently associated. Drying at higher temperatures also increased the bound water content due to dissociation of the polymer chains. Therefore, a modified two-step process in which free water is removed by convection and bound freezing water is removed by lyphophilization has been suggested for optimal retention of the functionality of the polysaccharide. The technique could also be adopted for polysaccharides from other sources.[122]

Extensive chemical analyses suggest that cyanobacterial polysaccharides contain various acidic and neutral sugars. Glucose and galactose are the neutral sugars widely present in the microalgal polysaccharides, although other sugars such as mannose and xylose are plentiful in some species. Glucosamine and uronic acids are also found in the extracellular polysaccharides of some cyanobacteria. Most polymers are anionic in nature due to the presence of uronic acids or other charged groups such as sulfate or pyruvate. Cyanobacterial isolates recovered from Polynesian microbial mats consisted of 7 to 10 different monosaccharides with neutral sugars predominating; four isolates exhibited sulfate contents ranging from 6 to 19%. They also contained uronic acids.[116,123,124] The microalga *Chlorella pyrenoidosa* produced two polysaccharides having molecular weights of 81,877 Da and 1749 Da. The major monosaccharide in both of them was mannose (76 to 78%) followed by glucose (8 to 13%).[117]

Analysis of the chemical composition, structure, and physicochemical properties of polysaccharides from red microalgae (*Porphyridium aerugineum* and *Rhodella reticulata*) revealed differences in the composition of the extracellular polysaccharides, although their interesting rheological characteristics were comparable.[125] Glucose (81%) was found to be the most abundant monosaccharide in exopolysaccharides produced by the diatom *Amphora coffeaeformis*. The products were acidic sulfated polysaccharides containing high concentrations of pyruvate (22%) and uronic acids (18%).[121]

Rheological studies indicate that almost all of the cyanobacterial polysaccharides show a pseudoplastic behavior, with marked differences in both viscosity values and shear thinning. Some of these polysaccharides may possess unique properties for special applications not fulfilled by the polymers currently available. The heteropolysaccharide produced by the cyanobacterium *Anabaena* in aqueous dispersions at concentrations ranging from 0.2 to 0.6% (w/w) showed marked shear-thinning properties (power-law behavior). Linear dynamic viscoelastic properties showed that the elastic component was always higher than the viscous component. Stress–growth experiments revealed that 0.4% and 0.6% (w/w) EPS dispersions showed thixotropic properties. Viscoelastic spectra demonstrated that the EPS dispersion turned out to be more fluid like. The flow indexes indicated that the EPS dispersion was less shear sensitive than that of xanthan gum, but they showed essentially the same viscosity.[119]

Polysaccharides of the red microalgae *Porphyridium aerugineum* and *Rhodella reticulata* showed interesting rheological properties. Aqueous solutions of the biopolymers were stable over a wide range of pH values and temperatures. Mixtures of the algal polysaccharides with locust bean gum exhibited synergism.[125] These properties need to be carefully retained using appropriate process parameters during extraction and isolation procedures.[122] The polysaccharides of these microalgae are capable of wide industrial applications as thickening agents and food additives because of their high viscosities over a wide range of pH, temperature, and salinity.[125]

References

1. Kumar, C. S. et al., Seaweed as a source of nutritionally beneficial compounds: a review, *J. Food Sci. Technol.*, 45, 1, 2008.
2. Plaza, M. et al., Innovative natural functional ingredients from microalgae, *J. Agric. Food Chem.*, 57, 7159, 2009.
3. Rudolph, B., Seaweed products: red algae of economic importance, in *Marine & Freshwater Products Handbook*, Martin, R. E., Carter, E. P., Flick, Jr., G. J., and Davis, L. M., Eds., CRC Press, Boca Raton, FL, 2000, pp. 515–530.
4. McLachlan, J. L., *Chondrus crispus* (Irish moss), an ecologically important and commercially valuable species of red seaweed of the North Atlantic Ocean, in *Marine Biology: Its Accomplishment and Future Prospect*, Mauchline, T. and Nemoto, T., Eds., Elsevier, Amsterdam, 1991, p. 217.
5. Ito, K. and Hori, K., Seaweed: chemical composition and potential food uses, *Food Rev. Int.*, 5, 101, 1989.
6. Ruperez, P., Mineral contents of some edible marine seaweed, *Food Chem.*, 79, 21, 2002.
7. Fleurence, J., Seaweed proteins: biochemical, nutritional aspects and potential uses, *Trends Food Sci. Technol.*, 10, 25, 1999.
8. Dawczynski, C. et al., Amino acids, fatty acids and dietary fibre in edible seaweed products, *Food Chem.*, 103, 891, 2007.
9. Ortiz, J. et al., Dietary fiber, amino acid, fatty acid and tocopherol contents of the edible seaweed *Ulva lactuca* and *Durvillaea antarctica*, *Food Chem.*, 99, 98, 2006.
10. Sun, H. H. et al., Polysaccharides from marine green seaweed *Ulva* species and their characteristics, *Agro Food Industry Hi-Tech.*, 18, 4, 2007.
11. Dembitsky, B. M. et al., Glycolipids, phospholipids and fatty acids of brown algae species, *Phytochemistry*, 29, 3417, 1990.
12. Sánchez-Machado, D. L. et al., Fatty acids, total lipid, protein and ash contents of processed edible seaweed, *Food Chem.*, 85, 439, 2004.
13. Marsham, S. et al., Comparison of nutrition chemistry of a range of temperate seaweed, *Food Chem.*, 100, 1331, 2007.
14. Nagai, T., Suzuki, N., and Nagaashima, T., Angiotensin l-converting enzyme inhibitory activities of beverages made from sea algae and commercially available tea extracts, *J. Food Agric. Environ.*, 4, 17, 2006.
15. Venugopal, V., *Marine Products for Healthcare: Functional and Bioactive Nutraceuticals from the Ocean*, CRC Press, Boca Raton, FL, 2008, chap. 9.
16. Akhilender, S. R. et al., Evaluation of nutritional quality and safety of seaweed in India, *J. Food Safety*, 13, 77, 1990.
17. Hoebler, C. et al., Supplementation of pig diet with algal fiber changes the chemical and physicochemical characteristics of digesta, *J. Sci. Food Agric.*, 80, 1357, 2000.
18. Cardozo, K. H. M. et al., Metabolites from algae with economical impact, *Comp. Biochem. Physiol. C*, 146, 60, 2007.
19. Mabean, S. and Fleurence, J., Seaweed in food products: biochemical and nutritional aspects, *Trends Food Sci. Technol.*, 4, 103, 1993.

20. Taboada, C., Millán, R., and Míguez, I., Composition, nutritional aspects and effect on serum parameters of marine algae *Ulva rigida*, *J. Sci. Food Agric.*, 90, 445, 2010.
21. Wong, K. H. and Cheung, P.C. K., Effect of fiber-rich brown seaweed on protein bioavailability of casein in growing rats, *Int. J. Food Sci. Nutr.*, 54, 269, 2003.
22. Urbani, M. G. and Goni, I., Bioavailability of nutrients in rats fed on edible seaweed nori (*Porphyra tenera*) and wakame (*U. pinnatifida*), as a source of dietary fibre, *Food Chem.*, 76, 281, 2002.
23. Miyashita, K., Edible Seaweed and Its Multibiological Functions, Paper No. AGFD 39, 232nd American Chemical Society National Meeting, September 10–14, 2006, San Francisco, CA.
24. Norziah, M. H. and Chio, Y. C., Nutritional composition of edible seaweed *Gracilaria changii*, *Food Chem.*, 68, 69, 2001.
25. Chapman, S. and Baek, I., Gums and thickeners: review of food hydrocolloids, *Campden & Chorleywood Food Res. Assoc. Rev.*, 34, 73, 2002.
26. McHugh, D. J., *Seaweed Production and Markets*, FAO/GLOBEFISH Research Programme, Vol. 48, Food and Agricultural Organization, Rome, 1996, 73 pp.
27. McHugh, D. J., *A Guide to the Seaweed Industry*, FAO Fisheries Technical Paper No. 441, Food and Agricultural Organization, Rome, 2003, 105 pp.
28. Rouxel, C. et al., Species identification by SDS–PAGE of red algae used as seafood or a food ingredient, *Food Chem.*, 74, 349, 2001.
29. Haines, J. and Patel, P., New assays for rapid analysis of food grade gums and thickeners, *New Food*, 6, 81, 2003.
30. Joubert, Y. et al., Development of a molecular method for the rapid discrimination of red seaweed used for agar production, *Food Chem.*, 113, 1384, 2009.
31. Bixler, H. J. and Porse, H., A decade of change in the seaweed hydrocolloids industry, *J. Appl. Phycol.*, 2010 (DOI 10.1007/s10811-010-9529-3).
32. Wheaton, F. W. and Lawson, T. B., *Processing of Aquatic Food Products*, John Wiley & Sons, New York, 1985.
33. Freile-Pelegrín, Y. and Murano, E., Agars from three species of *Gracilaria* (Rhodophyta) from Yucatán Peninsula, *Bioresour. Technol.*, 96, 295, 2005.
34. Naidu, A. S., Agar, in *Natural Food Antimicrobial Systems*, Naidu, A. S., Ed., CRC Press, Boca Raton, FL, 2000, p. 417–427.
35. Li, H. et al., Development of an eco-friendly agar extraction technique from the red seaweed *Gracilaria lemaneiformis*, *Bioresour. Technol.*, 99, 3301, 2008.
36. Romero, J. B. et al., Stability of agar in the seaweed *Gracilaria eucheumatoides* (Gracilariales, Rhodophyta) during post-harvest storage, *Bioresour. Technol.*, 99, 8151, 2008.
37. Higuera, D. L. A., Effect of alkali treatment time and extraction time on agar from *Gracilaria vermiculophylla*, *J. Appl. Phycol.*, 20, 515, 2008.
38. Marinho-Soriano, E. and Bourret, E., Polysaccharides from the red seaweed *Gracilaria dura* (Gracilariales, Rhodophyta), *Bioresour. Technol.*, 96, 379, 2005.
39. Rodriguez, M. C. et al., Agar from *Gracilaria gracilis* (Gracilariales, Rhodophyta) of the Patagonic coast of Argentina: content, structure and physical properties, *Bioresour. Technol.*, 100, 1435, 2009.
40. Marinho-Soriano, E., Silva, T. S. F., and Moreira, W. S. C., Seasonal variation in the biomass and agar yield from *Gracilaria cervicornis* and *Hydropuntia cornea* from Brazil, *Bioresour. Technol.*, 77, 115, 2001.

41. Chattopadhyay, K., Sulphated polysaccharides from Indian samples of *E. compressa* (Ulvales, Chlorophyta): isolation and structural features, *Food Chem.*, 104, 928, 2007.
42. Ray, B., Polysaccharides from *Enteromorpha compressa*: isolation, purification and structural features, *Carbohydr. Polym.*, 66, 408, 2006.
43. Prasad, K. et al., Agars of *Gelidiella acerosa* of west and southeast coasts of India, *Bioresour. Technol.*, 97, 1907, 2007.
44. Lahaye, M. and Rochas, C., Chemical structure and physicochemical properties of agar, *Hydrobiologia*, 221, 137, 1991.
45. Usov, A. I., Structural analysis of red seaweed galactans of agar and carrageenan groups, *Food Hydrocoll.*, 12, 301, 1998.
46. Morris, V. J., Gelation of polysaccharides, in *Functional Properties of Food Macromolecules*, 2nd ed., Hill, S. E., Ledward, D. A., and Mitchell, J. R., Eds., Aspen Publishers, New York, 1998, pp. 143–226.
47. Labuza, T. P. and Busk, G. C., An analysis of the water binding in gels, *J. Food Sci.*, 44, 1379, 1979.
48. Mitsuiki, M. et al., Glass transition properties as a function of water content for various low-moisture galacatans, *J. Agric. Food Chem.*, 46, 5638, 1998.
49. Kavanagh, G. M. and Ross-Murphy, S. B., Rheological characterization of polymer gels, *Prog. Polym. Sci.*, 23, 533, 1998
50. Barrangou, L. M., Daubert, C. R., and Foegeding, E. A., Textural properties of agarose gels. I. Rheological and fracture properties, *Food Hydrocoll.*, 20, 184, 2006.
51. Lai, M.-F. and Lii, C.-Y., Rheological and thermal characteristics of gel structures from various agar fractions, *Int. J. Biol. Macromol.*, 21, 123, 1997.
52. Roller, S. and Dea, I. C. M., Biotechnology in the production and modification of biopolymers for foods, *Crit. Rev. Biotechnol.*, 12, 261, 2002.
53. Ellis, A. and Jacquier, J. C., Manufacture and characterisation of agarose microparticles, *J. Food Eng.*, 90, 141, 2009.
54. Meena, R., Prasad, K., and Siddhanta, A. K., Studies on "sugar reactivity" of agars extracted from some Indian agarophytes, *Food Hydrocoll.*, 20, 1206, 2006.
55. Norziah, M. H., Foo, S. L., and Karim, A. A., Rheological studies on mixtures of agar (*Gracilaria changii*) and κ-carrageenan, *Food Hydrocoll.*, 20, 204, 2006.
56. Draget, K. I., Smidsrød, O., and Skjåk-Brœk, G., Alginates from algae, in *Polysaccharides and Polyamides in the Food Industry*, Steinbüchel, A. and Rhee, S. K., Eds., Wiley, New York, 2005, pp. 1–30.
57. Rioux, L.-E. et al., Rheological characterization of polysaccharides extracted from brown seaweed, *J. Sci. Food Agric.*, 87, 1630, 2007.
58. Zvyagintseva, T. N. et al., A new procedure for the separation of water-soluble polysaccharides from brown seaweed, *Carbohydr. Res.*, 322, 32, 1999.
59. Owusu-Apenten, R. K., *Introduction to Food Chemistry*, CRC Press, Boca Raton, FL, 2004, p. 55.
60. Clementi, F. et al., Production and characterisation of alginate from *Azotobacter vinelandii*, *J. Sci. Food Agric.*, 79, 602,1999.
61. Braccini, I. and Pérez, S., Molecular basis of Ca^{2+}-induced gelation in alginates and pectins: the egg-box model revisited, *Biomacromolecules*, 2, 1089, 2001.
62. Quignard, F. et al., Aerogel materials from marine polysaccharides, *New J. Chem.*, 32, 1300, 2008.
63. Mancini, M., Moresi, M., and Sappino, F., Rheological behaviour of aqueous dispersions of algal sodium alginate, *J. Food Eng.*, 28, 283, 1996.

64. Zhang, J. et al., Additive effects on the rheological behavior of alginate gels, *J. Text. Stud.*, 39(5), 582, 2008.
65. Nagasawa, N. et al., Radiation-induced degradation of sodium alginate, *Polym. Degrad. Stabil.*, 69, 279, 2000.
66. Harnsilawat, T., Pongsawatmanit, R., and McClements, D. J., Characterization of β-lactoglobulin–sodium alginate interactions in aqueous solutions: a calorimetry, light scattering, electrophoretic mobility and solubility study, *Food Hydrocoll.*, 20, 577, 2006.
67. Panouillé, M. and Larreta-Garde, V., Gelation behavior of gelatin and alginate mixtures, *Food Hydrocoll.*, 23, 1074, 2009.
68. Walkenström, P. et al., Microstructure and rheological behavior of alginate/pectin mixed gels, *Food Hydrocoll.*, 17, 593, 2003.
69. BeMiller, J. N., Carbohydrates, in *Kirk–Othmer Encyclopedia of Chemical Technology*, 4th ed., Vol. 4, Howe-Grant, M., Ed., Wiley, New York, 2004, pp. 911–948.
70. Phillips, G. O. and Williams, P. A., Eds., *Handbook of Hydrocolloids*, 2nd ed., Woodhead Publishing, Cambridge, U.K., 2009.
71. Sen, D. P., *Advances in Fish Processing*, Allied Publishers, Mumbai, 2005, chap. 18.
72. Montolalu, R. I. et al., Effects of extraction parameters on gel properties of carrageenan from *Kappaphycus alvarezii* (Rhodophyta), *J. Appl. Phycol.*, 20, 521, 2008.
73. Uy, S. F. et al., Seaweed processing using industrial single-mode cavity microwave heating: a preliminary investigation, *Carbohydr. Res.*, 340, 1357, 2005.
74. Hillou, L. et al., Effect of extraction parameters on the chemical structure and gel properties of κ-/ι-hybrid carrageenans obtained from *Mastocarpus stellatus*, *Biomol. Eng.* 23, 201, 2006.
75. Zhi, Q.O., Take, M., and Toyama, S., Chemical characterization of κ-carrageenan of ibaranori (*Hypnea charoides* Lamoroux), *J. Appl. Glycosci.*, 44(2), 137, 1998.
76. Hoffmann, R. A. et al., Effect of isolation procedures on the molecular composition and physical properties of *Eucheuma cottonii* carrageenan, *Food Hydrocoll.*, 9, 281, 1995.
77. Angelin, T. S. et al., Physicochemical properties of carrageenans extracted from *Sarconema filiforme* and *Hypnea valentiae*, *Seaweed Res. Util.*, 26, 197, 2004.
78. Parekh, R. G. et al., Polysaccharide from *Sarconema filiforme*, an Indian marine alga, *Phytochemistry*, 27, 933, 1988.
79. de Ruiter, G. A. and Rudolph, B., Carrageenan biotechnology, *Trends Food Sci. Technol.*, 8, 389, 1997.
80. Yuguchi, Y. et al., Structural characteristics of carrageenan gels: temperature and concentration dependence, *Food Hydrocoll.*, 16, 515, 2002.
81. Meunier, V., Nicolai,T., and Durand, D., Structure of aggregating κ-carrageenan fractions studied by light scattering, *Int. J. Biol. Macromol.*, 28, 157, 2001.
82. Ramakrishnan, S., Gerardin, C., and Prud'homme, R. K., Syneresis of carrageenan gels: NMR and rheology, *Soft Materials*, 2, 145, 2004.
83. Lin, L.-H., Tako, M., and Hongo, F., Molecular origin of the rheological characteristics of ι-carrageenan isolated from Togekirinsai (*Eucheuma serra*), *Food Sci. Technol. Res.*, 7, 176, 2001.
84. MacArtain, P., Jacquier, J. C., and Dawson, K. A., Physical characteristics of calcium induced κ-carrageenan networks, *Carbohydr. Polym.*, 53, 395, 2003.
85. Hilliou, L. et al., Thermal and viscoelastic properties of κ-/ι-hybrid carrageenan gels obtained from the Portuguese seaweed *Mastocarpus stellatus*, *J. Agric. Food Chem.*, 54, 7870, 2006.

86. Yamashita, S., Sugita, K. Y., and Shimizu, M., *In vitro* bacteriostatic effects of dietary polysaccharides, *Food Sci. Technol. Res.*, 7, 262, 2001.
87. Volery, P., Besson, R., and Schaffer-Lequart, C., Characterization of commercial carrageenans by Fourier transform infrared spectroscopy using single-reflection attenuated total reflection, *J. Agric. Food Chem.*, 52, 7457, 2004.
88. Yaphe, W. and Arsenault, G. P., Improved resorcinol reagent for the determination of fructose, and of 3,6-anhydrogalactose in polysaccharides, *Anal. Biochem.*, 13, 143, 1965.
89. Soedjak, H. S., Colorimetric determination of carrageenans and other anionic hydrocolloids with methylene blue, *Anal. Chem.*, 66, 4514, 1994.
90. Yabe, Y. et al., Simple colorimetric determination of carrageenan in jellies and salad dressings, *J. Assoc. Off. Anal. Chem.*, 74, 1019, 1991.
91. Graham, H. D., Quantitative determination of carrageenan in milk and milk products using papain and cetylpyridinium chloride, *J. Food Sci.*, 33, 390, 1968.
92. Roberts, M. A. and Quemener, B., Measurement of carrageenans in food: challenges, progress, and trends in analysis, *Trends Food Sci. Technol.*, 10, 169, 1999.
93. Ibanoglu, E., Effect of hydrocolloids on the thermal denaturation of proteins, *Food Chem.*, 90, 621, 2005.
94. Carp, D. J. et al., Impact of protein–κ-carrageenan interactions on foam properties, *Lebensm. Wiss. U. Technol.*, 37, 573, 2004.
95. Martinez, K. D. et al., Effect of limited hydrolysis of sunflower protein on the interactions with polysaccharides in foams, *Food Hydrocoll.*, 19, 361, 2005.
96. Chakraborty, B. K. and Randolph, H. E., Stabilization of calcium sensitive plant proteins by κ-carrageenan, *J. Food Sci.*, 37, 719, 1972.
97. Ercelebi, E. A. and Ibanoglu, E., Influence of hydrocolloids on phase separation and emulsion properties of whey protein isolate, *J. Food Eng.*, 80, 454, 2007.
98. Verbeken, D. et al., Influence of κ-carrageenan on the thermal gelation of salt-soluble meat proteins, *Meat Sci.*, 70, 161, 2005.
99. Hauga, I. J., Physical behaviour of fish gelatin kappa-carrageenan mixtures, *Carbohydr. Polym.*, 56, 11, 2004.
100. Gómez-Guillén, M. C. et al., Salt, nonmuscle proteins, and hydrocolloids affecting rigidity changes during gelation of giant squid (*Dosidicus gigas*), *J. Agric. Food Chem.*, 45, 616, 1997.
101. Techawipharat, J., Suphantharika, I. P., and BeMiller, J. N., Effects of cellulose derivatives and carrageenans on the pasting, paste, and gel properties of rice starches, *Carbohydr. Polym.*, 73, 417, 2008.
102. Funami, T. et al., Functions of ι-carrageenan on the gelatinization and retrogradation behaviors of corn starch in the presence or absence of various salts, *Food Hydrocoll.*, 22, 1273, 2008.
103. Murayama, A. et al., Sensory and rheological properties of κ-carrageenan gels mixed with locust bean gum, taragum or guar gum, *J. Tex. Stud.*, 26, 239, 1995.
104. Arda, E., Kara, S., and Pekcan, O., Synergistic effect of locust bean gum on the thermal phase transitions of κ-carrageenan gels, *Food Hydrocoll.*, 23, 451, 2009.
105. Bayarri, S., Duran, L., and Costell, E., Influence of sweeteners on the viscoelasticity of hydrocolloids gelled systems, *Food Hydrocoll.*, 18, 611, 2004.
106. Lahaye, M. and Robic, A., Structure and function properties of Ulvan, a polysaccharide from green seaweed, *Biomacromolecules*, 8, 1765, 2007.
107. Muffler, K. and Ulber, R., Downstream processing in marine biotechnology, *Adv. Biochem. Eng. Biotechnol.*, 96, 85, 2005.

108. Da Bilan, M. I. et al., Structure of a fucoidan from the brown seaweed *Fucus serratus* L., *Carbohydrate Res.*, 341, 238, 2006.
109. Kusaykin, M. et al., Structure, biological activity, and enzymatic transformation of fucoidans from the brown seaweed, *Biotechnol. J.*, 3, 904, 2008.
110. Holtkamp, A. D. et al., Fucoidans and fucoidanases—focus on techniques for molecular structure elucidation and modification of marine polysaccharides, *Appl. Microbiol. Biotechnol.*, 82, 1, 2009.
111. Takedo, M., Nakada, T., and Hongo, F., Chemical characterization of fucoidan from commercially cultured *Nemacystus deciphens*, *Biosci. Biotechnol. Biochem.*, 63, 1813, 1999.
112. Belitz, H. D., Grosch, W., and Schieberle, P., *Food Chemistry*, 3rd ed., Springer-Verlag, Heidelberg, 2004, chap. 4.
113. Yu, S. et al., Physicochemical characterization of Floridean starch of red algae, *Starch/Stärke*, 54, 66, 2002.
114. Rippka, R., Isolation and purification of cyanobacteria, in *Methods in Enzymology*, Packer, L. and Glazer, A. N., Eds., Academic Press, Orlando, FL, 1988, pp. 3–27.
115. Nicaulas, B. et al., Chemical composition and production of EPS from representative members of heterocystous and non-heterocystous cyanobacteria, *Phytochemistry*, 52, 639, 1999.
116. De Philippis, R. and Vincenzini, M., Exocellular polysaccharides from cyanobacteria and their possible applications, *FEMS Microbiol.*, 22, 151, 1998.
117. Ying Shi, Y. et al., Purification and identification of polysaccharide derived from *Chlorella pyrenoidosa*, *Food Chem.*, 103, 101, 2007.
118. Parikh, A. and Madamwar, D., Partial characterization of extracellular polysaccharides from cyanobacteria, *Bioresour. Technol.*, 97, 1822, 2006.
119. Moreno, J. et al., Chemical and rheological properties of an extracellular polysaccharide produced by the cyanobacterium *Anabaena* sp. ATCC 33047, *Biotechnol. Bioeng.*, 67, 283, 2000.
120. Otero, A. and Vincenzini, M., Extracellular polysaccharide synthesis by Nostoc strains as affected by N source and light intensity, *J. Biotechnol.*, 102, 143, 2003.
121. Bhosle, N. B. et al., Chemical characterization of exopolysaccharides from the marine fouling diatom *Amphora coffeaeformis*, *Biofouling*, 10, 301, 1996.
122. Ginzberg, A., Korin, E., and, Arad, S., Effect of drying on the biological activities of a red microalgal polysaccharide, *Biotechnol. Bioeng.*, 99, 411, 2008.
123. Rougeaux, H. et al., Microbial communities and exopolysaccharides from Polynesian mats, *Mar. Biotechnol.*, 3, 181, 2001.
124. Richert, L. et al., Characterization of exopolysaccharides produced by cyanobacteria isolated from Polynesian microbial mats, *Curr. Microbiol.*, 51, 379, 2005.
125. Geresh, S. and Arad, S., The extracellular polysaccharides of the red microalgae: chemistry and rheology, *Bioresour. Technol.*, 38, 195, 1991.
126. Hermansson, A. M. et al., Effects of potassium, sodium and calcium on the microstructure and rheological behaviour of kappa-carrageenan gels, *Carbohydr. Polym.*, 3, 297, 1991.
127. NAAS, *Seaweed Cultivation and Utilization*, NAAS Documents on Policy Issues, National Academy of Agricultural Sciences, New Delhi, India, 2003.

5

Extracellular Polysaccharides from Marine Microorganisms

5.1 Introduction

Microbial polysaccharides are a class of water-soluble polymers that have grown to industrial importance over the past 40 years. These compounds are produced by different types of microorganisms and are recognized to assist microbial communities in their survival. These polymeric compounds, because of their novel and unique properties, are rapidly emerging as materials that can offer a wide range of applications in such diverse fields as food, pharmaceutical, and other industries.[1-5] Over the last few decades, there has been an increasing interest in the isolation and identification of microbial extracellular polysaccharides to better understand their functional properties and compare them with those of traditional polysaccharides. This interest has also encouraged efforts to elucidate their composition and structure, evaluate their biosynthetic routes, identify applications of their molecular biology, determine their functionality, develop the technology necessary for their production, and identify potential uses in food, medicine, biotechnology, and other industries.[1-6] This chapter discusses extracellular polysaccharides from marine microorganisms and their characteristics. Many related areas, such as biosynthetic pathways, polymer secretion, microbial biofilms, genetic engineering, and chemical modifications of polysaccharides, are not within the scope of this discussion and can be further studied elsewhere.[1-6] In order to understand the production and properties of extracellular polysaccharides from marine microorganisms, it is pertinent to briefly present a few examples of polysaccharides from some *non-marine* microorganisms that are important food additives, including dextran, xanthan, gellan, cellulose, bacterial alginate, and curdlan. This chapter, then, begins with a general discussion on extracellular polysaccharides from non-marine microorganisms before moving on to a discussion of marine polysaccharides and their potential applications.

5.2 Functions of Exopolysaccharides in Microbial Cells

Microbial polysaccharides can be divided into intracellular, structural, and extracellular polysaccharides (exopolysaccharides, or EPSs). EPSs belong to extracellular polymeric substances that also contain proteins, neutral hexoses, lipids, DNA, humic acid substances, and slime. EPSs may occur in two basic forms: as *capsular polysaccharides* (CPSs), where the polysaccharide is associated with the cell surface, and as *slime polysaccharides*, which are loosely bound to the cell structure.[1] The ability of a microorganism to surround itself with a hydrated EPS layer protects it against desiccation and predation by protozoans and controls diffusion of substances such as antibiotics, toxic metals, and pathogenic organisms into the cell, thereby facilitating survival of these organisms under adverse conditions. This is particularly so in the case of marine bacteria, where EPSs allow microbes to endure extremes of temperature, salinity, and nutrient availability (see Section 5.6). EPSs may also be involved in pathogenic and symbiotic interactions between bacteria and plants and microbial aggregates such as biofilms and biological sludge. Biofilms are usually complex assemblages of microorganisms embedded within a matrix composed of water together with extracellular polymeric substances. Biofilms are of particular interest in the context of food spoilage and hygiene, with regard to the role of biofilms in the attachment of microorganisms on food and food–contact surfaces and their increased resistance to cleaning and disinfection processes.[1,7,8] EPSs are comparatively simple to isolate, as they are produced during microbial fermentation, as is discussed in Section 5.4.

5.3 Examples of Exopolysaccharides Produced by Microorganisms from Non-Marine Sources

Dextran, xanthan, gellan, cellulose, bacterial alginate, and curdlan are some of the commercially important microbial EPSs. The ability to produce EPSs is widespread among microorganisms. The lactic acid bacteria (LAB) produce both neutral and charged EPSs, having quite distinct functional properties. *Lactobacillus* EPSs have attracted increasing attention because of their importance in food product development. The chemical composition, molecular size, charge, type of side chains, and rigidity of the EPSs from these organisms greatly depend on the conditions employed for their culture. These characteristics, in turn, determine the intrinsic properties of EPSs, their functionality, and interactions with food compounds, including those of milk. In a survey, out of 182 *Lactobacillus* strains screened, 60 EPS-positive strains were identified, with 17 strains producing more than 100 mg soluble

EPSs per liter.[9,11] The probiotic *Bacillus coagulans* produces a heteropolymer EPS composed of galactose, mannose, fucose, glucose, and glucosamine during exponential and stationary growth phases.[10] Because of their significant commercial potential, the biosynthesis of EPSs from *Lactobacillus*, their genetics and molecular organization, strain improvement, nutritional and physiological aspects, and uses as food additives have been the subject of detailed studies.[11–13]

A number of *Pseudomonas* organisms have been found to produce EPSs.[14–17] *P. mendocina* cells grown at room temperature in sodium benzoate as the sole source of carbon produce an EPS characterized by significant viscosity of the medium. The EPS was found to be associated with the cells and not released into the supernatant fluid; however, a combination of sodium dodecyl sulfate, sodium citrate buffer, and homogenization was effective in releasing the EPS. This EPS is a heteropolysaccharide, consisting of rhamnose, fucose, glucose, ribose, arabinose, and mannose and having good emulsifying activity.[14]

Exopolysaccharides are also produced by yeasts and fungi. Pullulan is a commercially important water-soluble glucan gum produced extracellularly and aerobically by the yeast-like fungus *Aureobasidium pullulans* (see Section 5.5.6). Two types of EPSs are produced by the fungi *Sclerotium rolfsii* after 48 and 72 hr of cultivation.[18] Ammonium sulfate was the most favorable nitrogen source for an optimal expolysaccharide yield of 69% by the yeast *Rhodotorula acheniorum*, at an acidic pH of 2.0 and sugar concentration of 5%. The EPS contained mannose as the main monosaccharide component.[19] The lactose-negative yeast *R. glutinis* produced EPS with a yield of about 9 g/L when cultured in a cheese whey ultrafiltrate containing lactose. The yeast also produced β-carotene.[20]

Five yeast strains and one yeast-like fungus produced EPSs when grown on glucose, ethanol, or methanol. These polysaccharides were comparable with commercial xanthan in rheological properties.[21] In liquid culture conditions, the yeast-like fungus *Tremella mesenterica* synthesizes an EPS capsule, which is eventually released into the culture fluid. It is composed of an α-(1,3)-D-mannan backbone to which β-(1,2) side chains are attached, consisting of D-xylose and D-glucuronic acid.[21] It is possible to enhance EPS production by metabolic engineering, as observed in the case of *Streptococcus thermophilus*.[22]

The bioemulsifier V2-7 is an EPS synthesized by *Halomonas eurihalina* strain F2-7. It is capable of emulsifying a wide range of hydrocarbons, including heavy oils, petrol, and crude oil. The EPS could be considered highly beneficial for its application as bioemulsifier for bioremediation of oil pollutants.[23] *Penicillium citrinum* produced an EPS with emulsifier properties during cultivation on mineral medium with 1% (v/v) olive oil as the carbon source. The EPS production reached maximal activity at 60 hr of cultivation. The EPS contained D-galactose, D-glucose, and D-xylose in a ratio of 8.2:1.0:5.3, with a total carbohydrate content of 43%. It showed maximum emulsifying activity for xylene and diesel oil and was stable over a wide range of pH and

TABLE 5.1

Important Microbial Polysaccharides and Their Major Sources

Polysaccharide	Microorganism
Alginate	*Azotobacter, Pseudomonas*
Cellulose	*Acetobacter xylinum, Agrobacterium*
Curdlan	*Agrobacterium, Grifola frondosa* (fungus), *Lentinus ecodes*
Gellan	*Sphingomonas*
Rhamsan	*Alcaligenes*
Welan	*Alcaligenes*
Hyaluronic acid	*Streptococcus*
Xanthan	*Xanthomonas*
Succinoglycan	*Alcaligenes, Agrobacterium*
Emulsan	*Acinetobacter*
Dextran	*Leuconostoc*
Levan	*Pseudomonas, Zymomonas*, lactic acid bacteria (e.g., *Bacillus, Leuconostoc, Pediococcus, Streptococcus*)
Pullulan	*Aureobasidium*
Sceleroglucan	*Sclerotium*
Schizophylan	*Schizophyllum*

temperatures values. The presence of salts stimulated the emulsification activity, suggesting its potential for industrial waste or marine remediation.[24]

Amylovoran, the acidic EPS of *Erwinia amylovora*, and stewartan, the capsular EPS of *E. stewartii*, were characterized by analytical ultracentrifugation and by size-exclusion chromatography with dual detection of light scattering and mass. The average molecular weights of amylovoran and stewartan were determined to be 1.0×10^6 and 1.7×10^6 Da, respectively, with polydispersity values (M_w/M_n) of 1.5 and 1.4. Based on the sugar composition and their molecular weight, both exopolysaccharides consist of approximately 1000 repeating units per molecule.[25] Table 5.1 lists some important microbial polysaccharides and their major sources.

5.4 Fermentation of Microorganisms for Exopolysaccharides

Microorganisms are better suited for the production of polysaccharides than plants or algae, because they are capable of higher growth rates and are more amenable to manipulation of fermentation conditions for enhancing growth and production. Commercially important EPSs, such as dextran, xanthan, gellan, and curdlan, can be produced by microbial fermentation.[26,27] Successful commercial production of microbial EPSs requires one or more of the following techniques: (1) cultivating the organism in an appropriate medium for optimal EPS synthesis, (2) applying a suitable method to promote cell separation during downstream processing, (3) modifying the EPSs during or after production, (4) preventing loss of functionality of the EPSs by unwanted enzymes, and (5) transferring genetic determinants of EPS to more efficient host producers.[28] These aspects are discussed briefly below.

5.4.1 Cultivation

Organisms differ in their carbon and nitrogen utilization and mineral, temperature, and pH requirements for maximum exopolysaccharide production.[2] The use of organic nitrogen sources often results in a higher specific growth rate and greater EPS production. Microorganisms usually reach their optimal growth within the initial 24 hr, and maximal EPS production occurs in the later stages of growth (i.e., during the stationary phase).[15,26–28] The wide variety of carbon sources used to produce microbial EPSs includes sucrose, glucose, lactose, maltose, mannitol, sorbitol, whey, starch, commercial sugar concentrates, methanol, and *n*-alkanes. The nitrogen sources include ammonium sulfate, peptone, sodium nitrate, urea, and yeast extract. EPS production is usually favored by a high carbon to nitrogen ratio—ideally, 10:1. Limitations of one or more of the nutrients carbon, nitrogen, phosphate, and oxygen in the media can enhance production of polysaccharides. For example, nitrogen, carbon, and oxygen limitations affected the conversion of glucose into alginate and the proportion of mannuronate to glucuronate residues in *Pseudomonas mendocina*.[14,29] Phosphate was found to have a specific influence on EPS production by a *Pseudomonas* sp. During aerobic submerged fermentation, the exopolysaccharide synthesis by a *Pseudomonas* strain increased when the pH was maintained at 7 during fermentation. The polymer exhibited a pseudoplastic nature, had good thermostability, and was not affected by pH or high concentrations of salt.[17] Oxygen, pH, temperature, medium constituents such as orotic acid, and carbon source influenced EPS production by a lactic acid bacterium in a chemically defined medium; EPS production was greatest during the stationary phase. The relative proportions of the individual monosaccharides in the EPS varied according to specific medium alterations.[30]

The high cost of the carbon sources used, mainly sugars such as glucose, sucrose, and fructose, has a direct impact on production costs; therefore, it is advisable to search for less expensive carbon sources in order to reduce the production costs. Food processing generates large amounts of wastes and creates environmental problems. Common agrowaste components, such as corn starch, corn gluten meal, and corn steep liquor, could be used as fermentation media.[31] For example, *Agrobacterium* produces a curdlan-like EPS when grown in coconut water containing up to 4% sugar. After fermentation in shake flasks, the organism produced an EPS comparable to that produced in a sucrose medium. The optimal pH and temperature were 6 and 30°C, respectively. Instead of coconut water, molasses could also be used for fermentation.[32] It is also possible to increase EPS production through genetic modification by altering the levels of enzymes in the central carbohydrate metabolism.[22]

Apart from nutrients, temperature also influences the production of EPSs. An incubation temperature below the optimum growth temperature results in greater production of EPSs; a lower incubation temperature (e.g., 32°C instead of 37°C) can cause a reduction in growth rate and cell mass, which in turn results in an extended logarithmic growth phase and higher EPS production. A constant pH is ideal for the maximum production of EPS.[17] Overproduction of EPSs by an *Escherichia coli* K-12 mutant in response to osmotic stress has been reported.[33] In some cases, the presence of detergents such as Triton® X-100 may enhance the production of EPSs.[2] The degree of aeration also influences EPS production.[2] Maximum EPSs were produced at a low dilution rate of continuous cultures of *Pseudomonas, Alcaligenes,* and *Klebsiella.*[5]

Fermentation of microorganisms for EPS production can occur via either batch or fed-batch processes.[21] The agitation rate and dissolved oxygen tension (DOT) influence the growth and gellan production of *Sphingomonas paucimobilis.* A cell growth rate of 5.4 g/L was obtained at an agitation rate of 700 rpm, but maximum gellan (15 g/L) was produced at 500 rpm. DOT levels above 20% had no effect on cell growth, but gellan yield was increased to a maximum of 23 g/L with an increase in DOT, which was also reflected in enhanced viscosity and molecular weight of the polymer along with changes in its acetate and glycerate contents.[34] Shear influenced EPS yields in *Aureobasidium pullulans.* The yield dramatically reduced when the organism was grown in an airlift reactor. This fall in production could be reversed by improving fluid circulation through the placement of impellers within the draft tube, a strategy that resulted in the highest EPS concentration of 13 g/L.[35] A novel bioreactor was constructed for optimal production of EPSs by *Methylobacterium organophilum.*[36]

Exopolysaccharides are generally present in low concentrations in the fermentation broth; their presence is indicated by a high broth viscosity. During EPS production, the broth usually develops non-Newtonian characteristics and acts as a pseudoplastic fluid, with the measured viscosity decreasing with increasing shear rate. This is due to the increased secretion of EPSs having a pseudoplastic character to the medium, while contribution of cells to viscosity is negligible.[17] Production of the polysaccharide methylan increased gradually with increasing shear stress up to 30 Pa and remained constant beyond this shear. A fermentation broth viscosity of 127 Pa·s corresponded to a xanthan concentration of 68 g/L.[36] *Lactobacillus* produced EPSs when grown in whey medium supplemented with lactose and other nutrients at an optimal pH of 6.2 and 30°C. The production of EPSs was indicated by an increase in the viscosity of the medium;[37,38] therefore, viscosity can be used to monitor EPS production.[2,15,17,37–40] Because of the pseudoplastic nature of

EPSs, the power required for mixing viscous non-Newtonian systems, mass transfer, and scale-up problems are additional factors that need to be considered in polysaccharide fermentations.[1,15,17]

5.4.2 Postfermentation Recovery of Exopolysaccharides

Recovery of EPSs from the broth involves concentration, isolation, and purification, which also determine the total production costs. These steps need to be carried out without affecting the functional properties of the EPSs. EPSs are generally recovered from cell-free culture supernatant by solvent precipitation of the broth. For this, excess organic solvents that are miscible with water (e.g., alcohols or acetone) are added. These solvents favor EPS separation by lowering their solubility in water. During solvent treatment, proteins and salts of the medium may also precipitate along with the EPSs, which may be removed by dialysis or other suitable methods. EPSs from capsular polysaccharides and slime can be separated by centrifugation, the speed and duration of centrifugation depending on the nature and viscosity of the polysaccharide. If the capsular EPS is strongly associated to the cells, additional measures such as alkaline treatment prior to centrifugation and alcohol precipitation may be needed. In cases where the EPSs are thermally stable, heat treatment can be used, which lowers the viscosity and inactivates the contaminant microorganisms as well as the enzymes present in the broth.[2,5] A new assay system for EPSs has been developed that will not interfere with proteins and lactose in the growth medium. The method involved initial hydrolysis of contaminating protein and optimizing ethanol concentration to prevent lactose crystallization allowing complete EPS precipitation.[40]

A typical cultivation of a *Pseuodomonas* and its isolation should be highlighted here. *P. oleovorans* NRRL B-14682 was grown in a 10-L bioreactor operated in a fed-batch mode at a controlled pH of 6.7 to 6.8, a temperature of 30°C, and an air flow rate of 1 L/min. The growth medium was supplemented with a solution containing 25 g/L glycerol as the carbon source and 3.3 g/L $(NH4)_2HPO_4$ as the nitrogen source. During fermentation, the dissolved oxygen concentration was controlled below 10% by maintaining the stirrer speed between 400 and 800 rpm. When the culture entered an exponential growth phase, the ammonium concentration in the culture broth became limiting. Significant EPS production was observed after the culture entered the stationary phase, as shown in Figure 5.1. Concomitant with EPS production was a drastic increase in viscosity of the culture. After 96 hr of fermentation, the EPS attained a concentration of about 8 g/L. The shear-rate-dependent apparent viscosity of culture broth during EPS production is shown in Figure 5.2. After the experiment, culture broth was diluted with deionized water to reduce viscosity. The cell-free supernatant was collected by centrifugation, the protein was denatured with trichloroacetic acid, and the polymer was precipitated by the addition of cold 96% ethanol followed

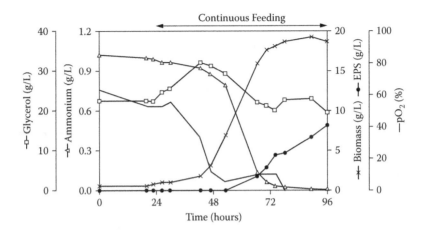

FIGURE 5.1
Growth and exopolysaccharide production by *Pseudomonas oleovorans* cultivated in a 10-L bio-reactor operated at 30°C, pH 6.75 to 6.85, and constant air flow rate. (From Freitas, V.D. et al., *Biores. Technol.*, 100, 859, 2009. With permission from Elsevier/Rightslink.)

by centrifugation. The pellet was washed with ethanol, redissolved in deion-ized water, and freeze-dried. The average molecular weight of the EPSs pro-duced was in the range of 1.0 to 5.0×10^6 Da.[15,37]

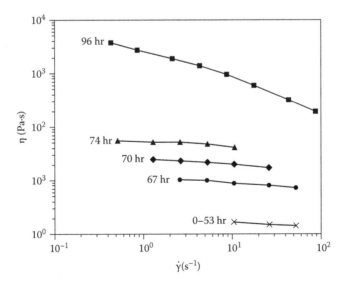

FIGURE 5.2
Shear-rate-dependent apparent viscosity of culture broth during exopolysaccharide produc-tion by a *Pseudomonas* strain grown on glycerol. (From Freitas, V.D. et al., *Biores. Technol.*, 100, 859, 2009. With permission from Elsevier/Rightslink.)

5.4.3 Concentration of Exopolysaccharides

After isolation of EPS from the fermentation broth, it is usually dried to a powder. It may be noted that conventional methods of drying tend to shrink the polymers; as their water content evaporates, a crust develops on the surface that results in a loss of porosity. This crust reduces the drying rate as well as the rehydration capacity of the dried product. Further, with the decline of moisture content in the material, the glass transition temperature (see Chapter 2) rises, causing transformation of the material from a rubbery to a glassy state.[41]

A recent study showed that conventional drying of EPSs produced by *Porphyridium* above 90°C resulted in reduced elasticity and intrinsic viscosity, and the biological activities (e.g., antiviral and anticell proliferation) were adversely affected. The loss of functionality was attributed to alteration of the hydration layer surrounding the charged groups of the polysaccharide.[42] It is important, therefore, that the drying process should be optimized to prevent the loss of functional properties of the material. A two-step drying process has been suggested as an ideal method of concentrating EPS isolates. The free water is removed by convection drying of the isolate followed by its freeze-drying to remove the bound water,[42] and the dry EPS is milled to the desired mesh size. Drying by supercritical CO_2 is an effective method for retaining the structure of hydrogels. The microspheres are dehydrated by immersion in a series of ethanol–water baths of increasing alcohol concentration (10 to 100%) for several minutes. The microspheres are then dried under supercritical CO_2 conditions (74 bar, 32°C); however, supercritical drying may cause some degree of shrinkage, as noted in the case of some marine polysaccharides.[43]

5.4.4 Structure and Properties

Microbial EPSs represent a wide range of chemical structures. The EPS produced by *Pseudomonas fluorescens*, for example, consists of the monosaccharides rhamnose, fucose, ribose, arabinose, xylose, mannose, galactose, and glucose. The acidic groups in the EPS are mainly composed of carboxylic acid, followed by sulfate and phosphate. Up to 70% of total carbohydrates are uronic acids, and total carbohydrates make up 26 to 31% of the organic carbon. In addition to the neutral and acidic sugars, the EPS also contains some proteins.[16]

Chemical structure strongly determines the functional properties of EPSs, and modifying the chemical structure may allow the preparation of derivatives with new functional properties and a wider variety of applications. Chemical modifications usually result in the alteration of side chains by the removal of unwanted groups, such as acyl groups. Hydrolysis of native dextran at elevated temperatures with hydrochloric acid gives dextrans of varying molecular mass. Treating xanthan with mild alkali removes β-D-mannosyl residues.[13] For the biosynthesis of bacterial alginate, GDP-mannuronic acid is polymerized to polymannuronate; this homopolymer can be modified by acetylation at positions O-2 and/or O-3, leading to a variable content of acetyl groups and G-residues, which strongly affects the gel-forming capacity of alginates.[44] Biotechnology provides tools to develop EPSs having novel structures and functional properties.[2]

5.5 Characteristics of Some Typical Commercial Microbial Exopolysaccharides

The characteristics of some commercial microbial EPSs are discussed below to facilitate comparison of their properties with those from marine microorganisms.

5.5.1 Xanthan

Xanthan is an anionic polysaccharide, produced from commercial fermentation of *Xanthomonas*, typically *X. campestris*. It is a white to cream colored, free flowing, powder soluble in hot and cold water but insoluble in most organic solvents.[45,46] Xanthan has a backbone of glucose units linked with β-(1,4)-glycoside bonds with branching at carbon 3 atoms. The branches contain D-mannopyranose-(2,1)-β-D-glucuronic acid-(4,1)-β-D-mannopyranose. The presence of side chains prevents strong polymer–polymer interactions, leading to enhanced solubility compared with cellulose. Aqueous solutions of xanthan exhibit pseudoplasticity and high yield values; relatively high stress is needed to produce a shear rate of 0.01 s[-1]. Because of their pseudoplastic nature, the viscosity of xanthan solutions decreases with higher shear rates but it increases again when the shear forces are removed. The relatively low viscosity at high shear rates makes it easy to mix, pour, and swallow. Its high viscosity at low shear rates gives it good suspension properties and stability in colloidal suspensions. Viscosity varies slightly with temperature, and the solutions remain viscous even at 100°C. Salts do not usually reduce the viscosity.

In its native state, xanthan has a single-helix conformation; however, heating dilute solutions causes an irreversible conformational change, leading to gelation and the polysaccharide adopting a double-helix conformation. Gelation of xanthan is promoted by cations in the order of $Ca^{2+} > K^+ > Na^+$. The gel is stable to freezing; however, the chains separate easily under shearing, which allows the solution to flow and accentuates the shear thinning behavior. These properties make the biopolymer a valuable thickening and texturizing agent over a wide pH range.[46]

Xanthan interacts with other food ingredients such as whey proteins. These interactions lead to a synergistic effect on the emulsifying capabilities and an increase in the stability of emulsion droplets against coalescence and flocculation. Salt has synergistic and antagonist effects, depending on the pH and salt combination.[47] Xanthan also shows synergistic interactions with guar gum and locust bean gum (LBG). The mixed polysaccharides reveal enhanced viscosity or gelation. Mixtures of xanthan or LBG produce elastic and cohesive gels at about a total concentration of 0.03%. The melting temperature of the gel is in the range of 43 to 60°C and is independent of ionic concentrations.[48]

5.5.2 Gellan

Gellan is a polysaccharide well known for its strong gel characteristics.[49–51] Gellan is elaborated by the Gram-negative bacteria belonging to the genus *Sphingomonas*, specifically by *S. paucimobilis* and *S. elodea*. These organisms secrete structurally related EPSs, collectively called *sphingans*, which include gellan, welan, and rhamsan. Of these, gellan is predominant. Production of gellan by the fermentation of *S. paucimobilis* and *Pseudomonas elodea* (now designated as *S. elodea*) is influenced by temperature, pH, stirring rate, oxygen transfer, and composition of the production medium. Aeration exerts a critical influence on the average molecular mass of the gellan produced. Cheese whey and molasses containing high sugar concentrations can be used as raw materials for low-cost fermentative production of gellan.[49]

In its native form, gellan is a linear anionic EPS based on a tetrasaccharide repeat unit composed of two molecules of D-glucose, one of L-rhamnose, and one of D-glucuronic acid. The native gellan is partially esterified with acyl substituents. The viscosity of aqueous gellan solution at a concentration of 0.4% (w/v) is more than 4000 cP (more than five times that of xanthan gum), and it is stable over a wide range of pH and temperatures. Native gellan has acetyl and glyceric side groups. Alkaline treatment under mild conditions removes the acetyl groups of gellan, without affecting the double-helix structure. Severe alkaline treatment removes all of the substituents. Commercial gellan produced from *Pseudomonas*, particularly *P. elodea*, is marketed under the trade names Gelrite® and Kelcogel®.

Deacetylated gellan forms a double helix whose transition temperature is influenced by external salt. The gel is brittle, firm, and optically clear. The solution is viscoelastic, with a molecular weight of 1×10^6 Da. The EPS is insoluble in cold water but gives hard gels when heated in the presence of calcium.[48] Acylated gellan can produce hard, brittle gels. Deacetylated polymer produces a clear, agar-like, thermoreversible gel in the presence of cations. The gel strength of the modified polymer is four times that of agar and can withstand autoclaving. Compared with other polysaccharides, gellan gel has many advantages, including an excellent thermal and acid stability, adjustable gel elasticity and rigidity, high transparency, and good flavor release.[50,51,52]

5.5.3 Dextrans from Lactic Acid Bacteria

Several lactic acid bacteria, including *Streptococcus* and *Leuconostoc*, provide dextrans, which are highly water-soluble polysaccharides. Different types of dextrans of varying size and structure are produced extracellularly by these bacteria by the action of secreted enzymes (e.g., dextran sucrase or sucrose).[53] The glucan is designated as such because dextran is comprised of polymeric chains of glucosyl units. Dextran produced by the bacterial strain *L. dextranicum* has been purified to homogeneity and characterized.[53] Dextran is a homopolymer containing (1,6) glucosidic linkages in the main linear chain, with little branching (<5%). This lack of branching makes the dextran highly water soluble. The biopolymer has unique rheological properties because of its potential to form very viscous solutions at low concentrations. The viscosity decreases with an increase in shear rate, exhibiting a typical non-Newtonian pseudoplastic behavior. The rheological properties favor its use as a thickening or gelling agent in food.[38,39]

5.5.4 Levan

Levan is a homopolysaccharide produced from a wide group of bacteria, yeasts, and fungi. They are formed by the fermentation of cane sugar and the action of polysaccharide-producing bacteria present in the cane sugar. These microorganisms include *Bacillus subtilis* and *B. circulans*, *Zymomonas mobilis*, *Biopolymyxa*, *Aerobacter levanicum*, *Acetobacter xylinum*, *Actinomyces naeslundii*, *Erwinia amylovora*, *Rhanella aquatilis*, *Lactobacillus reuteri*, *Microbacterium laveaniformans*, and *Serratia levanicum*. The organism *B. subtilis* (Natto) Takahashi could be the most efficient strain for levan production. When cultivated in 20% (w/w) sucrose for 21 hr, it yielded 40 to 50 mg/mL of levan. The product consisted of two fractions with different molecular weights of 1794 Da and 11 kDa, which can be easily separated by fractionation under a gradient of ethanol.[53] Levan is a β-(2,6)-linked fructose polymer, with about 12% branching through β-(2,1) linkages. The polysaccharide has a molecular weight of about 2×10^6 Da and is readily soluble in water.[53]

5.5.5 Curdlan

Curdlan is an insoluble microbial exopolymer. The name derives from its ability to curdle when heated. *Alcaligenes* bacteria (now taxonomically reclassified as *Agrobacterium*) and a few fungi produce the EPS. Factors such as initial pH and the concentrations of urea and sucrose have a significant influence on curdlan production. Curdlan from *Agrobacterium* is a linear polysaccharide homopolymer composed of exclusively β-(1,3) glucosidic linkages with a molecular weight of 1×10^5 Da, with an average degree of polymerization of approximately 450 units. Some curdlans show branched structures, and the average molecular weight ranges between 3×10^4 and 2×10^6 Da. The EPS is characterized by low crystallinity and is insoluble in cold water but soluble in hot water. When aqueous suspensions of curdlan were heated to 100°C or higher, they formed a gel, and gel strength increased with temperature.[53]

5.5.6 Pullulan

Pullulan is a water-soluble glucan produced extracellularly and aerobically by the yeast-like fungus *Aureobasidium pullulans*. Agrowastes such as grape skin pulp extract, starch waste, olive oil waste effluents, and molasses can all serve as substrates for the fermentation of *A. pullulans*. Fermentation of the ideal substrate (namely, grape skin pulp extract) produced pullulan at a concentration of 22.3 g/L. Microbial sources, structural geometry, upstream processing, downstream processing, and the distinctive characteristics and applications of pullulan have been discussed. Jaggery (a traditional byproduct of sugar manufacture from sugar cane) is a promising carbon source for the economical production of pullulan. Pullulan is a linear polysaccharide consisting of maltotriose units. It is an odorless, white powder that is soluble in water and produces clear, viscous solutions. It also has significant adhesion, sticking, lubrication, and film-forming abilities. The EPS products, in general, are usually off-white to white in color, depending on their purity.[54]

5.5.7 Bacterial Alginate

Although all commercial alginates are of algal origin, there is interest in the production of alginate-like polymers from bacteria.[44] The bacterium *Azotobacter vinelandii* is an ideal candidate for microbial production of alginate when grown aerobically. Alginate is produced when this organism is grown in glucose in a shaken flask. It is interesting that, although this alginate had a molecular weight of 220 kDa after 40 to 42 hours of fermentation, it decreased to 60 kDa after an additional fermentation for 20 to 22 hr. Certain *Pseuodomonas* species, such as *P. aeruginosa*, *P. fluorescens*, *P. putida*, and *P. mendocina* also produce alginate. In *A. vinelandii*, alginate helps maintain structural integrity, but the alginate in *P. aeruginosa*, a well-characterized human pathogen, seems to be an important virulence factor during the infection process of human epithelia. Unlike algal alginates, bacterial alginates are additionally *O*-acetylated on the 2 and 3 positions of the D-mannuronic acid residues. The high degree of acetylation in bacterial alginates can provide higher gel strength, as compared with seaweed alginate. The genetic modification of alginate-producing microorganisms could allow biotechnological production of new alginates with unique properties, suitable for medical and industrial applications.[55]

5.5.8 Bacterial Cellulose

To date, bacterial cellulose is rather unexploited, but it represents a polymeric material with major potential. Bacterial strains of *Acetobacter xylinum* and *A. pasteurianus* are able to produce an almost pure form of cellulose, homo-β-(1,4)-glucan. Its chemical and physical structure is identical to the cellulose formed in plants. Plant cellulose, however, has to undergo harsh chemical treatment to remove lignin and hemicellulose. Bacterial cellulose has been produced by fermentation of waste from beer culture fermentation. Static cultivation was found to be superior to shaking cultivation on the basis of production.[56]

5.5.9 Others

Clevan is another exopolysaccharide produced from microorganisms.[57] The polysaccharide is a heptasaccharide composed of rhamnose, mannose, glucose, and glucuronic acid. XM6 is a polysaccharide with interesting gelation properties that is produced by *Enterobacter* strain XM6. The polysaccharide forms gels with both monovalent and divalent cations and has a melting point at about 30°C.[5]

5.5.10 Interactions of Exopolysaccharides with Food Components

Interactions between bacterial exopolysaccharides and food compound proteins can influence the texture of foods. Interactions of bacterial EPSs with caseins and with whey protein isolates (WPIs) have been reported at an acidic pH in the range of 5 to 6. Charge density is the main factor influencing the amount of EPSs complexed onto casein molecules; molecular weight, chain stiffness, and charge density are all structural features influencing their binding with WPIs.[58,59] The tendency of starch to retrograde is common in processed starch-based foods (see Chapter 2). This can be prevented by the addition of xanthan gum, which is more effective than locust bean gum and konjac glucomannan, whereas guar gum was found to have no effect in retarding the retrogradation of tapioca starch gel. Syneresis of a 4% curdlan gel after freezing and thawing was reduced from 20.6% to 2.1% by the addition of 5% waxy corn starch and to 8.9% by the addition of 20% sucrose.[60] Curdlan conjugated with soy protein through naturally occurring Maillard reactions can be used as a functional food additive that has excellent gel-forming, emulsifying, and antioxidative properties.[61] The EPSs produced by lactic acid bacteria, being negatively charged, can interact with whey proteins at an acidic pH, leading to modified textural properties. Isolated EPSs demonstrate a rheological behavior different from that of the whole fermented media.[61]

5.6 Exopolysaccharides from Marine Microorganisms

Organisms that inhabit the deep sea are invariably adapted to three types of extreme environments—namely, low temperature, high pressure, and low nutrient levels (see Chapter 1). An increasing number of microorganisms isolated from marine extreme environments, such as Arctic and Antarctic ecosystems and deep-sea hydrothermal vents, are capable of producing EPSs (see Table 5.2).[62–69] In the ocean, these polysaccharides help microbial communities endure the extremes of temperature, salinity, and nutrient availability; however, unlike algae, production of polysaccharides by marine microorganisms is less sensitive to factors such as marine pollution or climatic changes. A number of studies on marine microbial EPSs have been reported. Ten bacterial strains isolated from Southern Ocean particulate material or from sea ice were representatives of the genera *Pseudoalteromonas, Shewanella, Polaribacter,* and *Flavobacterium,* including one strain belonging to a new

TABLE 5.2

Marine Bacteria that Secrete Exopolysaccharides

Organism	Source
Shewanella colwelliana	Eastern oyster
Vibrio alginolyticus	Marine fouling material
Vibrio parahaemolyticus	Marine water
Vibrio diabolicus	East Pacific Rise deep-sea hydrothermal vents
Pseudomonas sp. strain NCMB 2021	Halifax, Nova Scotia
Pseudoalteromonas	Marine sediment, seawater, and sea ice in the Southern Ocean
Pseudoalteromonas ruthenica	Marine biofilm-forming bacteria
Hyphomonas sp. strain MHS-3	Shallow water segment of Puget Sound
Alteromonas sp. strains KMM 156 and 2MM6	Halifax, Nova Scotia
Alteromonas sp. strains 1544 and 1644	East Pacific Rise
Alteromonas macleodii	East Pacific Rise
Geobacillus thermodenitrificans	Vulcano Island
Bacillus licheniformis	Vulcano Island
Thermophilic bacterial strains	Gulf of Naples
Oscillatoria (marine cyanobacteria)	Marine stromatolites, Bahamas
Thermococcus titoralis (marine archibacteria)	Mediterranean sea
Halomonas	Antarctic region, also hot spring and hypersaline habitats

Source: Adapted from Manusco-Nichols, C.A. et al., *Mar. Biotechnol.,* 7, 253, 2005; Kennedy, A.F.D. and Sutherland, I.W., *Biotechnol. Appl. Biochem.,* 9, 12, 1987; Nichols, M.C. et al., *J. Appl. Microbiol.,* 96, 1057, 2004; Muralidharan, J. and Jayachandran, S., *Proc. Biochem.,* 38, 841, 2003.

bacterial genus in the family Flavobacteriaceae. A novel EPS provides cryo-protection for the deep-sea psychrotolerant *Pseudoalteromonas* by enhancing the stability of cold-adapted protease by preventing its autolysis, presumably through a chelating action.[66] The EPS synthesis in these organisms is stimulated primarily by low temperature.[66] An *Alteromonas* sp. that produces polysaccharide has been isolated from deep-sea hydrothermal vents.[69]

Many species of halophilic bacteria belonging to the genus *Halomonas* secrete EPSs, with the quantity of EPS produced, its chemical composition, and its physical properties all depending on the bacterial strain. These organisms have been isolated from marine hot springs and saline lakes.[70] A highly anionic EPS bound to protein was isolated recently when a *Halomonas* sp. was grown on 1% glucose (w/v) and a peptone/yeast extract. This EPS, in addition to exhibiting high anionic activity, also had significant emulsifying activity. The combined effect of protein and anionic residues of the EPS contributed to its emulsification activity. The EPS was a heterogeneous polymer that displayed specific rheological properties in the presence or absence of monovalent and divalent ions. Also, it exhibited novel biological activities, metal binding capabilities, and unique chemical compositions that may find interesting applications in the near future, including biotechnological and food applications.[71–73]

5.6.1 Cultivation of Marine Microorganisms for Exopolysaccharides

Because of stringent growth requirements, only a minority of marine micro-organisms have been cultivated so far to isolate their metabolites, including EPSs. Marine microorganisms generally require salt in the medium for optimal growth. The cultivation of some marine bacteria under laboratory conditions has led to the production of EPSs. Currently, most production is at the shake-flask level, but investigations in bioreactor engineering and fermentation protocol design are in progress.[79] Secretion of EPS by a marine *Alteromonas* strain was stimulated by restricted growth conditions under limitation of nitrogen in fed-batch cultures.[74] A *Halomonas* sp. produced EPSs at a maximum of 2.8 g/L.[75] The degree of aeration influences EPS production, as observed in the case of the marine bacterium *Hahella chejuensis*.[75] *Bacillus thermantarcticus*, a thermophilic organism, produces two exopolysaccharides, with a total yield of 400 mg/mL when grown in mannose as the carbon and energy source.[53,76–79] Another thermophilic strain produced three EPSs with a total yield of 90 mg/L. The organism grew well at an optimal temperature of 60°C at pH 7 using sucrose or trehalose as the sole carbon source. EPS3 had an average molecular mass of 1×10^6 Da and contained mannose, glucose, and galactose in various proportions.[78]

Exopolysaccharides secreted by *Pseudoalteromonas*, a psychrotolerant bacterium isolated from deep-sea sediment, increased as the culture temperature was decreased from 30°C to 10°C, reaching a maximum value of 5.25 g/L. EPS production was optimal at 15°C when grown for 52 hr.[67] In batch culture, two EPS-producing bacterial strains belonging to the genus *Pseudoalteromonas* produced EPS. The yield of EPS produced by one strain was 30-fold higher at −2° and 10°C than at 20°C. Chemical analyses showed that the EPS molecules were composed primarily of neutral sugars and uronic acids with sulfates.[80] Comparison of the abilities of 19 strains belonging to the genus *Halomonas* to produce EPSs showed that one strain had a maximum EPS production of 1.6 g/L. The chemical composition of the polysaccharides was affected by the culture medium.[80] Production of EPSs by the moderately halophilic bacterium *Enterobacter cloacae* has been reported by the Central Salt and Marine Chemicals Research Institute, India.[81] A marine bacterium, *Antarctobacter*, was isolated and selected for its production of an extracellular emulsifying agent, AE22. Production of AE22 commenced toward the late exponential phase of growth, with maximum emulsifying activity detected after approximately 4 days of the cells entering the death phase.[82]

The extremely halophilic archaebacterium *Haloferax mediterranei* produces a heteropolysaccharide that gives the colonies a typical mucous character and is responsible for the appearance of a superficial layer in unshaken liquid medium. This exocellular polymeric substance can be obtained from the supernatant of shaken liquid cultures by cold ethanol precipitation, and yields as high as 3 mg/mL have been detected. The substance was produced under all the conditions tested and with all substrates assayed, although higher yields were obtained with sugars, particularly glucose, as the carbon and energy source. The polymer contains mannose as the major component. Glucose, galactose, and another unidentified sugar were also present, as well as amino sugars, uronic acids, and a considerable amount of sulfate, which accounts for the acidic nature of the polymer. Infrared spectrum and specific assays showed the absence of acyl groups.[89]

5.6.2 Chemical Nature of Marine Exopolysaccharides

Table 5.3 provides the chemical composition of EPSs produced by various marine bacteria. Most of these EPSs are linear heteropolysaccharides consisting of three or four different monosaccharides arranged in groups of 10 or fewer to form repeating units. The monosaccharides may be pentoses, hexoses, amino sugars, or uronic acids. EPSs possess different types of functional groups; most EPSs are sulfated and high in uronic content, and this gives them a net negative charge and acidic properties at the pH (near a value of 8.0) of seawater. These polysaccharides are very diverse, even among closely related organisms. Some strains produced unusually large polymers (molecular weights up to 5.7 MDa).[83] *Pseudoalteromonas ruthenica* isolated from a marine environment produced copious amount of EPSs, which contained eight individual sugars.[84] An EPS having a molecular weight of about 6.39 × 10^6 Da from a marine biofouling bacterium, *Vibrio alginolyticus*, showed the presence of glucose, aminoarabinose, aminoribose, and xylose in the molar ratio of 2:1:9:1.[85]

TABLE 5.3

Chemical Composition of Exopolysaccharide Produced by Marine Bacteria

Organism	Chemical Composition	Ref.
Alteromonas sp. strain 1545	Glc/Gal/4,6-Pyr–Gal/GlcA/GalA	Helene and Jean[86]
Alteromonas infernas	Glc/Gal/GlcA/Gal	Maugeri et al.[69]
Vibrio alginolyticus	Glc/amino-Ara/amino-Rib/Xyl in a molar ratio of 2:1:9:1	Muralidharan and Jayachandran[85]
Alteromonas macleodii	Glc/Gal/4,6-Pyr–Man/GlcA/GalA	Helene and Jean[86]
Thermophilic microorganism	Man/Glu/Gal in a relative ratio of 0.5:1.0:0.3 in EPS1, Man/Glu/Gal in a ratio of 1.0:0.3:trace in EPS2, and Gal/Man/GlcA/Ara in a ratio of 1.0:0.8:0.4:0.2 in EPS3	Moriello et al.[78]
Pseudoalteromonas sp. strain SM9913	Glu/Ara/4,6-Pyr–Glu	Quin et al.[66]

Abbreviations: 4,6-Py–Man, 4,6-O-{1-carboxyethylidene}-mannose residue; 4,6-Pyr–Gal, 4,6-O-carboxyethilidene-galactose; Ara, arabinose; Gal, galactose; GalA, galacturonic acid; Glc, glucose; GlcA, glucuronic acid; Man, mannose; Rib, ribose; Xyl, xylose.

An *Alteromonas* sp. producing an EPS has been isolated from marine environments. The EPS consists of residues of glucose, galactose, glucuronic acid, galacturonic acid, and pyruvated mannose associated into a repeating hexasaccharidic unit.[86] A marine bacterium, *Antarctobacter*, produced an emulsifier that is a high-molecular-weight (>2000 kDa) glycoprotein with high uronic acid content.[82] EPSs from Polynesian mats were primarily composed of carbohydrate, heavy metals, and humic acid, plus small quantities of proteins and DNA. Neutral hexose concentrations corresponded to more than 90% of the total EPS dry weight. The proportions of metals in the EPSs are similar to the proportions present in the water from each locality where the biofilms were collected.[87]

Bacillus thermantarcticus, a thermophilic organism, produces two EPSs, which have complex primary structures composed of different repetitive units, including a galactomannan type and mannan type. EPS1 seems to be close to some xanthan polymers, whereas EPS2 is a mannan.[88,89] EPS of *Pseudoalteromonas* sp. strain SM9913, a psychrotolerant bacterium isolated from deep-sea sediment, has a structure different from that of EPSs reported for other marine bacteria. The major sugar unit of the EPS was (1,6)-linked glucose.[66,90] An EPS produced by *Halomonas* is a heteropolysaccharide composed of glucose, galactose, mannose, and glucuronic acid in an equimolar ratio.[74] A comparison of the chemical structures of 19 strains belonging to the genus *Halomonas* showed that all of the EPSs had unusually high sulfate contents, and one EPS had a significant amount of uronic acid.[80,91] Purification and characterization of an extracellular polysaccharide from haloalkalophilic *Bacillus* sp. I-450 has also been reported.[92]

Two bacterial isolates from the intertidal zone produced significant quantities of two EPSs with interesting properties: PS 3a24 and PS 3a35. The relative proportion of sugars in PS 3a35 was 51.6% glucose, 39.0% galactose, 3.1% mannose, and 6.3% rhamnose, with a trace of an unidentified sugar. PS 3a24 was composed of 40.2% glucose, 57.2% galactose, and 2.6% mannose. PS 3a35 contained 6% pyruvate, whereas PS 3a24 contained no pyruvate.[93] A process to prepare marine fungi polysaccharide was reported in a recent patent.[94] *Haloferax mediterranei* EPS contains mannose as the major component. Glucose, galactose, and another unidentified sugar were also present, as well as amino sugars, uronic acids, and a considerable amount of sulfate, which accounts for the acidic nature of the polymer. Infrared spectrum and specific assays showed an absence of acyl groups.[89]

5.6.3 Functional Properties

The extreme environmental habitats of marine microorganisms give their EPS molecules novel functional properties such as enhanced stabilities to heat and pH as well as interesting rheological characteristics. These properties may be more pronounced in comparison with those of traditional polymers of nonmicrobial origin such as algae (alginates, carrageenans), crustaceans (chitin), or plants (starch and cellulose). The physical and functional properties of marine polysaccharides are influenced not only by their chemical nature but also by the method of their preparation and isolation. A relationship exists between the glycoside linkages, the geometry of polysaccharides, and their conformation; however, studies in these fields are in their infancy. Unlike EPSs from terrestrial microorganisms or aquatic microalgae, marine bacterial EPSs are often highly polyanionic, a property attributed to their relatively high uronic acid content.[72] Hence, these compounds are often highly active and represent a potential source of commercially valuable compounds.[71]

The main properties of marine EPSs—their rheology, flocculating and emulsifying properties, and film-forming capacity—make them good alternatives to other natural polysaccharides. Their rheological properties, particularly, support their functionality. *Pseudomonas ruthenica* isolated from marine environments has good shearing properties,[84] and the EPS secreted by *P. oleovorans* has a pseudoplastic fluid behavior in aqueous media.[15] The apparent viscosity is stable in the pH range of 2.9 to 7.1 and NaCl concentrations up to 1.0 *M*. The marine bacterium *Antarctobacter* produces a high-molecular-weight glycoprotein that is an emulsifier.[83] Though its apparent viscosity decreased at high temperatures, under alkaline conditions, and at NaCl concentrations of 2.0 *M*, its pseudoplastic fluid behavior was retained. The EPS was capable of stabilizing water emulsions with several hydrophobic compounds, including vegetable oils and hydrocarbons. The emulsifying activity was stable at temperatures ranging from 30 to 50°C and at both acidic and basic pH values.[15]

Available data on the functional properties of a few polysaccharides from marine microorganisms indicate that these EPSs undergo gelation comparable to their non-marine counterparts. Gelation, in general, involves a transition from an ordered state at a lower temperature in the presence of ions to a disordered state at an elevated temperature in a low ionic environment. Slight changes may induce considerable differences in the physical properties of these gels. The side chains found on many linear polysaccharides promote conformational disorder and inhibit any ordered assembly. These properties have a direct influence on water-holding, viscosity, and emulsifying properties. In addition, these properties are also influenced by the presence of cations, pH, and sometimes the presence of proteins. The extent of water binding by the hydrocolloids is important in the functional properties of their gels (see Chapter 1), because their structural changes can affect water binding at any given polymer concentration. Information on water-binding properties also provides information on the flexibility of the polysaccharides in a given situation.[95,96]

5.7 Marine Biotechnology

Marine biotechnology encompasses those efforts that help harness marine resources of the world. Recent work in this field includes the development of marine ingredients such as EPSs for food and other industries, seaweed farming, and monitoring ocean pollution, among others. Attention is focused on culturing EPSs under commercial conditions. Bioprocessing strategies in microbial cultivation include solid-state and submerged-state fermentation, the latter being popular for fungi in the industry. The growth and production of EPSs are affected by a wide range of parameters, including cultivation media, inoculums, pH, temperature, aeration agitation, and shear stress. Compared to many unicellular microbes, fermentation of filamentous fungi presents special challenges in optimization and scale-up because of the varying fungal morphological forms.[97–99] There is a potential for using the gene pools of marine bacteria for recombinant DNA technology to increase polysaccharide yield.[72]

5.8 Summary

Several marine microorganisms are capable of producing exopolysaccharides, which could be isolated by the cultivation of such microorganisms under conditions that simulate their marine habitats. The EPSs of these organisms have several features comparable to those produced by terrestrial microorganisms. Their functionality, particularly with respect to their possible food applications, is covered in Chapter 8, which presents some case studies.

References

1. Kumar, A. S., Mody, K., and Jha, B., Bacterial exopolysaccharides: a perception, *J. Basic Microbiol.*, 47, 103, 2007.
2. Morin, A., Screening of polysaccharide-producing microorganisms, factors influencing the production and recovery of microbial polysaccharides, in *Polysaccharides: Structural Diversity and Functional Versatility*, Dumitriu, S., Ed., Marcel Dekker, New York, 1998, pp. 275–296.
3. Song, S. F. et al., Progress in researches of microbial polysaccharides, *Oilfield Chem.*, 21, 91, 2004.
4. Crescenzi, V., Microbial polysaccharides of applied interest: ongoing research activities in Europe, *Biotechnol. Prog.*, 11, 251, 1995.
5. Sutherland, I. W., Polysaccharides from microorganisms, plants and animals, in *Biopolymers. Vol. 5. Polysaccharides I: Polysaccharides from Prokaryotes*, Vandamme, E., De Baets, S., and Steinbuchel, A., Eds., Wiley–VCH, Weinheim, 2002, pp. 1–19.
6. Harrah, T., Panilaitis, B., and Kaplan, D., Microbial exopolysaccharides, in *The Prokaryotes. Vol. 1. Symbiotic Associations, Biotechnology, Applied Microbiology*, Dworkin, M. et al., Eds., Springer, New York, 2006, pp. 766–776.
7. Brooks, J. D. and Flint, S. H., Biofilms in the food industry: problems and potential solutions, *Int. J. Food Sci. Technol.*, 43, 2163, 2008.
8. Kumar, C. G. and Anand, S. K., Significance of microbial biofilms in food industry: a review, *Int. J. Food Microbiol.*, 42, 9, 1998.
9. Majumder, A. and Goyal, A., Rheological and gelling properties of a novel glucan from *Leuconostoc dextranicum* NRRL B-1146, *Food Res. Int.*, 42, 525, 2009.
10. Kodali, V. P., Das, S., and Sen, R., An exopolysaccharide from a probiotic: biosynthesis dynamics, composition and emulsifying activity, *Food Res. Int.*, 42, 695, 2009.
11. De Vuysta, L. and Degeesta, B., Heteropolysaccharides from lactic acid bacteria, *FEMS Microbiol. Rev.*, 23, 153, 1999.
12. Welman, A. D. and Maddox, I. S., Exopolysaccharides from lactic acid bacteria: perspectives and challenges, *Trends Biotechnol.*, 21, 269, 2003.
13. Sutherland, I. W., Microbial polysaccharides from Gram-negative bacteria, *Int. Dairy J.*, 11, 663, 2001.
14. Royan, S., Parulekar, C., and Mavinkurve, S., Exopolysaccharides of *Pseudomonas mendocina* P_2d, *Lett. Appl. Microbiol.*, 29, 342, 1999.
15. Freitas, F. et al., Emulsifying behavior and rheological properties of the extracellular polysaccharide produced by *Pseudomonas oleovorans* grown on glycerol byproduct, *Carb. Polym.*, 78, 549, 2009.
16. Hung, C. C., Santschi, P. H., and Gillow, J. B., Isolation and characterization of extracellular polysaccharides produced by *Pseudomonas fluorescens* Biovar II, *Carb. Poly.*, 61, 141, 2005.

17. Marqués, A. M. et al., Production and rheological properties of the extracellular polysaccharide synthesized by *Pseudomonas* sp. strain EPS-5028, *Appl. Environ. Microbiol.*, 52, 1221, 1986.

18. Vinarta, S. C. et al., A further insight into the practical applications of exopolysaccharides from *Sclerotium rolfsii*, *Food Hydrocoll.*, 20, 5619, 2006.

19. Grigorova, D. et al., Preparation and preliminary characterization of exopolysaccharides by yeast *Rhodotorula acheniorum* MC, *Appl. Biochem. Biotechnol.*, 81, 181, 1999.

20. Frengova, G., Simova, E., and Beshkova, D., Caroteno-protein and exopolysaccharide production by co-cultures of *Rhodotorula glutinis* and *Lactobacillus helveticus*, *J. Ind. Microbiol. Biotechnol.*, 18, 272, 1997.

21. De Baets, S. et al., Optimization of exopolysaccharide production by *Tremella mesenterica* NRRL Y-6158 through implementation of fed-batch fermentation *J. Ind. Microbiol. Technol.*, 29, 181, 2002.

22. Levander, F. et al., Enhanced exopolysaccharide production by metabolic engineering of *Streptococcus thermophilus*, *Appl. Environ. Microbiol.*, 68, 784, 2002.

23. Martiínez-Checa, F. et al., Yield production, chemical composition, and functional properties of emulsifier H28 synthesized by *Halomonas eurihalina* strain H-28 in media containing various hydrocarbons, *Appl. Environ. Microbiol.*, 58, 358, 2008.

24. Camargo-de-Morais, M. M. et al., Production of an extracellular polysaccharide with emulsifier properties by *Penicillium citrinum*, *World J. Microbiol.*, 19, 191, 2003.

25. Jumel, K., Geider, K., and Harding, S. E., The solution molecular weight and shape of the bacterial exopolysaccharides amylovoran and stewartan, *Int. J. Biol. Macromol.*, 20, 251, 2007.

26. Fett, W. F., Biopolymers from fermentation, *ACS Symp. Ser.*, 647, 74, 1996.

27. Morris, J., Fermentation-derived polysaccharides for use in foods, *J. Chem. Technol. Biotechnol.*, 58, 193, 1993.

28. Vanhooren, P. and Vandamme, E. J., Biosynthesis, physiological role, use and fermentation process characteristics of bacterial exopolysaccharides, *Recent Res. Dev. Ferment. Bioeng.*, 1, 253, 1998.

29. Govan, J. R. W. et al., Isolation of alginate-producing mutants of *Pseudomonas fluorescens*, *P. putida* and *P. mendocina*, *J. Gen. Microbiol.*, 125, 217, 1981.

30. Petry, S. et al. Factors affecting exocellular polysaccharide production by *Lactobacillus delbrueckii* subsp. *Bulgaricus* grown in a chemically defined medium, *Appl. Environ. Microbiol.*, 66, 3427, 2000.

31. Digman, B. and Kim, D. S., Review: Alternative energy from food processing wastes, *Environ. Prog.*, 27, 524, 2008;.

32. .Lee, I.Y. et al., Production of curdlan using sucrose or sugarcane molasses by two-step fed-batch cultivation of *Agrobacterium* species, *J. Ind. Microbiol. Biotechnol.*, 18, 255, 1997.

33. Ionescu, M. and Belkin, S., Overproduction of exopolysaccharides by an *Escherichia coli* K-12 *rpoS* mutant in response to osmotic stress, *Appl. Environ. Microbiol.*, 75, 483, 2009.

34. Banik, R. M. and Santhiagu, A., Improvement in production and quality of gellan gum by *Sphingomonas paucimobilis* under high dissolved oxygen tension levels, *Biotechnol. Lett.*, 28, 1347, 2006.

35. Gibbs, P. A. and Seviour, R. J., The production of EPSs by *Aureobasidium pullulans* in fermenters with low-shear configurations, *Appl. Microbiol. Biotechnol.*, 49, 168, 1998.
36. Oh, D.-K., Kim, J.-H., and Yoshida, T., Production of a high viscosity polysaccharide, methylan, in a novel bioreactor, *Biotechnol. Bioeng.*, 54, 115, 1997.
37. Freitas, V. D. et al., Characterization of an extracellular polysaccharide produced by a *Pseudomonas* strain grown on glycerol, *Biores. Technol.*, 100, 859, 2009.
38. Majumder, A., Singh, A., and Goyal, A., Application of response surface methodology for glucan production from *Leuconostoc dextranicum* and its structural characterization, *Carb. Polym.*, 75, 150, 2009.
39. Purama, R. K. et al., Structural analysis and properties of dextran produced by *Leuconostoc mesenteroides* NRRL B-640, *Carb. Polym.*, 76, 30, 2009.
40. Goh, K. K. T. et al., Evaluation and modification of existing methods for the quantification of exopolysaccharides in milk-based media, *Food Res. Int.*, 38, 605, 2005.
41. Achanta, S. and Okos, M. R., Predicting the quality of dehydrated foods and biopolymers—research needs and opportunities, *Dry Technol.*, 14, 1329, 1996.
42. Ginzberg, A. et al., Effect of drying on the biological activities of a red microalgal polysaccharide, *Biotechnol. Bioeng.*, 99, 411, 2008.
43. Quignard, F. et al., Aerogel materials from marine polysaccharides, *New J. Chem.*, 32, 1300, 2008.
44. Remminghorst, U. and Rehm, R. H. A., Bacterial alginates: from biosynthesis to applications, *Biotech. Lett.*, 28, 1701, 2006.
45. Yang, S.-T., Lo, Y.-M., and Chattopadhyay, D., Production of cell-free xanthan fermentation broth by cell adsorption on fibers, *Biotechnol. Progr.*, 14, 259, 1998.
46. Born, K., Langendorff, V., and Boulenguer, P., Xanthan, in *Polysaccharides and Polyamides in the Food Industry*, Vol. 1, Steinbuchel, A. and Rhee, S. K., Eds., John Wiley & Sons, New York, 2005, pp. 481–518.
47. Bertrand, M. E. and Turgeon, S. L., Improved gelling properties of whey protein isolate by addition of xanthan gum, *Food Hydrocoll.*, 21, 159, 2007.
48. Rinaudo, M., Relation between the molecular structure of some polysaccharides and original properties in sol and gel states, *Food Hydrocoll.*, 15, 433, 2001.
49. Wang, X. et al., Modeling for gellan-gum production by *Sphingomonas paucimobilis* ATCC 31461 in a simplified medium, *Appl. Environ. Microbiol.*, 72, 3367, 2006.
50. Ashtaputre, A. A. and Shah, A. K., Studies on a viscous, gel-forming exopolysaccharide from *Sphingomonas paucimobilis* GS1, *Appl. Environ. Microbiol.*, 61, 1159, 1995.
51. Fialho, A. M. et al., Occurrence, production, and applications of gellan: current state and perspectives, *Appl. Microbiol. Biotechnol.*, 79, 889, 2008.
52. Naessens, M. et al., Leuconostoc dextransucrase and dextran: production, properties and applications, *J. Chem. Technol. Biotechnol.*, 80, 845, 2005.
53. McIntosh, M., Stone, B. A., and Stanisich, V. A., Curdlan and other bacterial (1→3)-beta-D-glucans, *Appl. Microbiol. Biotechnol.*, 68, 163, 2005.
54. Singh, R. S. et al., Pullulan: microbial sources, production and applications, *Carb. Polym.*, 73, 515, 2008.
55. Clementi, F. et al., Production and characterisation of alginate from *Azotobacter vinelandii*, *J. Sci. Food Agric.*, 79, 602, 1999.

56. Ha, J. H. et al., Production of bacterial cellulose by a static cultivation using the waste from beer culture broth. *Korean J. Chem. Eng.*, 25, 812, 2008.
57. Vanhooren, P. T. and Vandamme, E. J., Microbial production of clavan, an L-fucose rich exopolysaccharide, in *Food Biotechnology*, Bielecki, S. et al., Eds., Elsevier, Amsterdam, 2000, pp. 109–114.
58. Girard, M. D. and Schaffer-Lequart, C., Attractive interactions between selected anionic exopolysaccharides and milk proteins, *Food Hydrocoll.*, 22, 1425, 2008.
59. Muadklay, J. and Charoenrein, S., Effects of hydrocolloids and freezing rates on freeze–thaw stability of tapioca starch gels, *Food Hydrocoll.*, 22, 1268, 2008.
60. Junfeng, F. et al., Improving functional of soy protein hydrolysate by conjugation with curdlan, *J. Food Sci.*, 71, C285, 2006.
61. Ayala-Hernandaez, I. et al., Production, isolation and characterization of exopolysaccharides produced by *Lactococcus lactis* subsp. *cremoris* JFR1 and their interaction with milk proteins: effect of pH and media composition. *Int. Dairy J.*, 18, 1109, 2008.
62. Zhenming, C. and Yan, F., Exopolysaccharides from marine bacteria, *J. Ocean Univ. China*, 4, 67, 2005.
63. Manusco-Nichols, C. A. et al., Bacterial exopolysaccharides from extreme marine environments with special consideration of the Southern Ocean, sea ice, and deep-sea hydrothermal vents: a review, *Mar. Biotechnol.*, 7, 253, 2005.
64. Guezennec, J., Deep-sea hydrothermal vents: a new source of innovative bacterial exopolysaccharides of biotechnological interest?, *J. Ind. Microbiol. Biotechnol.*, 29, 204, 2002.
65. Decho, A. W., Microbial exopolymer secretions in ocean environments: their role(s) in food webs and marine processes, *Oceanogr. Mar. Biol. Ann. Rev.*, 28, 73, 1990.
66. Quin, G. et al., Structural characterization and ecological roles of a novel exopolysaccharide from the deep-sea psychrotolerant bacterium *Pseudoalteromonas* sp. SM9913, *Microbiology*, 153, 1566, 2007.
67. Nichols, C. M. et al., Chemical characterization of exopolysaccharides from Antarctic marine bacteria, *Microbial Ecol.*, 49, 578, 2005.
68. Raguenes, G. et al., *Alteromonas infernos* spp. Nov., a new polysaccharide-producing bacterium isolated from deep-sea hydrothermal vent, *J. Appl. Bacteriol.*, 82, 422, 1997.
69. Maugeri, T. L. et al., A halophilic thermotolerant *Bacillus* isolated from a marine hot spring able to produce a new exopolysaccharide, *Biotechnol. Lett.*, 24, 515, 2002.
70. Gutierrez, T., Morris, G., and Green, D. H., Yield and physicochemical properties of exopolysaccharide from *Halomonas* sp. strain TG39 identifies a role for protein and anionic residues (sulfate and phosphate) in emulsification of *n*-hexadecane, *Biotechnol. Bioeng.*, 103, 207, 2009.
71. Kennedy, A. F. D. and Sutherland, I. W., Analysis of bacterial exopolysaccharides, *Biotechnol. Appl. Biochem.*, 9, 12, 1987.
72. Weiner, R. M. et al., Applications of biotechnology to the production, recovery and use of marine polysaccharides, *Bio/Technology*, 3, 899, 1985.
73. Samain, E. et al., Simultaneous production of two different gel-forming exopolysaccharides by an *Alteromonas* strain originating from deep sea hydrothermal vents, *Carb. Polym.*, 34, 235, 1998.

74. Bouchotroch, S. et al., Bacterial exopolysaccharides produced by newly discovered bacteria belonging to the genus *Halomonas*, isolated from hypersaline habitats in Morocco, *J. Ind. Microbiol. Biotechnol.*, 24, 374, 2000.
75. Lee, H. S. et al., Effect of aeration rates on production of extracellular polysaccharide, EPS-R, by marine bacterium *Hahella chejuensis*, *Biotechnol. Bioprod. Eng.*, 6, 359, 2001.
76. Nicoulas, B. et al., Polysaccharides from extremophilic microorganisms, *Orig. Life Evol. Biosph.*, 34, 159, 2004.
77. Nicolaus, B. et al., A thermophilic *Bacillus* isolated from an Eolian shallow hydrothermal vent, able to produce exopolysaccharides, *System. Appl. Microbiol.*, 23, 426, 2000.
78. Moriello, S.V. et al., Production of exopolysaccharides from a thermophilic microorganism isolated from a marine hot spring in Flegrean areas, *J. Ind. Microbiol. Biotechnol.*, 30, 95, 2003.
79. Nichols, C. M., Bowman, J. P., and Guezennec, J., Effects of incubation temperature on growth and production of exopolysaccharides by an Antarctic sea ice bacterium grown in batch culture, *Appl. Environ. Microbiol.*, 71, 3519, 2005.
80. Bejar, V. et al., Characterization of exopolysaccharides produced by 19 halophilic strains of the species *Halomonas eurihalina*, *J. Biotechnol.*, 61, 135, 1998.
81. Iyer, A. and Mody, K. H., Production of exopolysaccharides by a moderately halophilic bacterium, *Trends Carbohydr. Chem.*, 7, 77, 2001.
82. Gutierrez, T. et al., Partial purification and chemical characterization of a glycoprotein (putative hydrocolloid) emulsifier produced by a marine bacterium, *Antarctobacter*, *Appl. Microbiol. Biotechnol.*, 76, 1017, 2007.
83. Nichols, M. C. et al., Production of exopolysaccharides by Antarctic marine bacterial isolates, *J. Appl. Microbiol.*, 96, 1057, 2004.
84. Saravanan, S. and Jayachandran, P., Preliminary characterization of exopolysaccharides produced by a marine biofilm-forming bacterium, *Pseudoalteromonas ruthenica* (SBT 033), *Lett. Appl. Microbiol.*, 46, 1, 2008.
85. Muralidharan, J. and Jayachandran, S., Physicochemical analyses of the exopolysaccharides produced by a marine biofouling bacterium, *Vibrio alginolyticus*, *Proc. Biochem.*, 38, 841, 2003.
86. Helene, R. and Jean, G., Novel Polysaccharide Obtained from Marine Bacteria *Altermonas macleodii*, European Patent No. FR2780063 (A1), 1999.
87. Aguilera, A. et al., Microbial communities and exopolysaccharides from Polynesian mats, *Appl. Microbiol. Biotechnol.*, 78, 1079, 2008.
88. Nicolaus, B. et al., A thermophilic *Bacillus* isolated from an Eolian shallow hydrothermal vent, able to produce exopolysaccharides, *System. Appl. Microbiol.*, 23, 426, 2000.
89. Anton, J. et al., Production of an extracellular polysaccharide by *Haloferax mediterranei*, *Appl. Environ. Microbiol.*, 54, 2381, 1988.
90. Talmont, F. et al., Structural investigation of an acidic exopolysaccharide from a deep-sea hydrothermal vent marine bacteria, *Food Hydrocoll.*, 5, 171, 1991.
91. Arias, S., Mauran, an exopolysaccharide produced by the halophilic bacterium *Halomonas maura*, with a novel composition and interesting properties for biotechnology, *Extremophiles*, 7, 319, 2003.
92. Kumar, C. G. et al., Purification and characterization of an extracellular polysaccharide from haloalkalophilic *Bacillus* spp. I-450, *Enz. Microbiol. Technol.*, 34, 673, 2004.

93. Boyle, C. D. and Reade, A. E., Characterization of two extracellular polysaccharides from marine bacteria, *Appl. Environ. Microbiol.*, 46, 392, 1983.
94. Bing, W., Lijuan, Z., and Rui, Y., Marine Fungi Polysaccharide and Its Extraction Method and Use, European Patent No. CN1657542 (A), 2005.
95. Hart, T. D. et al., A stray field magnetic resonance study of water diffusion in bacterial exopolysaccharides, *Enz. Microbiol. Technol.*, 24, 339, 2004.
96. Calvo, C. et al., Effect of cations, pH, and sulfate content on the viscosity and emulsifying activity of the *Halomonas eurihalina* exopolysaccharide, *J. Ind. Microbiol. Biotechnol.*, 20, 205, 1998.
97. Mikhailov, V. V., Kuznetsova, T. A., and Belyakov, G. B., Marine microorganisms and their biologically active secondary metabolites, *J. Appl. Biochem. Microbiol.*, 36, 613, 2000.
98. Blunt, J. W. et al., Marine natural products, *Nat. Prod. Rep.*, 22, 15, 2005.
99. Liebezeit, G., Aquaculture of "non-food" organisms for natural substance production, *Adv. Biotechnol. Eng. Biotechnol.*, 97, 1, 2005.

Section II

Food Applications

6

Crustacean Polysaccharides: Food Applications

6.1 Introduction

Chitin and its deacetylated product chitosan, as well as their derivatives, have found varied applications in agriculture, food processing, biotechnology, chemistry, cosmetics, dentistry, medicine, textiles, veterinary medicine, and environmental sciences. Their food uses cover a wide range of applications, including control of microbial deterioration, inhibition of lipid oxidation, emulsification, thickening, stabilization of color, and as dietary supplements. These benefits are derived due to their polyelectrolyte nature and the presence of reactive functional groups that offer gel-forming ability, high adsorption capacity, biodegradability, barrier properties, and bacteriostatic, fungistatic, and physiological advantages. In addition to their renewable and biocompatible nature, other characteristics of chitin and chitosan that make them useful for food applications are their deacetylated amino group reactivity, chelating action, complex formation, and ability to form films that have excellent barrier properties, as indicated in Table 6.1 (see also Chapter 3). There has been great interest in recent years in the use of chitosan to improve the quality and shelf life of diverse foods (e.g., fruits and vegetables, poultry, and seafood).[1–8] This chapter discusses the properties of chitin and chitosan that are relevant to food product development, as well as their applications in various food products. Uses of crustacean polysaccharides for the development of edible films and their applications for food preservation are discussed in Chapter 9.

6.2 Properties Important to Food Applications

6.2.1 Antimicrobial Activity

Interest in chitin and chitosan arises from their antimicrobial activities. Chitosan displays a broad spectrum of antimicrobial activities against both Gram-positive and Gram-negative microorganisms, including *Aeromonas*

TABLE 6.1

Features of Chitosan Useful in Food Applications

Properties	Applications
Renewable resource	Abundantly available from marine sources and hence renewable
Bioactivity	Antimicrobial activity (useful in food packaging); stimulation of immunological system; anticholesterolemic activity; obesity control
Biodegradability	Substrate for single cell production; biodegradable packaging material; controlled release of drugs and nutrients
Deacetylated amino group reactivity	Moisture control; thickening agent
Chelating capacity	Water treatment (e.g., removal of metals); antioxidant activity
Complex formation with other macromolecules	Complexes with proteins (useful for removal of hypoallergenic β-lactoglobulin from whey, clarification of wines); removal of protein from seafood industry wastewater by chitosan–alginate
Biocompatibility	Nontoxic; biological tolerance
Film-forming properties	Edible packaging; encapsulation materials; delivery of nutraceuticals

hydrophila, Bacillus cereus, B. licheniformis, B. subtilis, Clostridium perfringens, Brochothrix spp., *Enterobacter sakazakii, Lactobacillus* spp., *Listeria monocytogenes, Pseudomonas* spp., *Salmonella typhimurium, S. enteritidis, Serratia liquefaciens, Staphylococcus aureus,* and *Escherichia coli* O157; the yeasts *Candida, Saccharomyces,* and *Rhodotorula;* and the molds *Aspergillus, Penicillium,* and *Rhizopus.* The antibacterial activities of chitin and chitosan against these organisms have been screened in a variety of foods, including infant milk formula. These studies indicated that most of these organisms are sensitive to low concentrations of chitosan, ranging from about 4 to 75 mg%. Generally, Gram-negative bacteria are more sensitive to chitosan, with a minimum inhibitory concentration (MIC) as low as 0.006% (w/v). The sensitivity of Gram-positive bacteria is highly variable. Yeasts require a slightly higher MIC value of 0.01% (w/v). Storage of chitosan does not affect its antibacterial activities; in fact, storage may enhance this property.[9–13]

Chitosan oligosaccharides have also been shown to possess antimicrobial activity, generally depending on their average molecular weight. Oligomers of low molecular weight (2000 to 4000 Da) appeared to be more effective than those having molecular weights of 8000 to 12,000 Da.[14] A 5% hydrolyzate showed appreciable inhibition on pure cultures of *Bacillus cereus, Lactobacillus brevis, Leuconostoc mesenteroides, Micrococcus varians, Staphylococcus aureus, Acinetobacter* spp., *Escherichia coli, Pseudomonas aeruginosa, Salmonella typhimurium*, and *Serratia liquefaciens*. High-molecular-weight chitosan oligosaccharides inhibited the growth of yeasts (*Candida albicans*, *Rhodotorula glutinis*, *Saccharomyces cerevisiae*, and *Trichosporon pullulans*), while the products showed only weak inhibition of the mold *Aspergillus niger*. Treatment with 3000 ppm of chitosan hydrolyzates resulted in 75% inhibition of *Mucor mucedo* and 95% inhibition of *Rhizopus stolonifer*.[15] The antimicrobial effect on *S. aureus* was enhanced when the molecular weight of chitosan was below 300 kDa. In contrast, the effect on *E. coli* was weakened.[16]

Antimicrobial activities of chitosan oligosaccharides with different molecular weights (1333, 432, 201, 131, and 104 kDa) prepared by ozone treatment were examined against two Gram-positive bacteria (*Listeria monocytogenes* and *Staphylococcus aureus*) and two Gram-negative bacteria (*Escherichia coli* and *Pseudomonas fluorescens*). The antimicrobial activity varied depending on the molecular weight and concentration of the oligosaccharides and type of microorganism. Generally, the effectiveness significantly increased with increasing concentration, regardless of molecular size and types of bacteria. Chitosan oligosaccharides with molecular weights ranging from 104 to 1333 kDa showed relatively greater antimicrobial activity against *L. monocytogenes*, *S. aureus*, and *P. fluorescens*; whereas, for *E. coli*, chitosan of an intermediate molecular weight was more effective in growth inhibition than lower or higher molecular weight chitosan, particularly at 0.1% concentration.[17]

A wide spectrum of diverse derivatives of chitin and chitosan is also reported to have antibacterial activities. These include water-soluble chitin derivatives, partially deacetylated chitin (DAC), N-trimethyl derivatives of DAC (NTM–DAC), chitose oligomers, sulfuryl and phosphoryl chitins, and sulfonated chitosan, among others. These compounds, at concentrations ranging from 0.6 to 2.5% (w/v), inhibited the growth of *Bacillus subtilis*, *Pseudomonas aeruginosa*, *Staphylococcus aureus*, *S. epidermidis*, *Klebsiella pneumoniae*, and *Proteus vulgaris* to different extents. Chitin deacetylated up to 80% (i.e., chitosan with 20% acetylation) inhibited these organisms at concentrations of 0.13 to 0.5% (w/v), suggesting that the number of amino groups in the DAC determined the degree of inhibition. NTM–DAC was more potent in its antibacterial effect than DAC, requiring only 0.03 to 0.5% (w/v) to inhibit the organisms. Similarly, sulfuryl chitin, phosphoryl chitin, and the thiourea derivative of chitosan inhibited these organisms at varying concentrations. Certain derivatives of chitosan also possessed antimicrobial activity. The antimicrobial activity of the Schiff base of chitosan synthesized by the reaction of chitosan with citral was greater than that of native chitosan, and the activity increased against *Escherichia coli*, *S. aureus*, and *Aspergillus niger* with an increase in the concentration.[18]

Microcrystalline chitosan and its derivatives, especially salts, demonstrate significant antiviral activity. The high sulfur content in sulfonated chitosan adversely influences its antibacterial effect. Minimal inhibitory concentrations of sulfonated chitosan (0.63% sulfur content) against *Shigella dysenteriae*,

Aeromonas hydrophila, Salmonella typhimurium, and *Bacillus cereus* were found to be lower than those of 69% deacetylated chitosan. Sulfobenzoyl chitosan (SBC) has excellent water solubility and an antibacterial effect comparable to that of sulfonated chitosan. The growth of coliforms and *Pseudomonas, Aeromonas*, and *Vibrio* species on oysters was retarded by the addition of 69% deacetylated chitosan or SBC. SBC at 1000 and 2000 ppm extended the shelf life of oysters at 5°C by 4 and 7 days, respectively. Sulfobenzoyl chitosan increased the storage life of oysters at 5°C.[19–23]

Another chitosan derivative, chitosan glutamate, suppressed the growth of *Lactobacillus fructivorans* in mayonnaise and mayonnaise-based shrimp stored for 8 days at 5°C when the compound was incorporated at 0.3% in the products that also contained 0.16% acetic acid or 2.6% lemon juice. The counts of *Zygosaccharomyces bailii* were also reduced after one day of storage at 5°C from an initial inoculated level of log 5 to 6 colony forming units (CFU) per gram. At 25°C, chitosan was ineffective in controlling these organisms.[22,23] Sulfuryl chitin, phosphoryl chitin, and the NTM derivative of deacetylated chitin were found to have higher bacterial inhibition activity than carboxymethyl chitosan.[76] The susceptibility of *Staphylococcus aureus* to a water-soluble lactose chitosan derivative was influenced by pH, temperature, and age of culture. The lactose chitosan derivative exerted a higher antibacterial activity at 37°C than at 22°C and 5°C. The effect was also maximum at pH 6.5 to 7.0. The organism was most susceptible to the chitosan derivative in the late exponential stage of growth. The lactose chitosan derivative also inhibited enterotoxin production.[23] The thiourea derivative of chitosan is more active against the gray mold fungus *Botrytis aurea* and the rice leaf blast fungus *Pyricularia oryzae*. Schiff bases of chitosan, N-substituted chitosan, and quaternized chitosan had better antifungal activities against *B. cinerea* and *Colletotrichum lagenarium* than normal chitosan.[24] Twenty kinds of 2-(α-arylamino phosphonate)-chitosan (2-α-AAPCS) had antifungal activities at concentrations ranging from 50 to 500 μg mL.[25]

The antimicrobial effect of chitosan can be attributed to its cationic nature. In an acid environment, the NH_2 group in the C2 position of chitosan protonates to yield NH_3^+, which binds to anionic sites on bacterial and fungal cell wall surfaces, resulting in disruption of the barrier properties of the outer membranes of the microorganisms. In addition, migration of glucosamine from the biopolymer into microbial cultures is also responsible for the antimicrobial effect, as observed in the case of *Staphylococcus aureus* and *Salmonella* spp.[26,27] In addition, chitin and chitosan are capable of forming complexes with many of the transition metals and some of those from Groups 3 to 7 of the periodic table, thereby interfering with microbial growth and toxin formation.[1,9] The antifungal action of chitosan occurs via the formation of gas-permeable coats, interference with fungal growth, and stimulation of many defense processes, including accumulation of chitinases, production of proteinase inhibitors, and stimulators of callous synthesis.[12]

6.2.2 Antioxidant Activity

Reactive oxygen species or free radicals, such as the superoxide anion (O^{-2}), hydroxyl ($HO\cdot$), peroxy ($ROO\cdot$), alkoxy ($RO\cdot$), and hydroperoxy ($HOO\cdot$) radicals, are generated during metabolism and have detrimental effects on human health. Antioxidants can inhibit or retard oxidation either by scavenging the free radicals that initiate oxidation or by breaking the oxidative chain reactions. Antioxidant activity is determined in terms of assay of 1,1-diphenyl-2-picrylhydrazyl (DPPH) radicals, hydrogen peroxide and superoxide anion radicals, and metal ion chelating capacity.[28] The mechanisms of action involve binding metal ions, scavenging oxygen, converting hydroperoxides to nonradical species, and deactivating singlet oxygen, thereby reducing the rate of oxidation. Compounds such as butylhydroxyanisole (BHA), α-tocopherol, flavonoids, butylhydroxytoluene (BHT), *tert*-butylhydroquinone (TBHQ), and esters of gallic acid (e.g., propyl gallate) are the major synthetic antioxidants. They are used generally in concentrations as low as 0.02% of the fat or oil content and sometimes in combination for synergistic effects. A database released by the U.S. Department of Agriculture provides the antioxidant values of 277 selected foods.[29]

Growing consumer demand for food devoid of synthetic antioxidants has focused efforts on the discovery of natural antioxidants, which are presumed to be safe as they occur normally in foods. These compounds include ascorbic acid, carotenoids, α-tocopherol, and flavonoids.[28] Chitosan and chitosan oligosaccharides have the advantages of being natural antioxidants. They control lipid oxidation by scavenging free radicals, which can be attributed to their ability to chelate metals and combine with lipids.[30–32] The antioxidant effects of chitin and chitosan are dependent on their molecular weight, viscosity, and degree of deacetylation. Highly deacetylated chitosan oligosaccharide is preferable for scavenging radicals such as DPPH, hydroxyl, and carbon-centered radicals. Low-molecular-weight chitosan (LMWC) (12 kDa) exhibited stronger scavenging activity toward DPPH radicals, superoxide anion radicals, and hydrogen peroxide, compared to either medium-molecular-weight chitosan (MMWC) (95 kDa) or high-molecular-weight chitosan (HMWC) (318 kDa).[1,32,33] The antioxidant activity of chitosan could be compared with that of propyl gallate;[1,34] however, comprehensive studies need to be conducted to ascertain the *in vivo* safety of LMWC (see Chapter 11).

Improvement of antioxidant activity of chitosan could be achieved by techniques such as derivatization or subjecting chitosan to ionizing radiation. Water-soluble chitosan derivatives that varied in the degree of substitution (DS) ranging from 20 to 70% were prepared by reductive alkylation of α-chitosan with lactose, maltose, or cellobiose. Antioxidant activities were determined, including radical scavenging effect, for α,α-diphenyl-β-picrylhydrazyl (DPPH) radicals, superoxide anion radicals, and hydrogen peroxide, as well as the copper ion chelating ability of these chitosan derivatives. It

TABLE 6.2

Influence of Gamma Irradiation on Chitosan Characteristics

Properties	Non-Irradiated Chitosan	Irradiated (25 kGy) Chitosan
Molecular weight	1.8×10^6 Da	1.75×10^4 Da
Viscosity	1200 cP	16 cP
β-Carotene bleaching assay[a]	24.0	681.7
2,4-Dinitrophenylhydrazine (DNPH) scavenging activity[b]	9.5	60.8

[a] Expressed as antioxidant activity coefficient.
[b] Expressed as percentage 1,1-diphenyl-2-picrylhydrazyl (DPPH) scavenging.
Source: Adapted from Kanatt, S. et al., *Int. J. Food Sci. Technol.*, 39, 997, 2004.
         With permission.

was found that the chitosan derivatives exhibited multiple antioxidant activities that varied with the concentration, degree of substitution with disaccharides, and the kind of disaccharide present in the derivative molecule. The antioxidant activities increased as the concentration of these derivatives increased up to a certain extent. A stronger scavenging effect of superoxide anion radicals, DPPH radicals, and H_2O_2 was noted with the chitosan derivatives having lower DS with disaccharide than those with higher DS. Among the various compounds, the disaccharide chitosan derivatives were found to show the highest hydrogen peroxide scavenging activity. At a concentration of 400 µg/mL, all of the chitosan derivatives exhibited 60% or greater scavenging activity.[35] Derivatives of chitosan—namely, *N,O*-carboxymethyl chitosan (NOCC) and its lactate, acetate, and pyrrolidine carboxylate salts— inhibited lipid oxidation in the range of 46 to 69%. NOCC inhibited iron-activated autooxidation by chelating action.[4,21]

Gamma irradiation of chitosan at 25 kGy resulted in a sixfold increase in its antioxidant activity, as measured by the β-carotene bleaching assay and DPPH scavenging activity. The antioxidant activity measured by the β-carotene bleaching assay increased from 24 to 681, while that measured by DPPH scavenging activity increased from 9.5 to 60.8. It should be noted, however, that irradiated chitosan has only limited potential for other roles, such as texturizer, because irradiation drastically reduced the chitosan viscosity from 1200 cPs to 16 cPs and the molecular weight from 1.8×10^6 Da to 1.75×10^4 Da.[36] Table 6.2 shows the influence of gamma irradiation of chitosan on its characteristics, including antioxidant activity.

Chitosan undergoes a loss of molecular weight in the presence of hydrogen peroxide. The molecular weight of H_2O_2-treated chitosan decreased with concentration, temperature, pH, and time of treatment. The dissolution of chitosan at pH 5.5 enhanced the degradation, but excessive hydrogen ion potentially inhibited the degradation. There was no significant chemical change in the backbone of chitosan having molecular weight of 51×10^3, but chitosan having lower molecular weights of 3.5×10^3 and 1.2×10^3 lost about 15% and 40% of amino groups, respectively. Further degradation led to more oxidation associated with ring opening, the formation of carboxyl groups, and faster deamination.[37] Irradiation from 10 to 100 kGy in the presence of H_2O_2 can also lead to loss of crystallinity of chitosan.[38]

6.2.3 Emulsification Capacity

Chitosan interacts with lipid, a characteristic that could be applied toward the use of the polysaccharide as an emulsifier; however, the interactions are pH sensitive. In acidic conditions, chitosan stabilizes an emulsion due to its capacity to bind with anionic lipid molecules.[39] The fat-binding capacity of chitin, chitosan, and microcrystalline chitin ranges from 170 to 315% (w/w), with microcrystalline chitin exhibiting superior emulsifying properties.[40] The influence of chitosan content (0 to 0.5%) on emulsifying properties, particle size distribution in the emulsion, creaming stability, apparent viscosity, and microstructure of oil-in-water emulsions (40% rapeseed oil) containing 4% whey protein isolate (WPI) at pH 3 was investigated. A WPI and chitosan mixture exhibited a slightly higher emulsifying activity than whey protein did alone. An increase in chitosan content resulted in a decreased average particle size, higher viscosity, and increased creaming stability of emulsions. Increasing concentration of chitosan resulted in the formation of a flocculated droplet network.[41]

Chitosan can stabilize flavor compounds. At present, emulsified flavor oils are usually stabilized by gum arabic, which is a naturally occurring polysaccharide–protein complex. In a recent study, it was shown that citral and limonene, the major flavor components of citrus oils, undergo rapid chemical degradation leading to a loss of flavor. The combination of sodium dodecyl sulfate (SDS) and chitosan more effectively stabilized emulsions and retarded formation of the citral oxidation product *p*-cymene than gum arabic. The ability of chitosan could be due to the formation of a cationic and thick emulsion droplet interface that could repel prooxidative metals.[42] Emulsification of sunflower oil by chitosan solutions with a degree of deacetylation between 75 and 95% showed that the polysaccharide produced stable water-in-oil-in-water (W/O/W) emulsions. The droplet size distribution was independent of both chitosan solution viscosity and emulsion viscosity. Emulsion viscosity, emulsion stability, and aging were proportional to chitosan concentration.[39]

The influence of low-molecular-weight chitosan on the physicochemical properties and stability of a low-acid (pH 6) tuna oil-in-water emulsion stabilized by non-ionic surfactant (Tween® 80) was studied. The added chitosan was adsorbed on the surface of oil droplets through electrostatic interactions. Although the droplet diameter was not affected, the impact of chitosan on the strength of the colloidal interaction between the emulsion droplets increased with increasing chitosan concentration. It was concluded that the addition of low-molecular-weight chitosan is helpful to create tuna oil emulsions with a low-acid to neutral character, as well as to impart required physicochemical and stability properties to the emulsion.[43]

6.3 Food Applications of Chitin and Chitosan

From a food technological point of view, chitin and chitosan could serve as versatile additives. Chitin has low toxicity and is inert in the gastrointestinal tract of mammals. It is biodegradable due to its sensitivity to chitinases, which are widely distributed in bacteria, fungi, plants, and the digestive systems of many animals. It is sensitive to degradation by lysozymes found in egg white, fig, and papaya plants. Chitosan, unlike chitin, is highly soluble under mild acidic conditions and has interesting cationic properties. The functional role of chitosan as a food additive stems from its ability to function as an antimicrobial and antioxidant agent. The antimicrobial properties could be advantageous in combination with, for example, chilling, modified atmosphere packaging, or high pressure to extend the shelf life of food products. In addition, its ability to interact with food macromolecules such as lipids, proteins, and starch enables it to function as a texturizing and emulsifying agent. Other advantages are its potential to function as fiber and its ability to lower cholesterol. Furthermore, it is amenable to formation of films having attractive barrier properties, making it a biocompatible coating that can prevent quality losses in many foods, including fruits, seafood, and vegetables, among others.[4,44,45]

6.3.1 Fruits and Vegetables

A number of benefits, particularly antimicrobial and antioxidant, can be derived from chitosan with regard to fruits and vegetables. Also, the polysaccharide can control plant pathogens and pathogenic nematodes and enhance host-plant resistance against these pathogens. The antimicrobial activity of chitosan is achieved by dipping fruits and vegetables in a solution of the polysaccharide to coat them. Usually, the products are dipped in a chitosan lactic acid/sodium lactate solution, the pH of which is adjusted to the pH of the products. For better antimicrobial activity, the treated products may be stored under modified atmosphere and at chilled temperatures. The microbiological loads on the chitosan-coated samples are usually lower in comparison with uncoated products, and the effect depends on the type of fruit and vegetables.

A chitosan coating inhibited the growth of microorganisms in sliced strawberries and significantly enhanced the stability of the products, particularly when the samples were packaged in a modified atmosphere with high (80%) and low (5%) percentages of oxygen followed by chilled storage at a maximum temperature of 15°C. The treatment also helped maintain the color of processed strawberries.[46] Preharvest chitosan sprays can prevent postharvest infection of strawberries by such pathogenic organisms as *Botrytis cinerea*. Small bunches of berries dipped in chitosan solutions (0.1, 0.5, and 1.0%) and inoculated with *B. cinerea* showed a reduction of incidence, severity, and nesting of gray mold in comparison with the control. Single berries artificially wounded, treated with the polymer, and inoculated with *B. cinerea* showed a reduced percentage of infected berries. Higher chitosan concentrations demonstrated greater reduction in decay. All preharvest treatments significantly reduced the incidence of gray mold, as compared to the control.[47]

Chitosan added to pickled vegetables and soy sauces inhibits the growth of molds at low levels of sodium chloride.[3] A combination of chitosan and high-pressure treatment has been recently shown to enhance the storage life of apple juice and apple cider. These products were inoculated with *Escherichia coli* K-12 and processed using a high-pressure homogenizer at pressures in the range of 100 to 200 MPa, which resulted in significant inactivation of the bacterium. Inactivation of the bacterium at the same homogenizing pressure was higher in apple juice than apple cider. While chitosan alone did not inactivate the bacterium, there was synergistic inactivation of the organism when the pressure treatment was done in the presence of up to 0.1% chitosan.[48] When edible coatings contain hydrophilic plasticizers such as glycerol, sorbitol, and Tween® 80 as surfactants, the surface properties, particularly wettability of skins of the coated products, may be adversely affected; however, incorporation of 1.5% chitosan in the coating solution enhanced wettability of coated tomato and carrot.[49] Chitosan was also effective in reducing the turbidity of apple juice and improving its color.[50]

Coatings of low-, medium-, and high-molecular-weight chitosan at concentrations of 0.01 and 0.02 g/mL were applied to fresh-cut papaya cubes. The treated cubes were stored at 5°C and changes in quality were evaluated. The chitosan coatings suppressed mesophilic plate count and the growth of molds and yeast compared to controls. The medium-molecular-weight chitosan maintained the highest color values (L^* and b^*) and firmness. The coating, because of its antimicrobial activities, also resulted in control of microbial deteriorative process. Also, it decreased the activity of the enzymes polygalacturonase and pectin methylesterase in the papaya.[51]

Coating by immersion in 1% chitosan solution containing 0.1% of Ca^{2+} limited changes in the sensory properties of stored bell pepper fruits and cucumbers. Chitosan coatings also controlled growth of the pathogen *Botrytis cinerea* in the fruit. In untreated control tissue, massive fungal colonization was followed by extensive degradation of the pectin component of host walls. In chitosan-treated tissue, the preservation of pectin binding sites and the intense and regular cellulose distribution over host walls suggested that the polysaccharide prevented the maceration of host tissue by *B. cinerea*.[52]

The effectiveness of pre- and postharvest treatments with chitosan at 1.0% (w/v) to control *Botrytis cinerea* on table grapes has also been reported. In postharvest treatments, small bunches dipped in chitosan solutions and inoculated with the pathogen showed a reduction of incidence, severity, and nesting of gray mold in comparison with the control. Single berries artificially wounded, treated with the polymer, and inoculated with *B. cinerea* showed a reduced percentage of infection, which was also indicated by a significant increase in phenylalanine ammonia lyase activity.[53] Chitosan also controlled the pathogen in cucumber.[54]

Chitosan coating can control browning and help maintain the quality of fresh-cut Chinese water chestnut (CWC). After treatment with aqueous solutions of chitosan, 0.5 to 1% (w/v), the product was placed in trays that were then wrapped with plastic films and stored at 4°C. The coating delayed discoloration associated with reduced activities of phenylalanine ammonia lyase, polyphenol oxidase, and peroxidase as well as lower total phenolic content; it also slowed down the loss in eating quality associated with higher contents of total soluble solids, titratable acidity, and ascorbic acid of CWC. Disease development in the fresh-cut CWC with the chitosan coating was also inhibited compared to the control. Increasing the concentration of chitosan coating markedly enhanced the beneficial effects. The results showed that application of a chitosan coating can effectively extend shelf life and maintain the quality of fresh-cut CWC.[55]

Litchi pulp is very perishable and thus has an undesirably short shelf life. Manually peeled litchi fruits were treated with aqueous solutions of chitosan at a maximum concentration of 3%; they were placed in trays that were then wrapped with plastic film and stored at –1°C. The treatment retarded weight loss and declines in sensory quality due to higher contents of total soluble solids, titratable acid, and ascorbic acid, and it suppressed the increase in activities of polyphenol oxidase and peroxidase, suggesting that a chitosan coating can effectively maintain quality attributes and extend the shelf life of the peeled fruit.[56]

A coating of low-molecular-weight (15 kDa) chitosan at 0.1% concentration substantially slowed fungal decay of the citrus fruit hybrid Murcott tangor by over 20% when stored at 15°C in relation to the control sample. A concentration of 0.2% low-molecular-weight chitosan was more effective in controlling the growth of the fungi *Penicillium digitatum* and *P. italicum* on citrus fruits. The coating also improved firmness, titratable acidity, ascorbic acidity, and the water content of the fruit stored for 56 days at 15°C.[57]

The addition of chitin to soil is effective in the elimination of some plant diseases. Normally, fungi, arthropods, and nematodes are the major contributors of chitin in soil. The chitin limits the growth of plant pathogens both in soil and plant vascular systems through hydrolysis of fungal cell walls by chitinolytic enzymes secreted by antagonists. When chitin decomposes, it produces ammonia, which takes part in the nitrogen cycle. The application of this polysaccharide was found to promote the growth of certain chitin-olytic microorganisms and make them dominant in the soil.[58] Chitosan has the potential to improve seed quality and enhance crop yields, as well as increase the value of stored grains intended for food and feed. Chitosan treatment (2 to 8 mg/mL) of wheat seeds significantly improved seed germination to recommended seed certification standards (>85%). The effect was due to chitosan controlling seedborne *Fusarium graminearum* infection and increasing resistance in seedlings by stimulating the accumulation of phenolics and lignin. The treatment can also stimulate plant root growth and enhance the strength of wheat stems.[59]

Soaking soybean seeds for 8 hr in a 0.5% (w/v) aqueous solution of chitosan having a molecular weight of 493 kDa increased total weight, vitamin C content, and hardness of soybean sprouts.[60] Optimal conditions selected for cultivation of sunflower sprouts involved soaking the seeds in 0.5% 28-kDa chitosan (dissolved in 0.5% lactic acid) for 18 hr. After cultivation for 6 days at 20°C, the treated seeds yielded sprouts with 12.9% higher total weight and a 16.0% higher germination rate compared to those of seeds soaked in water alone for 18 hr (control). Chitosan-treated sprouts also exhibited slightly improved DPPH radical scavenging activity and total isoflavone content.[61] The U.S. Environmental Protection Agency has approved chitosan for seed treatment and noted that it is not toxic to humans and animals, as it is naturally occurring in large concentrations and has been exempted from regulation by the U.S. FDA when used as a food or feed additive.[62] Table 6.3 summarizes the uses of chitosan in agricultural products.

6.3.2 Dairy Products

The flocculation property of chitosan and its interaction with whey proteins have been found useful in removing lipids and undenatured hypoallergenic β-lactoglobulin from cheese whey. The addition of chitosan caused selective precipitation of β-lactoglobulin that increased with pH. Adding a low concentration of chitosan at 0.01 to 0.02% to cheddar cheese whey at pH 4.5 resulted in the formation of a chitosan–fat globule membrane complex. The complex flocculated and precipitated when incubated at ambient temperature for 10 to 30 min. Centrifugation of the treated whey resulted in a clear supernatant that contained all of the whey protein, with a lipid content less than 0.2 g per 100 g protein. No residual chitosan could be detected. At pH 6.2, adding 1.9 to 3.0 mg/mL of chitosan led to complete removal of β-lactoglobulin, whereas at least 80% of the whey proteins remained in solution. A concentration of whey at about 4- to 5-fold volume only slightly increased the minimum concentration of chitosan required to flocculate the fat globule membrane. The production of cheese whey without β-lactoglobulin could help to expand the applications of dairy byproducts in food processing.[64]

TABLE 6.3

Uses of Chitosan in Agricultural Products

Product	Treatment	Effect	Refs.
Strawberries	Chitosan coating	Enhances chilled storage under 80% or 5% oxygen	Campaniello et al.[46]
Strawberries	Chitosan spray	Prevents postharvest infections by pathogenic organisms	Reddy et al.[47]
Table grapes	Chitosan dip	Controls decay	Ramanazzi et al.[53]
Carrot and tomato	Chitosan dip	Increases wettability	Casariego et al.[49]
Pickled vegetables and soy sauces	Chitosan dip	Inhibits molds	Prashanth and Tharanathan[3]
Papaya	Chitosan coating	Prevents deterioration and preserves quality	González-Aguilar et al.[51]
Apple juice	Chitosan added	Reduces turbidity and improves color	Soto-Peralta et al.[50]
Bell pepper fruits	Chitosan coating	Enhances storage stability; controls gray mold pathogen *Botrytis cinerea*	El Ghauth et al.[52]
Cucumber	Chitosan sprays	Controls gray mold pathogen *Botrytis cinerea*	Ben-Shalom et al.[54]
Fresh cut Chinese water chestnut	Dipping in dilute aqueous chitosan solution	Extends shelf life and quality	Pen and Jiang[55]
Litchi fruit	Dipping in aqueous chitosan solution	Extends shelf life	Dong et al.[56]
Citrus fruit (Murcott tangor)	Coating with low-molecular-weight chitosan	Improves firmness, titratable acidity, ascorbic acidity, and water content	Chien et al.[57]
Soybean sprouts	Soaking in chitosan solution	Increases total weight, vitamin C content, and hardness	No et al.[60]
Various seeds	Chitosan coating	Improves seed quality	Reddy et al.[59]

6.3.3 Muscle Foods

Advantages of chitosan as an additive in muscle products include control of flavor loss, antimicrobial and antioxidant properties, and increased storage stability. A "warmed-over" flavor develops in cooked poultry and uncured meat upon storage, resulting in a loss of perceived freshness. Chitosan is capable of preventing flavor deterioration due to its antioxidant activity. The addition of 1% chitosan to meat can result in a reduction in lipid oxidation of as much as 70% during storage at 4°C. N-carboxymethyl chitosan (NCMC) and its lactate, acetate, and pyrrolidine carboxylate derivatives were effective in controlling the oxidation and off-flavor development in cooked meat stored for 9 days at refrigerated temperatures. The polysaccharide can also be used as preservative for extended chilled storage of pork sausage.[65–67] Research by the U.S. Department of Agriculture has revealed that NCMC can be useful as a meat flavor preservative at a concentration of 0.1% (w/w) in meat. As a kitchen aid, the compound in granular form can be sprinkled on gravies or meat products. NCMC is very useful in preserving microwavable or quickly prepared foods as well as in preventing development of the "warmed-over" flavor of institutional foods. The compound itself is tasteless, blends well with foods as a colorless ingredient, and is nontoxic and nonallergenic. Application of NCMC with glaze formulation when flash-freezing many vegetable or muscle foods will inhibit surface oxidation and enhance shelf life. Meat and poultry processors can use NCMC as a postslaughter perfusion as a long-term flavor and storage preservative.[62]

Chitosan alone and in combination with either rosemary or α-tocopherol effectively controlled lipid oxidation in frozen ($-18°C$) beef burgers stored for 180 days. The effect was significantly higher compared to individual use of rosemary or α-tocopherol. The treatment also contributed to retention of the red color of the burger, thus improving its appearance.[63] Incorporation of 3% chitosan into ground beef or turkey may reduce the potential risk of *Clostridium perfringens* spore germination and outgrowth when heated products are cooled improperly. Chitosan was mixed into thawed beef (25% fat) or turkey (7% fat) at concentrations ranging from 0.5 to 3.0% (w/w). The meat was inoculated with heat-activated spores of the organism at a final concentration of 2 to 3 log CFU/g. Samples of the ground beef or turkey mixtures were then vacuum-packaged and cooked to 60°C for 1 hr, followed by cooling to 7.2°C at different rates (12 to 21 hr). Adding chitosan to beef or turkey resulted in a concentration- and time-dependent inhibition of *C. perfringens* spore germination and outgrowth. Chitosan at 3% reduced *C. perfringens* spore germination and outgrowth by 4 to 5 log CFU/g during slow (12 to 18 hr) cooling of the cooked beef or turkey, but the reduction was significantly lower at a chilling time of 21 hr (about 2 log CFU/g).[68]

Chitin oligosaccharides (COs) formed as a result of gamma irradiation of aqueous chitosan solutions exhibit antioxidant and antibacterial activities depending on their molecular weights. COs having a molecular weight of 8.3 kDa exhibited the highest antioxidant activity. A combination of COs and lysozyme was more effective against Gram-negative bacteria than when lysozyme was used alone. When tested in a meat model system, COs and lysozyme were effective in eliminating *Escherichia coli*, *Pseudomonas fluorescens*, and *Bacillus cereus* and reduced the load of *Staphylococcus aureus*. The shelf life of minced meat containing a CO–lysozyme mixture was extended up to 15 days at chilled temperatures.[69]

The enhanced antioxidant activity of gamma-irradiated chitosan could be used to preserve lamb meat. Adding irradiated chitosan to the meat before the radiation processing was found to suppress rancidity development during post-irradiation storage at 0 to 3°C. Rancidity of the irradiated meat containing irradiated chitosan was 88% lower in the leg portion and 54% in the rib portion as compared to corresponding samples devoid of chitosan. Further, after storage for a week, development of rancidity was reduced by 39 and 59% in the leg and rib portions, respectively, of the samples treated with chitosan.[36]

A chitosan–glucose complex (CGC), a Maillard reaction product, could also be used as an antioxidant. The CGC was prepared by heating chitosan with glucose. The product, similar to chitosan, also showed antimicrobial activity against *Escherichia coli*, *Pseudomonas* spp., *Staphylococcus aureus*, and *Bacillus cereus*, the common food spoilage and pathogenic bacteria. The antioxidant and antimicrobial activities of the complex suggest its use as a promising novel preservative for various food formulations. Addition of CGC to lamb meat increased its shelf life by more than 2 weeks during chilled storage. The complex also extended the shelf life of pork cocktail salami to 28 days.[70]

Another preparation that has both antioxidant and antibacterial activities is a mixture of chitosan and mint (CM). It efficiently scavenged superoxide and hydroxyl radicals. Also, the antimicrobial activities of CM and chitosan were effective against common Gram-negative food spoilage and pathogenic bacteria, the minimum inhibitory concentration being 0.05% (w/w). CM treatment extended the chilled shelf life of pork cocktail salami, as determined by total bacterial count and oxidative rancidity.[71] The inclusion of 1.5% (w/w) chitosan with a molecular weight of 1.84×10^5 Da and a degree of deacetylation of 94% as a cold-set binder improved the texture of salt-soluble proteins (SSPs) from chicken and porcine It also increased the disulfide content, rheological characteristics, and structure of cold-set SSP gels.[72,73]

6.3.4 Seafood

The benefits of chitosan for seafood products include its antimicrobial, antioxidant, and texturizing properties. Fishery products are highly perishable due primarily to microbial spoilage. Further, unlike red meat, fish contain significant amounts of unsaturated fatty acids and are highly sensitive to oxidation and associated flavor changes. Chitosan at a level of 1% (w/w) can control these changes to enhance the shelf life of fresh fishery products. Viscosity and molecular weights of the carrageenan influence these effects. Chitosans of various molecular weights (30, 90, and 120 kDa) exhibited antioxidative activities in salmon during 7 days of storage. At 0.2% (w/v) and 0.5% (w/v) concentrations, the thiobarbituric acid (TBA) reactive substances with chitosan added were decreased by 75% and 45%, respectively, over 15 days.[74] Chitosan reduced lipid oxidation in cod and herring muscle depending on the concentration (50 to 200 ppm). Various chitosan viscosities (14, 57, or 360 cP) can be used. Chitosan of varying molecular weight was also effective in controlling the oxidation of lipids of comminuted cooked cod (*Gadus morhua*). Both peroxide and 2-thiobarbituric acid reactive substance values were reduced as a result of the treatment of fish prior to cooking using chitosan.[75,76]

Chitosan could potentially be used to control microbial spoilage in fish in view of its antimicrobial properties. To study its effects, 3% (w/v) chitosan solutions were prepared incorporating 10% fish oil (w/w chitosan, containing 91.2% EPA and DHA) with or without the addition of 0.8% vitamin E. Fresh lingcod (*Ophiodon elongates*) fillets were vacuum-impregnated in a coating solution at 100 mmHg for 10 min followed by atmospheric restoration for 15 min, dried, and then stored at 2°C or –20°C for 3 weeks and 3 months, respectively. The chitosan and fish oil coating increased the total lipid and omega-3 fatty acid contents of fish by about threefold, reduced rancidity in both fresh and frozen samples, and decreased the drip loss of frozen samples by 14 to 28%. The coating also reduced total plate and psychrotrophic counts in cold-stored and frozen-stored samples, suggesting that the chitosan and fish oil coatings may be used to extend the shelf life and fortify omega-3 fatty acid in lean fish.[77]

Combination treatments involving high pressure and chitosan as an antibacterial additive have been reported. Chitosan can enhance the inhibitory effect of high pressure on microbial growth. Cod sausage was produced at a chilled temperature (7°C) and high pressure (350 MPa for 15 min). Incorporation of chitosan (1.5%) extended the shelf life of the product.[78] Chitosan has been found to control the black discoloration of crustaceans, including shrimp. The preferred concentration ranges from 0.1 to 2.0%, depending on particular requirements.

Interactions of chitosan with proteins help improve the texture of surimi products. The surimi gels of walleye pollock have poor gel strength, but the gel strength was nearly doubled by the addition of 1.5% chitosan. A combination of chitosan and transglutaminase has been especially beneficial in this respect. Chitosan at the 1.5% level alone did not substantially modify the rheological and microstructural properties of mackerel meat gels prepared under high-pressure conditions (300 MPa, 25°C, 15 min), although it reduced lipid oxidation. However, microbial transglutaminase (0.02%) in combination with chitosan caused an increase in hardness and a considerable decrease in elasticity and breaking deformation.[79] A blend of sucrose, sorbitol, and chitosan gave better textural properties to surimi than either chitosan alone or a blend of sucrose, sorbitol, and polyphosphate.[80]

Incorporation of chitosan and calcium chloride greatly improved the gelling properties of surimi from barred garfish in terms of breaking force and deformation of gel without changes in color. The effect could be due to its influence on the endogenous transglutaminase activity in cross-linking of protein–protein and protein–chitosan conjugates. Adding chitosan with a 65.6% degree of deacetylation at the level of 15 mg/g surimi favorably modified both breaking force and deformation of suwari and kamaboko gels compared to the control. A chitosan concentration of 10 mg/g was found to produce the highest breaking force of kamaboko gel.[81–83] Table 6.4 summarizes the various functional roles of chitosan in muscle food products.

TABLE 6.4

Functional Roles of Chitosan in Muscle Food Products

Product/System	Functional Activity	Refs.
Cod, herring, salmon, beef, pork, irradiated lamb meat	As edible coating or dip, chitosan provides protection against lipid oxidation (antioxidant activity).	No et al.[1]
Surimi from barred garfish	As texturizing agent, chitin and chitosan influence transglutaminase activity and cross-linking of surimi.	Benjakul et al.[81,82]
Fresh lingcod fillets	Chitosan–fish oil coating enhances omega-3 fatty acid content and reduces lipid oxidation.	Duan et al.[77]
Salmon	Chitosan provides antioxidant and antimicrobial activities.	Kyung et al.[74]
Kamaboko gel	Incorporation of chitosan modifies breaking strength and deformation.	Kataoka et al.[83]
Lamb meat	Gamma-irradiated chitosan enhances antioxidant activity; can be used as edible coating.	Kanatt et al.[36,70,71]
Sausages and patties	Antibacterial action reduces spoilage caused by bacteria; disrupts the barrier properties of the outer membrane of Gram-negative bacteria. The addition of chitosan reduces nitrite requirement as curing agent.	Lin and Chao,[66] Caballero et al.[78]
Various	N-Carboxymethyl chitosan (NCMC) and its lactate, acetate, and pyrrolidine carboxylate control oxidation and off-flavor.	Flick and Martin[62]
Beef burger	Chitosan in combination with rosemary extract and α-tocopherol controls lipid oxidation and provides color stability during frozen storage.	Georgamtelis et al.[63]

6.3.5 Bakery Products

Chitosan and chitin can also be used as food additives in cookies, noodles, and bread to improve texture. These effects are due to the antimicrobial properties of chitosan and its ability to control starch retrogradation. Microcrystalline chitin has a positive effect on emulsion stability, in addition to increasing the specific loaf volume of white bread and protein-fortified breads. Maillard reaction products (MRPs) prepared from chitosan and xylose extend the shelf life of fresh noodles.[84] Baking tests were performed with 0.5 to 2.0% (flour basis) microcrystalline chitin added to wheat flour bread or to potato-protein-fortified (8% potato protein concentrate) white bread. Chitosan and chitin did not produce emulsions, but microcrystalline chitin showed good emulsifying properties and was superior to microcrystalline cellulose. A chitin bread product has been reported that is 65% water, the remainder consisting of wheat flour and chitin.[40]

6.3.6 Wines and Vinegars

Browning and overoxidation are the most common defects affecting white wines. Reducing phenolic compounds by the use of adsorbents is most frequently employed to counter these problems. Chitosan can be useful for the clarification of wine and vinegars. Chitosan exhibits a high affinity to a number of phenolic compounds, particularly cinnamic acid, and prevents browning in a variety of white wines. It compared well with two conventional adsorbents being used for these applications.[85]

6.3.7 Nutritional Value and Use as Food Supplement

Chitin and chitosan lower plasma cholesterol and triglycerides and improve cholesterol ratios due to their ability to bind dietary lipids, thereby reducing intestinal lipid absorption. Chitosan reduces lipid absorption by trapping neutral lipids, such as cholesterol and other sterols, by means of hydrophobic interactions. Because of this inhibitory activity on fat absorption, these molecules act as fat scavengers in the digestive tract and remove fat and cholesterol via excretion. Apart from chitosan, chitosan oligomers having average molecular weights of 10,000 Da could significantly enhance fecal excretion of neutral steroids. The positive-charge nature of chitosan and its oligomers govern most of these biological activities. In this respect, chitosan satisfies the requirements of dietary fiber, including nondigestibility in the upper gastrointestinal tract, high viscosity, and high water binding ability in the lower gastrointestinal tract. From a physiological standpoint, the prime function of a dietary fiber is to lower cholesterol levels and to promote the loss of body weight through a reduction of intestinal lipid absorption.[86–89] In view of its functional properties, especially its role as a fiber, chitosan has been particularly recommended as a dietary supplement for the elderly. It is important that, for chitosan to be nutritionally active, it must be soluble in food or supplied as a powder that becomes soluble with an acid pH. Its prolonged use as fiber in diets should be monitored to ensure that it does not disturb the intestinal flora or interfere in the absorption of micronutrients, particularly lipid-soluble vitamins and minerals, and that it does not have any other negative effects. There are reports that chitosan may alter the normal flora of the intestinal tract which may result in the growth of resistant pathogens.[88,89]

Chitosan is not highly amenable to hydrolysis by digestive enzymes. *In vivo* toxicity studies indicate that the chitosan obtained from prawn shells with a molecular weight of 126 kDa is nontoxic and inert, neither causing hemolysis nor favoring microbial growth. Chitosan oligosaccharide functions as a stimulant of selective growth of lactobacilli and bifidobacteria. Studies with cells, tissues, and animals suggest that chitin and chitosan promote wound healing, improve immune responses, and possess antitumor activity. Certain medical precautions, however, should be observed with long-term ingestion of high doses of chitosan to avoid potential adverse metabolic consequences.[90]

6.3.8 Other Food-Related Applications

6.3.8.1 Treatment of Water

Chitosan and its derivatives carboxymethyl chitosan and cross-linked chitosan have been successfully used in water treatment to remove lead, copper, and cadmium from drinking water, due to complex formation between the amino group and heavy-metal ions. In comparison with activated charcoal, chitosan is more efficient in the removal of polychlorinated biphenyls from contaminated water. The hydroxamic acid derivatives of chitin and chitosan are most efficient at removing lead and copper.[91] These polysaccharides are also useful for the complete removal of mercury from water. Chitosan is currently employed in domestic sewage treatment systems in conjunction with other settling aids such as alum or bentonite clay to promote coagulation and settling of colloidal and other suspended solids. The polyelectrolyte is added at the rate of 1 to 2 ppm but can also be employed alone without alum when the concentration is raised to around 10 ppm. Being positively charged, it is very effective at agglomerating the negatively charged sludge particles.[92,93]

Chitosan can be applied as a coagulant in the treatment of wastewater from food industries. The production of surimi generates a large amount of washwater that contains sizeable amounts of proteins, indicated by high turbidity. Chitosan treatment of surimi washwater results in the recovery of soluble proteins. The protein recovery is further increased by adding a complex of chitosan and alginate. Flocculation at 20°C produced by agitation for 5 min at 130 rpm in the presence of 20 to 150 mg/L chitosan–alginate complex resulted in the recovery of as much as 83% protein, which was also associated with a 97% reduction in the turbidity of the washwater.[94]

Soluble proteins from surimi washwater (SWW) precipitated using a chitosan–alginate complex and recovered by centrifugation were freeze-dried. Analysis showed that SWW proteins had a crude protein content of 73.1% and a high concentration of essential amino acids. In a rat-feeding trial, SWW proteins as a single protein source showed higher modified protein efficiency ratios and net protein ratios than the casein control. Blood chemistry analysis revealed no deleterious effect from the full protein substitution or the chitosan in SWW proteins; therefore, proteins recovered from surimi washwater using the chitosan–alginate complex could be used in feed formulations.[95] The use of chitosan as a coagulant to treat wastewater from a milk processing plant has been reported.[96] The polysaccharide adsorbs metals from wastewaters.[97] Microcrystalline chitosan can be more effective than conventional chitosan for the treatment of industrial wastewater. It also has appreciable antimicrobial properties.[98]

6.3.8.2 Animal Feed

Chitin has a growth-promoting effect on broiler chickens. Increases in average live weight and dressed weight and decreases in wastage during dressing in broiler chickens fed a diet containing 0.5% chitin have been reported. No abnormal symptoms in broilers and hens were observed when chitosan was administered at a concentration of 1.4 g/kg of body weight per day up to 239 days nor in rabbits fed a chitosan concentration of <0.8 g/kg body weight for the same period. Both chitin and chitosan were digested at a rate of 35 to 83% and 88 to 98% by hen and broilers, respectively. Inclusion of chitin and chitosan in the feed also resulted in suppression of serum cholesterol, triglycerols, and free fatty acids; however, a higher rate of feeding of chitin and chitosan, at 3.6 to 4.2 g/kg body weight, resulted in a decrease in appetite and the egg-laying rate of hens.[99,100]

The use of chitin as a source of dietary fiber in chicken feed enhances the growth of bifidobacteria in the guts, which reduces other microorganisms and produces the β-galactosidase necessary for the digestion of feed supplemented by whey or other dairy byproducts. The effect has also been observed in the case of chitosan feeds for pigs and fish. Chitosan adipate and ascorbate administrated orally or intramuscularly to piglets stimulate immunity and decrease mortality of the animals through increased resistance against intestinal diseases.[101] Similarly, feeds containing chitin and glucosamine could also be used in aquaculture for improved growth of cultured fish.[102] In addition, chitin hydrolysates produced through the digestion of crustacean waste by chitinases could be utilized as a carbon source for the cultivation of yeast that can convert chitin oligosaccharides into single-cell proteins. The yeast could be utilized as feed component.[103]

6.3.8.3 Biotechnology

Chitin and chitosan have been found to be useful as a matrix for the immobilization of various enzymes for the processing of such products as wine and sugar, the synthesis of organic compounds, and the construction of sophisticated biosensors for *in situ* measurements of environmental pollutants and metabolite control in artificial organs.[104] In a recent interesting work, industrially important maltose from potato starch has been produced. Sweet potato β-amylase and pullulanase from *Bacillus brevis* were separately immobilized onto chitosan beads, which were held in 40% potato starch hydrolyzate for 14 days at 60°C and pH 6.0 to give a yield of maltose of 71%.[105]

Chitin could be used in cheese making as an immobilization matrix for seal gastric protease that would aid in the clotting of milk. The average degree of immobilization is 20%, and the immobilized enzyme matrix exhibits optimum performance at pH 2.0.[106] Krill chitin has been used to support the enzyme diastase. The enzyme was immobilized by simple adsorption or in the presence of 0.1% glutaraldehyde for 2 hr. The optimum pH for binding of the diastase on chitin preparations was 6.2 in the presence of glutaraldehyde or 6.7 without the cross-linking agent. Immobilization shifts the optimum pH for the activity of diastase by 0.5 units toward the acid side.[107] Chitosan can also be used as biosensors in food applications. A lactose biosensor using an enzyme-immobilized eggshell membrane has been developed for determination of lactate in dairy products and serum samples. The system consisted of L-lactate oxidase and chitosan, which were deposited on an eggshell membrane and used as an oxygen electrode.[108]

6.4 Glucosamine

A milk beverage supplemented with glucosamine and chondroitin sulfate is available commercially in Japan. Pasteurization of the supplemented milk at an ideal temperature of 80°C did not cause instantaneous aggregation and precipitation of the milk proteins. Further, the treatment had no adverse effect on the stability of chondroitin sulfate, suggesting that pasteurized, glucosamine- or chondroitin-sulfate-enriched beverages are feasible.[109]

6.5 Commercial Aspects

Potential annual global chitin production has been estimated at 118,000 t.[110] The chitin is primarily used to prepare chitosan, which is available in a wide variety of commercial products with varying deacetylation grades, molecular weights, viscosities, and, hence, functional properties. Chitosan is well positioned as a functional food ingredient, but it has not made major inroads into the American market, where annual sales are in the region of US$20 million per year.[33] Global Industry Report analyzes the worldwide markets for chitin and chitosan and other derived products, including glucosamine and chitosan.[112] The various chitosan applications include water treatment, cosmetics, healthcare, agrochemicals, and biotechnology, among others. More than 50 companies are currently involved in the business.[111] Japan is the major producer of chitin and chitosan from the shells of crabs and shrimp

TABLE 6.5

Future Research Needs Regarding Commercial Applications of Chitosan

Area	Brief Description
Process standardization with respect to deproteinization, demineralization, decolorization, and deacetylation	Traditional methods influence the molecular weight, degree of deacetylation, viscosity, fat and water absorption, and hydrophilic nature, parameters that should be standardized with respect to functionality.
Novel and simpler processes for chitosan and chitosan oligomers	Cost-effective processes can encourage applications.
Improvements in film-casting techniques, incorporation of plasticizers and antimicrobial additives	Better stability against humidity and antimicrobial properties is necessary.
Removal of astringent and bitter taste by techniques such as ozone technology	A wider range of food applications would be available.
Quality standards	Various applications reported so far have used chitosan having diverse properties; a need exists for common standards for universal applications.

Source: Adapted from No, H.K. et al., *J. Food Sci.*, 72, 87, 2007; No, H.K. and Meyers, S.P., *J. Aquat. Food Prod. Technol.*, 4, 27, 1999.

and is the largest market (20,000 t) for chitin-derived products.[111] In India, a few entrepreneurs are producing chitin and chitosan on a commercial scale under the technical guidance of Central Institute of Fisheries Technology, Cochin.

Chitosan is used as a food quality enhancer in several countries including Norway and Japan. Most of the chitosan preparations used in health-care are in tablet or capsule form, with a few preparations in powder form. In the European market, chitosan is sold in the form of dietary capsules to assist weight loss, and in some countries, such as Japan, it is added to various foods (e.g., noodles, potato crisps, biscuits). Some of the commercial chitosan products include Fat Absorb™, a product containing 250 mg of chitosan per capsule; Seaborne range of products such as Sea Essentials™ and Sea Essentials™ Plus, chitosan combined with other nutrients such as lecithin, vitamins C and E, garlic, and β-carotene; and MinFAT, a "fat trimmer" marketed in Malaysia that claims to absorb 21 times its weight of fat.

Chitosan-fortified fruit juices and chocolates are marketed in the United States. The role of chitosan as fiber is challenged by popular fiber products such as oats, soy, and bran; nevertheless, in spite of certain limitations, chitosan promises to offer innovative applications in diverse areas of food processing and other fields. Table 6.5 indicates future research needs with regard to commercial applications of chitosan.

References

1. No, H. K., Meyers, S. P., Prinyawiwatkul, W., and Xu, Z., Applications of chitosans for improvement of quality and shelf life of foods: a review, *J. Food Sci.*, 72, 87, 2007

2. Aguilo, E. et al., Present and future role of chitin and chitosan in food, *Macromol. Biosci.*, 3, 521, 2003.

3. Prashanth, K. V. H. and Tharanathan, R. N., Chitin/chitosan modifications and their unlimited application potential: an overview, *Trends Food Sci. Technol.*, 18, 117, 2007.

4. Kurita, K., Chitin and chitosan functional biopolymers from crustaceans, *Mar. Biotechnol.*, 8, 203, 2006.

5. Ravi Kumar, M. N. V., A review of chitin and chitosan applications, *Reactive and Functional Polymers*, 46, 1, 2000.

6. Rinaudo, M., Chitin and chitosan: properties and applications, *Prog. Polym. Sci.*, 31, 603, 2006.

7. Synowiecki, J. and Al-Khateeb, N. A., Production, properties, and some new applications of chitin and its derivatives, *Crit. Rev. Food Sci. Nutr.*, 43, 145, 2003.

8. Rudrapatnam, N., Kittur, T., and Kittur, F. S., Chitin: the undisputed biomolecule of great potential, *Crit. Rev. Food Sci. Nutr.*, 43, 61, 2002.

9. No, H. K. and Prinyawiwatkul, W., Stability of chitosan powder during long-term storage at room temperature, *J. Agric. Food Chem.*, 57, 8434, 2009.

10. Tsai, G.-J. and Su, W.-H., Antibacterial activity of shrimp chitosan against *Escherichia coli*, *J. Food Prot.*, 62, 239, 1999.

11. Gurtler, J. B., Kornacki, J. L., and Beuchat, L. R., *Enterobacter sakazakii*: a coliform of increased concern to infant health, *Int. J. Food Microbiol.*, 104, 1, 2005.

12. Faug, S. W. et al., Antifungal activity of chitosan and its use, *J. Food Prot.*, 56, 134, 1994.

13. Chen, M. C., Yeh, G. H. C., and Chiang, B. H., Antimicrobial and physicochemical properties of methylcellulose and chitosan, *Trends Food Sci. Technol.*, 12, 1, 2005.

14. Rhoades, J. and Roller, S., Antimicrobial action of degraded and native chitosan against spoilage organisms in lab media and foods, *Appl. Environ. Microbiol.*, 66, 80, 2000.

15. Barreteau, H., Delattre, C., and Michaud, P., Production of oligosaccharides as promising new food additive generation, *Food Technol. Biotechnol.*, 44, 323, 2006.

16. Zheng, L.-Y. and Zhu, J.-F., Study on antimicrobial activity of chitosan with different molecular weights, *Carb. Polym.*, 54, 527, 2003.

17. Seo, S. et al., Antibacterial activity of ozone-depolymerized crawfish chitosan, *J. Food Sci.*, 73, M400, 2008.

18. Jin, X., Wang, J., and Bar, J., Synthesis and antimicrobial activity of the Schiff base from chitosan and citral, *Carb. Res.*, 344, 825, 2009.

19. Chen, C. S. et al., Antibacterial effects of N-sulfonated and N-sulfobenzoyl chitosan and application to oyster preservation, *J. Food Prot.*, 61, 1124, 1998.

20. Yang, T. C., Chou, C. C., and Li, C. F., Antibacterial activity of N-alkylated disaccharide chitosan derivatives, *Int. J. Food Microbiol.*, 35, 707, 2004.

21. Je, J.-Y. and Kim, S.-K., Antimicrobial action of novel chitin derivative, *Biochim. Biophys. Acta*, 104, 1760, 2006.
22. Roller, S. and Kovill, N., The antimicrobial properties of chitosan in mayonnaise and mayonnaise-based shrimp salads, *J. Food Prot.*, 63, 202, 2000.
23. Chen, Y.-L. and Chou, C. C., Factors affecting the susceptibility of *S. aureus* CCRC 12657 to water soluble lactose chitosan derivative, *Food Microbiol.*, 22, 29, 2005.
24. Guo, Z. et al., Antifungal properties of Schiff bases of chitosan, *N*-substituted chitosan and quaternized chitosan, *Carb. Res.*, 342, 1329, 2007.
25. Zhang, Z. et al., Preparation, characterization and antifungal properties of 2-(alpha-arylamino phosphonate)-chitosan, *Int. J. Biol. Macromol.*, 45, 255, 2009.
26. Fernandez-Saiz, P., Optimization of the biocide properties of chitosan for its application in the design of active films of interest in the food area, *Food Hydrocoll.*, 23, 913, 2009.
27. Helander, I. M. et al., Chitosan disrupts the barrier properties of the outer membrane of Gram-negative bacteria, *Int. J. Food Microbiol.*, 71, 235, 2001.
28. Pokorny, J., Natural antioxidants for food use, *Trends Food Sci. Technol.*, 2, 223, 1991.
29. FDA, *Oxygen Radical Absorbance Capacity (ORAC) of Selected Foods*, Release 2, U.S. Food and Drug Administration, Washington, D.C. (http://www.ars.usda.gov/Services/docs.htm?docid=15866).
30. Xue, C. et al., Anti-oxidant activities of several marine polysaccharides evaluated in a phosphatidylcholine–liposomal suspension and organic solvents, *Biosci. Biotechnol. Biochem.*, 62, 206, 1998.
31. Je, J. Y., Park, P. J., and Kim, S. K., Free radical scavenging properties of heterochito-oligosaccharides using ESR spectroscopy, *Food Chem. Toxicol.*, 42, 381, 2004.
32. Kim, S.-K. and Mendis, E., Bioactive compounds from marine processing byproducts: a review, *Food Res. Int.*, 39, 383, 2006.
33. Hayes, M. et al., Mining marine shellfish wastes for bioactive molecules: chitin and chitosan. Part B. Applications, *Biotechnol. J.*, 3, 878, 2008.
34. Krochta, J. M. and Mulder-Johnston, D. C., Edible and biodegradable polymer films. challenges and opportunities, *Food Technol.*, 51, 61, 1997.
35. Lin, H.-Y. and Chou, C.-C., Antioxidative activities of water-soluble disaccharide chitosan derivatives, *Food Res. Int.*, 37, 883, 2004.
36. Kanatt, S., Chander, R., and Sharma, A., Effect of irradiated chitosan on the rancidity of radiation-processed lamb meat, *Int. J. Food Sci. Technol.*, 39, 997, 2004.
37. Qin, C. Q., Du, Y. M., and Xiao, L., Effect of hydrogen peroxide treatment on the molecular weight and structure of chitosan, *Polym. Degrad. Stabil.*, 76, 211, 2002.
38. Kang, B. et al., Synergistic degradation of chitosan with gamma radiation and hydrogen peroxide, *Polym. Degrad. Stabil.*, 92, 359, 2007.
39. Rodriguez, M. S., Albertengo, L., and Agulló, E., Emulsification capacity of chitosan at different pH, in *Chitin and Chitosan in Life Science*, Uragami, T., Kurita, K., and Fukamizo, T., Eds., Kodansha Scientific, Tokyo, Japan, 2001, pp. 113–114.
40. Knorr, D., Functional properties of chitin and chitosan, *J. Food Sci.*, 47, 593, 1982.
41. Speiciene, V. et al., The effect of chitosan on the properties of emulsions stabilized by whey proteins, *Food Chem.*, 102, 1048, 2007.

42. Djordjevic, D. et al., Chemical and physical stability of citral and limonene in sodium dodecyl sulfate-chitosan and gum arabic-stabilized oil-in-water emulsions, *J. Agric. Food Chem.*, 55, 3585, 2007.
43. Klinkesorn, U. and Namatsila, Y., Influence of chitosan and NaCl on physicochemical properties of low acid tuna oil-in-water emulsions stabilized by nonionic surfactant, *Food Hydrocoll.*, 23, 1374, 2009.
44. Aider, M., Chitosan application for active bio-based films production and potential in the food industry: review, *LWT Food Sci. Technol.*, 43, 1, 2010.
45. Labuza, T. P. and Breene, W. M., Applications of active packaging for improvement of shelf life and nutritional quality of fresh and extended shelf life foods, *J. Food Proc. Preserv.*, 13, 1, 1988.
46. Campaniello, D. et al., Chitosan: antimicrobial activity and potential applications for preserving minimally processed strawberries, *Food Microbiol.*, 25, 992, 2008.
47. Reddy, B. M. V., Belkacemi, K., and Corcuff, K., Effect of pre-harvest chitosan sprays on post-harvest infection by *Botrytis cinerea* and quality of strawberry fruit, *Postharvest Biol. Technol.*, 20, 39, 2000.
48. Kumar, S. et al., Inactivation of *Escherichia coli* K-12 in apple juice using combination of high-pressure homogenization and chitosan, *J. Food Sci.*, 74, M8, 2009.
49. Casariego, A. et al., Chitosan coating surface properties as affected by plasticizer, surfactant and polymer concentrations in relation to the surface properties of tomato and carrot, *Food Hydrocoll.*, 22, 1452, 2008.
50. Soto-Peralta, N. V., Muller, H., and Knorr, D., Effects of chitosan treatments on the clarity and color of apple juice, *J. Food Sci.*, 54, 495, 1989.
51. González-Aguilar, G. A. et al., Effect of chitosan coating in preventing deterioration and preserving the quality of fresh-cut papaya "Maradol," *J. Sci. Food Agric.*, 89, 15, 2009.
52. El Ghauth, A. et al., Biochemical and cytochemical aspects of the interactions of chitosan and *Botrytis cinerea* in bell pepper fruit, *Postharvest Biol. Technol.*, 12, 183, 1997.
53. Ramanazzi, G. et al., Effects of pre- and postharvest chitosan treatments to control storage grey mold of table grapes, *J. Food Sci.*, 67, 1862, 2002.
54. Ben-Shalom, N. et al., Controlling gray mould caused by *Botrytis cinerea* in cucumber plants by means of chitosan, *Crop Protect.*, 23, 285, 2003.
55. Pen, L. T. and Jiang, Y. M., Effects of chitosan coating on shelf life and quality of fresh-cut Chinese water chestnut, *Lebensm. Wiss. u. Technol.*, 36, 359, 2003.
56. Dong, H. et al., Effects of chitosan coating on quality and shelf life of peeled litchi fruit, *Lebensm. Wiss. u. Technol.*, 64, 355, 2004.
57. Chien, P.-J., Sheu, F., and Lin, H.-R., Coating citrus (Murcott tangor) fruit with low molecular weight chitosan increases postharvest quality and shelf life, *Food Chem.*, 100, 1160, 2007.
58. Knorr, D., Use of chitinous polymers in food, *Food Technol.*, 38, 85, 1984.
59. Reddy, M. V. B. et al., Chitosan treatment of wheat seeds induces resistance to *Fusarium graminearum* and improves seed quality, *J. Agric. Food Chem.*, 47, 1208, 1999.
60. No, H. K. et al., Chitosan treatment affects yield, ascorbic acid content, and hardness of soybean sprouts, *J. Food Sci.*, 68, 680, 2003.
61. Cho, M. H., No, H. K., and Riinyawiwatkul, W., Chitosan treatments affect growth and selected quality of sunflower sprouts, *J. Food Sci.*, 73, S79, 2008.

62. Flick, G. J. and Martin, R. E., Chitin and chitosan, in *Marine & Freshwater Products Handbook*, Martin, R. E., Carter, E. P., Flick, G. J., and Davis, L.M., Eds., Technomic, Lancaster, PA, 2000, pp. 627–634.
63. Georgamtelis, D. et al., Effect of rosemary extract, chitosan and α-tocopherol on lipid oxidation and colour stability during frozen storage of beef burgers, *Meat Sci.*, 75, 256, 2007.
64. Caasal, E. et al., Use of chitosan for selective removal of beta-lactoglobulin from whey, *J. Dairy Sci.*, 89, 1384, 2006.
65. Sagoo, S., Board, R., and Roller, S., Chitosan inhibits growth of spoilage microorganisms in chilled pork products, *Food Microbiol.*, 19, 175, 2002.
66. Lin, K. W. and Chao, J. Y., Quality characteristics of reduced fat Chinese-style sausage as related to chitosan's molecular weight, *Meat Sci.*, 59, 343, 2001.
67. Jo, C. et al., Quality properties of pork sausage prepared with water-soluble chitosan oligomer, *Meat Sci.*, 59, 369, 2001.
68. Juneja, V. K. et al., Chitosan protects cooked ground beef turkey against *Clostridium perfringens* during chilling, *J. Food Sci.*, 71, M238, 2006.
69. Rao, M. S., Chander, R., and Sharma, A., Synergistic effect of chitooligosaccharides and lysozyme for meat preservation, *LWT—Food Sci. Technol.*, 41, 1995, 2008.
70. Kanatt, S. R., Chander R., and Sharma A., Chitosan glucose complex: a novel food preservative, *Food Chem.*, 106, 521, 2008.
71. Kanatt, S. R., Chander R., and Sharma A., Chitosan and mint mixture: a new preservative for meat and meat products, *Food Chem.*, 107, 845, 2008.
72. Kachanechai, T., Jantawat, P., and Pichyangkura, R., The influence of chitosan on physico-chemical properties of chicken salt-soluble protein gel, *Food Hydrocoll.*, 22, 74, 2008.
73. Chen, Y.-C. et al., Rheological properties of chitosan and its interaction with porcine myofibrillar proteins as influenced by chitosan's degree of deacetylation and concentration, *J. Food Sci.*, 68, 876, 2003.
74. Kyung, W. Kim, K. W., and Thomas, R. L., Antioxidative activity of chitosans with varying molecular weights, *Food Chem.*, 101, 308, 2007.
75. Jeon, Y. J., Kamil, J., and Shahidi, F., Chitosan as an edible invisible film for quality preservation of herring and Atlantic cod, *J. Agric. Food Chem.*, 50, 5167, 2002.
76. Shahidi, F. et al., Antioxidant role of chitosan in a cooked cod (*Gadus morhua*) model system, *J. Food Sci.*, 9, 57, 2002.
77. Duan, J., Cherian, G., and Zhao, Y., Quality enhancement in fresh and frozen lingcod (*Ophiodon elongates*) fillets by employment of fish oil incorporated chitosan coatings, *Food Chem.*, 119, 524, 2010.
78. Caballero, M. E. et al., A functional chitosan-enriched fish sausage treated by high pressure, *J. Food Sci.*, 70, M168, 2005.
79. Gómez-Guillién, M. C. et al., Effect of chitosan and microbial transglutaminase on the gel forming ability of horse mackerel (*Trachurus* spp.) muscle under high pressure, *Food Res. Int.*, 38(1), 103, 2005.
80. Rajalakshmi, M. and Mathew, P. T., Color and textural parameters of threadfin bream surimi during frozen storage as affected by cryoprotectants and chitosan, *Fishery Technol. (India)*, 44, 55, 2007.
81. Benjakul, S. et al., Effect of chitin and chitosan on gelling properties of surimi from barred garfish (*Hemiramphus far*), *J. Sci. Food Agric.*, 81, 102, 2001.

82. Benjakul, S. et al., Chitosan affects transglutaminase-induced surimi gelation, *J. Food Biochem.*, 27, 53, 2003.
83. Kataoka, J., Ishizaki, S., and Tanaka, M., Effects of chitosan on gelling properties of low quality surimi, *J. Muscle Foods*, 9, 209, 1998.
84. Huang, J.-R. et al., Shelf-life of fresh noodles as affected by chitosan and its Maillard reaction products, *Lebensm. Wiss. u. Technol.*, 40, 1287, 2007.
85. Spagna, G. et al., The stabilization of white wines by adsorption of phenolic compounds on chitin and chitosan, *Food Res. Int.*, 29, 241, 1996.
86. Je, J.-Y. et al., Antihypertensive activity of chitin derivatives, *Biopolymers*, 83, 250, 2006.
87. Muzarelli, R. A. A., Chitosan-based dietary foods, *Carbohydr. Polym.*, 29, 309, 1996.
88. Preuss, H. G. and Kaats, G. R., Chitosan as a dietary supplement for weight loss: a review, *Curr. Nutr. Food Sci.*, 2, 297, 2006.
89. Gooday, G. W., Chitinases, in *Enzymes in Biomass Conversion*, Leatham, G. F. and Himmel, M. E., Eds., American Chemical Society, Washington D.C., 1991, pp. 478–485.
90. Koide, S. S., Chitin–chitosan: properties, benefits and risks, *Nutr. Res.*, 18, 1091, 1998.
91. Hirotsu, T. et al., Synthesis of dihydroxamic acid chelating polymers, *J. Polym. Sci. A*, 24, 1953, 1986.
92. Prabhu, P. V., Radhakrishnan, A. G., and Iyer, T. S. G., Chitosan as a water clarifying agent, *Fishery Technol. (India)*, 13, 69, 1976.
93. No, H. K. and Meyers, S. P., Preparation and characterization of chitin and chitosan: a review, *J. Aquat. Food Prod. Technol.*, 4, 27, 1999.
94. Wibowo, S. et al., Surimi wash water treatment for protein recovery: effect of chitosan–alginate complex concentration and treatment time on protein adsorption, *Biores. Technol.*, 96, 665, 2005.
95. Wilsowo, S. et al., A feeding study to assess nutritional quality and safety of surimi wash water portions recovered by a chitosan–alginate complex, *J. Food Sci.*, 72(3), S179, 2007.
96. Chi, F. H. and Cheng, W. P., Use of chitosan as coagulant to treat wastewater from milk processing plant, *J. Polym. Environ.*, 14, 411, 2006.
97. Gerente, C. et al., Application of chitosan for the removal of metals from wastewaters by adsorption mechanisms and models review, *Crit. Rev. Environ. Sci. Technol.*, 37, 41, 2007.
98. Struszczyk, H. and Kivekas, O., Recent developments in microcrystalline chitosan applications, in *Advances in Chitin and Chitosan*, Brine, J., Standford, P. A., and Zikakis, J., Eds., Elsevier, London, 1991, pp. 580–585.
99. Hirano, S. et al., Chitosan as an ingredient for domestic animal feed, *J. Agric. Food Chem.*, 38, 1214, 1990.
100. Ramachandran Nair, K. G. et al., Chitin preparation from prawn shell, *Ind. J. Poultry Sci.*, 22, 40, 1987.
101. Ramisz, A. et al., The influence of chitosan on health and production in pigs, in *Chitin World*, Karnicki, Z. S. et al., Eds., Wirtschaftsverlag NW, Bremerhaven, pp. 612–616.
102. Kono, M., Matsui, T., and Shimizu, C., Effect of chitin, chitosan and cellulose as diet supplements on the growth of cultured fish, *Nippon Suisan Gakkaishi*, 53, 125, 1987.

103. Carroad, P. A. and Tom, R. A., Bioconversion of shellfish chitin waste: waste pretreatment, enzyme production, process design and economic analysis, *J. Food Sci.*, 43, 1158, 1978.

104. Krajewska, B., Application of chitin- and chitosan-based materials for enzyme immobilizations: a review, *Enzyme Microb. Technol.*, 35, 126, 2004.

105. Akoh, C. C. et al., Biocatalysis for the production of industrial products and functional foods from rice and other agricultural produce, *J. Agric. Food Chem.*, 56, 10445, 2008.

106. Han, X.-Q. and Shahidi, F., Extraction of harp seal gastric proteases and their immobilization on chitin, *Food Chem.*, 52, 71, 1995.

107. Synowiecki, J. et al., Immobilisation of amylases on krill chitin, *Food Chem.*, 8, 239, 1982.

108. Choi, M. M. F., Application of a long shelf-life biosensor for the analysis of L-lactate in dairy products and serum samples, *Food Chem.*, 92, 575, 2005.

109. Uzzan, M., Nechrebeki, J., and Labuza, T. P., Thermal and storage stability of nutraceuticals in a milk beverage dietary supplement, *J. Food Sci.*, 72, E109, 2007.

110. Arvanitoyannis, I. S. and Kassaveti, A., Fish industry waste treatments, environmental impacts, current and potential uses, *Int. J. Food Sci. Technol.*, 43, 726, 2008.

111. FAO, *State of Aquaculture and World Fisheries*, Food and Agriculture Organization of the United Nations, Geneva, 2008.

112. Global Industry Analysts, *Chitin & Chitosan (Specialty Biopolymers)*, Global Industry Analysts, San Jose, CA, 2007 (http//www.marketresearch.com).

7

Seaweed, Microalgae, and Their Polysaccharides: Food Applications

7.1 Introduction

Seaweeds have traditionally been used as food in several regions of the world, especially in East and Southeast Asia. Algae are eaten in various forms in coastal areas of Japan, China, Indonesia, Korea, and Philippines, with the earliest recorded use of seaweed dating back to 2700 B.C. The current average daily algal consumption per person in Asia varies between 3 and 13 g. In Japan, the average seaweed annual consumption is 1.4 kg per person. Seaweed species such as *Hydroclathrus, Caulerpa, Eucheuma*, and *Acanthophora* are used in many Asian countries as a green salad ingredient, whereas the coarser *Gracilaria* and *Eucheuma* are pickled. Other seaweed uses include salad, soups, pasta, and jellies. Seaweed has been a traditional food in many European societies, as well, especially in Ireland, Iceland, western Norway, Nova Scotia, Newfoundland, and some parts of the Atlantic coast of France. Nevertheless, in the west, seaweed is predominantly used as a source of hydrocolloids, including agar, algin, and carrageenan. Irish moss was used as a gelling agent for desserts in Ireland before gelatin was available. Later, large-scale commercial production of Irish moss extracts in purified and dehydrated form was initiated in the United States.[1-4] This chapter discusses recent food applications of seaweed, followed by uses of individual seaweed polysaccharides in food. Their isolation and functional properties were discussed in Chapter 4.

7.2 Functional Value of Seaweed as Dietary Supplement

Recent interest in seaweed as a food commodity is essentially due to recognition of their nutritional value and functional potential.[5] In addition to supplying diverse minerals and vitamins (see Chapter 4), seaweeds have a high content of polysaccharides, including alginate and carrageenans, which function as dietary fiber.[6] The crude fiber content of many seaweed species varies between 32 and 75% on a dry weight basis, most of which is water soluble.[7] Adequate intake of fiber is known to offer a number of health benefits, such as lowering the risk of coronary heart disease.[8] The potential of seaweed polysaccharides to lower serum cholesterol levels seems to be due to their ability to disperse in water, to retain cholesterol and related physiologically active compounds, and to inhibit lipid absorption in the gastrointestinal tract. In addition, due to their high water-holding capacity, these polysaccharides enhance a product's viscosity, binding ability, absorptive capacity, fecal bulking capacity, and fermentability in the alimentary canal.[9,10]

Many seaweed components, including polysaccharides, also have antioxidant, antibacterial, and antiviral properties that would be useful in a variety of food products. Their antioxidant properties can be superior to conventional antioxidants such as butylhydroxyanisol.[11] Their high antioxidant activities—measured in terms of hydroxyl and superoxide radical scavenging, erythrocyte hemolysis inhibition, and metal chelating activities—are due to the necessity of these algae to protect themselves against oxidative stress from ultraviolet light and desiccation during tidal fluctuations. The antioxidants include L-ascorbic acid, glutathione, carotenoids, tocopherols, chlorophyll derivatives, polyphenols such as the phlorotannins in brown kelp, and mycosporine-like amino acids in red algae.[12–16]

A *Sargassum* seaweed has been shown to prevent rancidity in fish oil, due to the presence of phlorotannins, at a rate about 2.6 times greater than 0.02% *tert*-butyl-4-hydroxytoluene (BHT).[17] Beverages prepared with added seaweed have significant antioxidant properties, as revealed by superoxide anion radical, 1,1-diphenyl-2-picrylhydrazyl (DPPH) radical, and hydroxyl radical scavenging tests. The antioxidant activities correlate with polyphenols from the added seaweed, suggesting that these beverages could serve as health drinks, particularly for patients with cancer, cardiovascular diseases, or diabetes.[18] In addition to whole fiber, the oligosaccharides derived from alginate and fucoidan also exhibit antioxidant properties. Oligosaccharides of alginate had the highest scavenging hydroxyl radical activity. Fucoidan oligosaccharides showed good chelation of Fe^{2+}, while alginate hardly had any activity in this respect.[19]

Seaweed can also have antimicrobial activities that are of some practical value when used as food or food supplement. These antimicrobial activities can be attributed to polysaccharides such as carrageenans (see Chapter 4). Crude extracts of six species of brown algae and ten species of red algae from the Black Sea were shown to exhibit antibacterial, antiviral, and cytotoxic properties. Most of the extracts showed pronounced effects against the Gram-positive bacterium *Staphylococcus aureus*, as well as influenza and herpes simplex viruses. This cytotoxic effect was identified in a significant proportion of the algae investigated. The biological activities may be attributed to the presence of volatile compounds.[20] A number of different types of seaweed from the southwest coast of India were found to have synergistic bioactivities. Methanol extracts of these seaweeds exhibited a varied range of ichthyotoxicity. The extract from one seaweed also exhibited cytotoxicity and larvicidal activity.[21]

Seaweed extracts may also display antifungal and acetylcholinesterase (AChE) inhibitory activities. *Dictyota humifusa* extracts from South Africa showed the highest antifungal and AChE activities and inhibited Gram-negative *Escherichia coli*. Other seaweed extracts inhibited the growth of the Gram-positive bacteria, including *Bacillus subtilis* and *Staphylococcus aureus*. Seasonal variations in the antibacterial but not antifungal properties of seaweed have been observed; the extracts generally have no activity in summer but do have antibacterial activity in the late winter and early spring.[22] Antimicrobial activities have also been detected among Indo-Pacific seaweed.[23] These results suggest that foods incorporating seaweed could offer a number of health benefits in terms of antioxidant and antimicrobial activities. Regular dietary intake of seaweed or seaweed-fortified food has other health benefits, such as stimulation of the immune system, purification of the blood, and proper functioning of the endocrine glands, especially the thyroid.[4,6,24] It should be pointed out, however, that although most marine algae are edible some freshwater algae can be toxic. Also, caution should also be exercised to avoid any hazard due to contamination of the algal species by metals such as arsenic.

7.2.1 Uses of Seaweed as Food and in Food Formulations

Seaweed enjoys wide use in diet foods, as it is calorie free and rich in fiber and minerals. Some of the popular types of edible seaweed (see Table 7.1) include nori (*Porphyra* spp.), laver (*Porphyra tenera*), kelp (*Macrocystis pyrifera*), wakame (*Undaria pinnatifida*), and dulse (*Eucheuma cottonii*). The red seaweed nori is the most widely utilized edible seaweed; Japan is the largest producer. Laver provides good amounts of vitamins A, B_2, and C; potassium; and magnesium. It is also a good source of iron. Kelp (kombu) is a good source of iodine, which also provides iron, magnesium, and folate (vitamin B_9). Tablets of dried kelp powder are sold in health food shops for their health benefits, including its iodine content (to prevent goiter). Wakame (brown seaweed) has been exploited in recent years as a valuable weight-loss aid. It contains good amounts of the essential fatty acid eicosapentaenoic acid (EPA). Wakame is sold salt cured and boiled, but other products such as kelp and laver are usually sold as dried foodstuff. Wakame has been shown to have many benefits, including protection against diabetes and fat-burning properties. Dulse is especially rich in protein (up to 20%) and contains magnesium, iron, and β-carotene. Other popular edible seaweed products are carola (South America), karengo (New Zealand), and ogonori (Japan).[28]

A number of species of seaweed can be used in the development of fortified foods that would benefit from incorporating the functional properties of algae. For example, blanched and salted seaweed prepared from wakame

TABLE 7.1

Common Edible Seaweed

Seaweed	Scientific Name
Arame	*Eisenia bicyclis*
Badderlocks	*Alaria esculenta*
Bladderwrack	*Fucus vesiculosus*
Carola	*Callophyllis variegata*
Carrageen moss	*Mastocarpus stellatus*
Dulse	*Eucheuma cottonii*
Gutweed	*Enteromorpha intestinalis*
Hijiki (hiziki)	*Sargassum fusiforme*
Irish moss	*Chondrus crispus*
Laver	*Porphyra laciniata/Porphyra umbilicalis*
Limu kala	*Sargassum echinocarpum*
Kombu	*Laminaria* spp.
Mozuku	*Cladosiphon okamuranus*
Nori	*Porphyra* spp.
Oarweed	*Laminaria digitata*
Ogonori	*Gracilaria*
Sea belt	*Laminaria saccharina*
Sea grapes (green caviar)	*Caulerpa lentillifera*
Sea lettuce	*Ulva* spp.
Wakame	*Undaria pinnatifida*
Thongweed	*Himanthalia elongata*

Source: Michael Guiry's Seaweed Site (http://www.seaweed.ie/uses_
general/humanfood.html).

(*Undaria pinnatifida*) is a popular product that has a high dietary fiber content. The product is made by blanching fresh wakame in water at 80°C for 1
minute followed by rapid cooling in cold water. About 30 kg of salt per 100
kg of the seaweed are mixed, and the mixture is stored for 24 hours, which
results in removal of excess of water. The dewatered product is stored frozen
at −10°C.[27] Nori (laver, purple laver, redware, sea tangle) is commonly eaten,
especially by the Japanese. Sheets of dried laver look somewhat like purple
cellophane. It is also popular in Wales, where it is used to make laver bread,
which is boiled laver mixed with oatmeal and deep fried. Laver bread is
used to thicken soups and in seafood stuffing. Nori is used to wrap rice into
rolls or small packages, in the same manner as Europeans use cabbage and
grape leaves.[29] The brown seaweed *Sargassum* (Gulfweed and sea holly) is a
free-floating brown alga with small berry-like air bladders that keep them
floating. It is used in soups and soy sauce (as well as in fertilizers).

    Seaweed polysaccharides can be incorporated in a variety of food items to
derive such benefits as enriched fiber content, modified texture, and antioxidant and antimicrobial activities. Due to their high affinity for water, these

polysaccharides in food products have a positive influence on syneresis and water activity in the foods.[30] For example, fishery products offer high nutritional properties but are low in fiber. Enrichment of these products, including surimi-based restructured products, with seaweed can increase the fiber content and improve such functional properties as water binding and gelling, leading to improved sensory attributes. Incorporation of seaweed can also enhance the emulsifying capacity of fish sausages.[31,32]

In meat products, incorporation of a popular seaweed such as sea spaghetti (*Himanthalia elongata*) or wakame (*Undaria pinnatifida*) at 2.5 to 5.0% (w/w) improved their water- and fat-binding properties and modified their texture. The hardness and chewiness of cooked products with seaweed added were increased while springiness and cohesiveness were reduced compared to control samples. Color changes in meat systems were affected by the type of seaweed. In general, products formulated with brown seaweed (sea spaghetti and wakame) exhibited comparable behavior, different from that of products made with the red seaweed.[33]

Microparticles (100 μm) of red seaweed (Rhodophyta) developed by high-speed shearing techniques could serve as low-cost fat replacers for food and as texturizers for beverages. Rheological studies showed that, even at a high solids content of around 30%, the viscosity was low and therefore did not affect the flow of the final product. At a solids content above 35%, a solid-like dispersion was observed that was characteristic of a fluid gel.[34]

7.2.1.1 Seaweed in Animal Nutrition

Seaweed can also be used for animal nutrition. In pigs, the extracts of brown seaweed can improve gut health and serve as a source of iodine. Feeding weaned piglets with dried iodine-rich intact marine seaweed (*Ascophyllum nodosum*) at 10 g/kg body weight revealed a significant depressive effect on pig gut flora, especially on *Escherichia coli*. The ratio of lactobacilli to *E. coli* was enhanced in the small intestine, indicating a beneficial shift in the microbial population. Increases in iodine content were noted for several tissues in piglets fed a diet incorporating seaweed (20 g/kg body weight, corresponding to 10 mg iodine per kilogram feed), suggesting that the seaweed may be introduced in pig nutrition as a feed material for improved gut health and iodine enrichment of porcine tissues.[25]

A seaweed-enriched diet has been reported to be beneficial for poultry in that it increased egg quality with regard to *n*-3 fatty acid (omega-3 fatty acid) and albumin content and yolk color, while the egg flavor was not affected. In this study, *Macrocystis pyrifera*, *Sargassum sinicola*, and *Enteromorpha* spp. were incorporated at a level of 10% in the diets of 35-week-old Leghorn hens for 8 weeks. The diet also contained 2% (w/w) sardine oil.[26]

Seaweed meal used an additive to animal feed are produced in Norway. The palatable meal is made from brown seaweed that is collected, dried, and milled. Seaweed meals prepared from *Gracilaria, Gelidiella, Hypnea,* and

Sargassum can increase fertility and the birth rate of animals, in addition to improving yolk color in eggs. Seaweed meal can also be used to enrich fish and prawn feed with minerals, amino acids, and carbohydrates. The feed also helps to maintain water quality in aquaculture.[2,6] Approximately 50,000 t of wet seaweed are harvested annually to yield 10,000 t of seaweed meal, valued at US$5 million.[27] Uses of seaweed in agriculture are discussed later in this chapter.

7.2.2 Some Seaweed-Based Food Products

7.2.2.1 Edible Powders

Conversion of seaweed into edible powders requires careful processing to make the nutrients bioavailable. Hand-harvested seaweed powders are available in many countries. These powders include kelp (kombu), green rockweed, rockweed, dulse (dilisk), nori, wakame, ulva, sea lettuce, sea spaghetti, and Irish moss, among others. From a nutritional point of view, they generally contain 10 to 30% minerals, comprised of the macroelements Ca, Cl, K, Na, P, Mg, and Fe, as well as the microelements Zn, Cu, Mn, I, Se, Mo, and Cr. The protein content is between 20 and 45%, with a good amino acid profile. Soluble fiber (agar, alginate, and carrageenan) and sugar and sugar alcohols represent up to 40% and 10 to 20%, respectively. They also contain β-carotene (pro-vitamin A) and vitamins of the B complex. The price for 85-g packets of these powders can vary from US$2 to 6.[35]

7.2.2.2 Processed Eucheuma Seaweed

Eucheuma seaweed, which is harvested around the Philippines and Indonesia, is directly treated with alkali to prepare a commercial product known by the various names of processed eucheuma seaweed (PES), semi-refined carrageenan (SRC), Philippines natural grade (PNG), semi-refined carrageenan (SRC), alternatively refined carrageenan (ARC), or alkali-modified flour (AMF). Two varieties of red seaweed (*E. cottonii* and *E. spinosum*) are used to prepare processed eucheuma seaweed. PES contains 23 to 31% ash and relatively low levels of lipid-soluble material, while water-soluble components of the flours comprise 70 to 90%. PES contains polysaccharides, including carrageenan, in the acid-insoluble matter. Acid-insoluble material in the product is usually high, in the range of 4.8 to 11.3%. The polysaccharides, particularly carrageenan, modify the hydration, appearance, and textural characteristics of the product. Because of carrageenan, the powder has significant water-holding capacity, as high as 17.7 g/g and an oil-holding capacity of 2 g/g. PES is predominately used as an additive in products such as processed meat, fish, and dairy products, at concentrations ranging from 8 to 10%.[2,36,37] Incorporation of the powder at 8% enhanced the fiber content, breaking strength, extensibility, and cooking yield of Chinese egg noodles by more than 400%. Also, higher water absorption by the seaweed powder imparted a softer and spongier texture to the noodles.[38]

7.2.2.3 Other Products

Modifilan® is a patented commercial extract of *Laminaria* that contains significant amounts of organic iodine, fucoxanthin, alginate, fucoidan, and laminarin. The manufacturer claims enhanced bioavailability of nutrients due to a low-temperature processing of the seaweed. The product is also claimed to enable the human body to detoxify heavy metals and toxins, to boost the immune system, to protect against thyroid cancer, and to decrease high blood sugar. Recommended daily usage is 2 to 3 g (4 to 6 capsules).[39] An anti-obesity diet has been developed recently for overweight people and diabetics based on seaweed and seaweed hydrocolloids. The formula includes natural ingredients, such as agar, carrageenans, alginate, and the microalgae *Chlorella* and *Spirulina*.[40]

Seaweed tea is made from fresh or dried rockweed (*Ascophyllum nodosum*). The blades from the tough main stalks of fresh seaweed are cut, washed well in freshwater, and dried under controlled conditions to make tea. A teaspoon of dried and crumbled rockweed is added to a cup of boiling water, steeped for about 5 minutes, and consumed after adding honey. It can be superior in its levels of essential minerals, phenols, and total vitamin C compared to green tea. A flavoring made from rockweed (Norwegian kelp) can add a seawater flavor to steamed seafood or soup products. The seaweed is enclosed in a cheesecloth bag, which is steamed along with the seafood; after cooking, the bag is removed. The *Seaweed Jelly-Diet Cookbook* provides seaweed recipes.[41] Recently, a salt product called Saloni K Salt (a mixture of approximately 30% potassium chloride and 60% sodium chloride), isolated from seaweed, has been shown to have the potential to alleviate hypertension. The salt is being commercially produced for Indian markets.[42] A number of seaweed species have been approved as sources of commercial polysaccharides, as indicated in Table 7.2. In general, these are used to impart a number of functional benefits to foods, as shown in Table 7.3.

TABLE 7.2

Seaweeds Approved as Sources of Commercial Gums

Danish agar (*Furcellaria fastigiata*)	Hypnean (*Hypnea* spp.)
Eucheuman (*Eucheuma* spp.)	Iridophycan (*Iridaea* spp.)
Furcellaran agar (*Furcellaria fastigiata*)	Irish moss (*Chondrus* spp.)

Source: FAO/WHO Codex Alimentarius Commission, Food Standards Programme, Food and Agriculture Organization, Rome, and World Health Organization, Geneva. With permission.

TABLE 7.3

Functional Benefits of Seaweed Polysaccharides in Foods

Syneresis control
Reduced dryness and toughness
Enhanced fiber content
Increased viscosity
Increased yield
Reduced product costs
Serve as water-binding agents, emulsifiers, texturizers, and stabilizers

7.3 Agar

Commercial agar is quite stable. The product contains less than 20% moisture and about 7% ash, and the remainder is fiber. In the Orient, natural agars in the form of strips and squares are used at home to prepare traditional dishes. Food-grade agar is used as a stabilizer in canned meat, in confectionery, and in glazing and icing for the baking industry. White and semitranslucent, it is sold in packages as washed and dried strips or in powdered form. Its nondigestible nature and colloidal and gelling properties make agar a popular thickener, gelling agent, stabilizer, lubricant, emulsifier, and absorbent. Some agars, especially those extracted from *Geochelone chilensis*, can be used in confectionery with a very high sugar content, such as fruit candies. It is also used to make jellies, puddings, and custards. Because of its bland taste, agar does not affect the flavors of foodstuffs.

Agar exhibits hysteresis, melting at 85°C and solidifying at 32 to 40°C. A popular Japanese sweet dish is *mitsumame*, which consists of cubes of agar gel containing fruit and added colors. Agar can be canned and sterilized without the cubes melting. In Indian cuisine, agar is known as *China grass* and is used for making desserts. Agar is preferred to gelatin because of its higher melting temperature and gel strength. It is used in vegetarian foods such as meat substitutes. It has been used to clarify wines, especially plum wine, which is difficult to clarify by traditional methods. Unlike starch, agar is not readily digested and so adds little caloric value to food. It improves the texture of dairy products such as cream cheese and yogurt. Interactions of agar with food components are important in determining the functional role of agar in food product development. The interaction of agar with sugar increases the strength of the gel through a phenomenon known as sugar reactivity (see Chapter 4). Although agar costs more than synthetic and other natural gelling agents, it is usually superior to such products because its gels have greater transparency, strength, and stability over a range of acidity and alkalinity.[7,43,44] Some specific applications of agar are discussed below (see Table 7.4).

TABLE 7.4

Applications of Agar in Food Products

Food Product	Applications
Bakery	Texture improvement, stabilization of doughs, reduction of pasting temperatures of starch, replacement of gluten and gelatin, prevention of product adherence to packaging, control of phase separation
Vegetable	Reduction of torque in extrusion cooked products
Miscellaneous	Enhanced texture (hydrogels of agar and carrageenan), enhanced satiety (useful as dietetic foods)

7.3.1 Bakery Products

In the bakery industry, hydrocolloids help to improve food texture (softer texture) and moisture retention, retard starch staling and retrogradation, and, finally, enhance the overall quality of the products in terms of specific loaf volume and viscoelastic properties of products.[45] A number of studies have identified uses of agar in baked products. Agar is usually added at 0.8% (w/w) in baked goods and baking mixes, 2.0% in confections and frostings, 1.2% in soft candy, and 0.25% in all other candy.[46] The ability of agar gels to withstand high temperatures makes the polysaccharide a useful stabilizer and thickener in a number of products, including pie fillings, icings, meringues, cakes, and buns. The polysaccharide largely modifies starch properties; for example, it reduces pasting temperatures (as measured by amylograph parameters). The effect depends on the chemical structure of the added hydrocolloid. Reducing the pasting temperature is important because it results in early starch gelatinization and, in turn, an increase in the availability of starch as an enzyme substrate during the baking period.[47] Agar also helps maintain crust texture characteristics, color, and moisture, which are important quality indicators for breads. The effect is due to better water retention by the polysaccharide leading to higher moisture content in the final baked product. The effect of agar is also manifested in the viscoelastic characteristics of crust of baked products, irrespective of duration of baking, whether semi-baked, full-baked, or bread products developed from cold stored dough.[45]

In addition, improvement in dough stability during fermentation has been noted after the incorporation of agar, which also results in an increase in the specific volume as well as enhanced moisture retention and water activity. The effects of the hydrocolloid, however, were highly dependent on the type of flour (white or whole-wheat flours) and the breadmaking process.[47] Algar also functioned as an emulsifier due to its softening effects and exhibited both synergistic and antagonistic interactions among anti-staling additives.[48–50]

7.3.2 Gluten-Free Products

The presence of gluten in wheat can cause health problems in predisposed individuals. The high incidence of gluten intolerance, reflected as celiac disease, among the Western population has given rise to increased need for gluten-free breads (see Chapter 2). Agar has been found to be a suitable replacement for gluten. It can also replace fat in such breads.[51–53]

7.3.3 Control of Syneresis

When starch pastes or gels are frozen, phase separation occurs due to the formation of ice crystals. Upon thawing, water is easily expressed from the gel network, a phenomenon known as *syneresis*. The extent of phase separation increases with additional freeze–thaw cycles. Freeze–thaw stability is an important criterion when evaluating the quality of a starch. The amount of syneresis can be used as an indicator of the tendency of a starch to retrograde. A fast freezing rate could prevent changes that occur due to retrogradation more than medium or slow freezing rates. Agar can prevent syneresis in starch products. In addition to their various applications in bakery products, hydrocolloids, including agar, keep products such as icings, toppings, and meringues from becoming sticky or adhering to packaging during storage and transport, especially during damp weather, and they control excessive drying and brittleness under conditions of low humidity.[49]

7.3.4 Other Applications

Gelatin jellies have long been favored because they melt at body temperature, resulting in a smooth mouth feel and easy release of flavors; however, if they are stored for a day or two, they toughen and are less pleasant to eat. Agar is an alternative to mammalian gelatin. With the appearance of bovine spongiform encephalopathy (BSE, or mad cow disease) and foot-and-mouth disease, efforts have been made to find suitable substitutes for gelatin. Agar and Irish moss extracts are used to replace conventional animal gelatin in many jelly candies and marshmallows, jellies, puddings, and fruit batters and jams. Being of vegetable origin, it is acceptable to vegetarians. Furthermore, compared to gelatin, agar possesses superior gel strength and high melting temperature. Other gelling agents are not as satisfactory, as they are more likely to melt. Agar is added to frozen desserts made with fruit juice, soy, water, or milk at about 0.1% (w/v), often in combination with gum tragacanth and locust bean gum. Agar in the amount of 0.1 to 1% (w/v) stabilizes yogurt, cheeses, and candy and pastry fillings. It can also be added to desserts and pretreated instant cereal products. Jelly-type candies are made with agar at concentrations ranging from 0.3 to 1.8% by weight, although starch and pectin are used whenever transparency and other characteristics of agar gel are not required.

A typical candy might be produced by soaking agar for 2 hours in water, followed by cooking the dissolved agar at 105°C in the presence of sugar. This is followed by the addition of sweetener (corn syrup or invert sugar), mixing, and finally the addition of color, flavor, and acid. When all of the ingredients have dissolved, the mixture is poured into molds and cooled.

In the extrusion cooking of corn grits, agar (and other hydrocolloids, including alginate, locust bean gum, guar, and gum arabic) has been found to reduce torque. The hydrocolloids were dry blended with corn grits at levels of 0.1 to 1.0% (w/w). Moisture at 20% was added, and each product was extruded in a Brabender® Model PL-V500 laboratory extruder at temperatures ranging from 50 to 150°C with a 1:1 screw operating at 100 rpm. The hydrocolloids reduced torque at 50°C but not at higher extrusion temperatures.[54] Agar and other hydrocolloids are used to improve the shelf life and handling properties of tortillas.[55]

A diet that features agar has been developed for obesity. The efficacy of the agar diet in combination with a conventional diet (traditional Japanese food) for obese patients with impaired glucose tolerance and type 2 diabetes has been reported. In one study, the agar diet resulted in marked weight loss due to the maintenance of reduced calorie intake and improvement in metabolic parameters. This diet has received some press coverage in the United States.[56]

7.3.5 Modification of Agar for Novel Uses

Efforts have been made to modify the gelling properties and solubility of agar so the polysaccharide can be put to additional food uses; for example, gel strength has been improved by increasing the average molecular weight and reducing the sulfate contents. A new type of agar, quick-soluble agar, has recently been developed that can dissolve in water at a lower temperature than ordinary agar. The product is obtained by drying agar directly from a solution instead of making the agar gel before drying. Microparticles of agar can be prepared by shearing bulk gels dispersed in cold water using a high-speed rotor/stator device. The unmixed gels had negligible porosity, but the mixed gels were porous and stronger than the unmixed gels. Mixing conditions influenced thermal transition temperatures, highlighting the importance of preparation methods with regard to the functionality of agar. Particular preparations can confer a range of textural functionalities to fluid gels, including beverages.[57]

Development of a stable hydrogel network of an agar–κ-carrageenan blend cross-linked with genipin for food applications has been reported recently. A mixture of agar and κ-carrageenan was treated with the natural cross-linker genipin in an aqueous medium. The blend at an optimum ratio of 25:75:0.8 (w/w) exhibited remarkable stability over the pH range of 1 to 12 and had excellent swelling properties. The cross-linked blend exhibited higher viscosity, thermal stability, and swelling ability, as well as lower weight loss compared to the unmodified blend, thus representing immense potential in food product development.[58]

Agar is also used in the silk and paper industries as a substitute for isinglass, for dying fabrics, as a lubricant for drawing wire, as a sizing for fabric, and for photographic films and plates. In the medical field, agar is used as a laxative and for the treatment of constipation. Its action in the intestinal tract is comparable to that of cellulose fiber in aiding bowel movement. It is also used to make dental casts in dentistry and as a surgical lubricant. Agar blocks streptococcal adhesion to biosurfaces and thus has the potential to be used in mouthwashes and spray washes for foods. In addition to its well-recognized uses in microbiology, agar is also used in gel electrophoresis, chromatography, immunology, biotechnology, and immobilization of enzymes.[1,4]

7.4 Alginic Acid and Alginates

Alginic acid (algin) and alginates are mainly used as thickening agents in a variety of food products such as salad dressings, sauces, syrups, milk shakes, ice cream toppings, pie fillings, cake mixes, and canned meat and vegetables. In some of these products, they help to retain moisture, while in some others they thicken the batter, in addition to aiding moisture retention, as per requirements. In canned meat and vegetables, they can give either temporary or delayed-action thickening. The remarkable gelling properties of alginic acid have also found unique applications in restructured foods, bakery fillings, dessert gels, and pet foods due to its interactions with proteins. Alginic acid improves and stabilizes the consistency of fillings for bakery products (cakes and pies), salad dressings, and milk chocolates, and it prevents the formation of large crystals in ice cream during storage. Alginates are used in a variety of gel products (e.g., cold instant puddings, fruit gels, dessert gels, onion rings, imitation caviar) and are used to stabilize fresh fruit juice and beer foams; for example, propylene glycol alginate (PGA) is used as a foam stabilizer in beer and cider. Green Manzanilla olives available in Spain are stuffed with flavored alginate-based pastes, such as garlic, herbs, hot pepper, lemon, and cheese. PGA is also used in high-oil salad dressings for its emulsifying properties; however, because it is required at high levels to stabilize the emulsion, its use may be cost prohibitive. To counter this problem, PGA is combined with less expensive xanthan.[59]

Compared to gelatin, alginate has several advantages with respect to its use in ice cream. Because of its thickening ability, sodium alginate is used as a stabilizer in ice cream. The whipping ability of mixes containing alginate is significantly greater than that of similar mixes containing gelatin. Furthermore, a lesser amount of alginate is required in ice cream, and it provides uniform viscosity during aging, lighter color, smoother and cleaner melt down, and better flavor. The only disadvantage is its insolubility when added to cold mixes. This can be overcome by warming the mixture to 68 to 70°C before adding the alginate. Alginate at 0.2% (w/w) is a good suspending agent for cocoa in chocolate milk. It is also used in soft cheese spreads at 0.1 to 0.2%. The colloid should be dissolved in hot water and added to the cream before pasteurization. As much as 0.8% of alginate may be used in cheese spreads. The biopolymer is also used in several bakery products such as icing, filling, marshmallow toppings, jellies, glazes, syrup, and bread. Alginate is also added in puddings and confectioneries.[1] Applications of alginate in individual product categories are discussed below.

7.4.1 Bakery Products

Alginate and other hydrocolloids (xanthan, κ-carrageenan) at 0.1% (w/w, flour basis) improve the properties of bread in terms of specific volume index, width/height ratio, crumb hardness, sensory properties (visual appearance, aroma, flavor, crunchiness), and overall acceptability. The hydrocolloids also prevent staling in bread stored for 24 hours, reduce the loss of moisture content during storage, and lower the crumb dehydration rate. Alginate was found to be effective in its anti-staling effect and the prevention of crumb hardening during storage.[60] The physical properties of fresh yellow layer cakes and changes that occurred during storage were notably influenced by the type of hydrocolloid used (i.e., carrageenan, pectin, hydroxypropylmethylcellulose, guar gum, and xanthan gum). In general, overall acceptability of the cakes was improved by all of the hydrocolloids except pectin, whereas xanthan was able to maintain an acceptable texture during storage.[61]

Alginate can be used to improve the physicochemical and rheological properties of wheat flour noodles. The hydration properties (water absorption index, water solubility, and swelling power) of wheat flour increase with increasing levels of alginate due to its high affinity to water. This is reflected in increased water absorption and dough development time of the flour and reduced tolerance of the dough to mixing. Further, the syneresis of the wheat flour gel is significantly reduced during freeze–thaw treatments when alginate is incorporated. Noodles containing alginate exhibited an increase in cooked weight and a decrease in cooking loss, in addition to a significant increase in the cutting and tensile forces.[49,62] Sodium alginate has a beneficial interaction with soy protein during the precipitation of soybean grits used in making textured vegetable protein products.[49,63]

7.4.2 Meat Products

Alginate gels can be used as binders in restructured meat products. These products are made by binding meat pieces together and shaping them to resemble usual cuts of meat, such as nuggets, roasts, meatloaf, even steaks.

A mixture of sodium alginate, calcium carbonate, lactic acid, and calcium lactate is used for the purpose. When mixed with the raw meat they form a calcium alginate gel that binds the meat pieces together. Similarly, shrimp substitutes can be made that incorporate alginate, proteins such as soy protein concentrate, and flavors. The mixture is extruded into a calcium chloride bath to form edible fibers, which are chopped, coated with sodium alginate, and shaped in a mold. Restructured fish fillets have been made using minced fish and a calcium alginate gel.[63]

Alginate along with carrageenan could be used to develop low-fat, precooked, beef patties that have higher yields and moisture contents but lower shear force values compared to either alginate or carrageenan treatment alone within the same fat level. Alginate appeared to improve texture slightly more than carrageenan, whereas carrageenan tended to release more free water after cooking and reheating. Patties with 10% fat were generally lower in shear value, cooking yield, and percentage free water released as compared to their 5% fat counterparts containing alginate. Low-fat, precooked, ground beef patties containing combinations of alginate and carrageenan were comparable to regular beef patties having 20% fat with respect to yields and textural properties.[64]

Buffalo meat patties were sequentially dipped in 2% solutions of alginate and calcium chloride for 30 seconds each, followed by draining. The coated patties were kept at 4°C in polyethylene pouches. Storage studies indicated that the coating significantly improved overall appearance and color, juiciness, flavor, texture, and overall palatability of the product. The growth of microorganisms in the product was also suppressed by the coating. The product was free of enterobacteria.[65] When beef cuts are coated with calcium alginate films before freezing, the meat juices released during thawing are reabsorbed into the meat, and the coating helps to protect the meat from bacterial contamination. If desired, the calcium alginate coating can be removed by redissolving it with sodium polyphosphate.[27]

7.4.3 Seafood

Alginate at 0.5% (w/w) helped retain the water-holding capacity of raw whiting muscle and protected against an increase in toughness of the minced fillets stored frozen at −18°C for 2 months. There was remarkable improvement in texture and water-holding capacity of the treated product, reflected in the extractable myosin, dimethylamine, and formaldehyde contents of the products during storage.[66] Alginate in combination with high pressure can modify characteristics of fish meat gel. Alginate incorporation gave a gel that was harder, more adhesive, less cohesive, and more yellow than pressure-induced samples. The gel properties were influenced by the pressure applied (200 or 375 MPa); lower pressure treatment showing significantly higher values for penetration values and cohesiveness and lower values for elasticity and lightness compared with higher pressure treated samples.[67]

Calcium alginate has been used to preserve frozen fish. It can function as a cryoprotectant in frozen fish products to control denaturation of proteins and to maintain texture.[10] Whereas sodium tripolyphosphate (STPP) is conventionally used as a cryoprotectant to control the problem, alginate (or ι-carrageenan) can be effective, as observed in the case of red hake (*Urophycis chuss*). The physicochemical and sensory properties of the fish mince were retained when stored at –20°C for 17 weeks when 0.4% alginate, 4% sorbitol, and 0.3% STPP were incorporated in the mince before freezing. The additives protected the mince from hardening and improved its dispersibility during mixing. Alginate appeared to be responsible for preventing muscle fiber interaction through electrostatic repulsion and chelating Ca^{2+}, thus improving dispersibility. Alginate can also be used to modify the texture of restructured shrimp or crab meat products.[68,69] The oils in fish such as herring and mackerel can become rancid through oxidation even when quick frozen and stored at low temperatures. If the fish is frozen in a calcium alginate jelly, the fish is protected from the air and rancidity from oxidation is limited.[27]

7.4.4 Vegetable Products

Onion is an economically important vegetable used in various food preparations. Fresh onion undergoes weight loss during ambient temperature storage, accompanied by a loss of skin layers. Coating the onion with alginate has been attempted to prevent this loss of quality during storage and to extend the shelf-life of onion and thus its export and domestic salability. The residual amount of mineral found in the skin layers suggests possible penetration of alginate into the skin.[70]

7.4.5 Miscellaneous Uses

Alginate gels in the form of sponges have been reported to be useful as carriers of vitamin A. These sponges were produced by preparing cold-set 1% alginate gels containing vitamin A. The sponges consisted of hydrocolloid matrices to which oil containing the vitamin had been added before the gelation process. After gelation and freeze-drying, a crunchy, chewable, cellular solid was produced. The product is devoid of flavor, odor, and color; these characteristics can be modified during processing to ensure broad acceptance by the targeted subjects. The edible sponges were tested as a means of supplementing preschool children who had endemic vitamin A deficiency. Administration of the sponges to children resulted in a significant increase in levels of vitamin A.[74] A low-viscosity soybean beverage was prepared through lactic acid fermentation of soy milk with *Lactobacillus casei*. During the fermentation process, an organoleptically undesirable powdery/gritty sensation developed. This off-taste could be effectively reduced by the addition of propylene glycol alginate (PGA); however, the emulsion stability of the fermented product was sometimes decreased when PGA was added. This defect could be overcome by adding calcium lactate with the PGA.[75]

Sodium alginate, when subjected to gamma irradiation, has been shown to promote the growth of red amaranth (*Amaranthus cruentus* L.). A 3% aqueous solution of the alginate was irradiated by ^{60}Co gamma radiation at a dose of 37.5 kGy. Irradiation decreased the viscosity of the alginate and a reduction of average molecular weight. The treated alginate, at an optimum concentration of 150 ppm, was applied to seedlings after 10 days at intervals of 6 days. The treatment resulted in a significant increase in plant height (17.8%), root length (12.7%), number of leaves (5.4%), and maximum leaf area (2%) compared to the control vegetative plant production.[76]

7.4.6 Nutritional Value of Alginate

Alginate also has an important use as dietary fiber, as it is not digestible (see Chapter 2). Consuming foods containing alginate can slow the absorption of fat and reduce serum cholesterol and triglycerides in the blood. This helps to prevent high blood pressure, diabetes, and adiposity and controls the accumulation of heavy metals, such as strontium, cadmium, and lead. Alginates are the basis of many weight-loss food products. Alginic acid swells in the stomach and promotes a feeling of satiety. Consuming alginate at a rate of 10 g once a day for 2 weeks was shown to have a beneficial effect on the levels of bifidobacteria, which increased significantly, while the levels of enterobacteria and frequency of occurrence of lecithinase-negative clostridial bacteria showed a tendency to decrease. Fecal sulfide, phenol, *p*-cresol, and indole were significantly decreased during alginate consumption. Fecal concentrations of ammonia and skatole were also significantly reduced, whereas the levels of acetic and propionic acids were increased. The water content and weight of the feces were slightly increased during consumption.[71] These benefits support the role of alginate as fiber in food.[72,73] The uses of alginate as dietary fiber and in various food products are shown in Table 7.5 and Table 7.6.

7.5 Carrageenan

Carrageenans are widely used as ingredients for diverse purposes, generally as natural thickeners, formulation stabilizers, or gelling agents at concentrations ranging from 0.005 to 3% depending on the food. The functional properties of carrageenans in food products depend on the source of the seaweed, type of carrageenan (κ, ι, or λ), and the isolation conditions. Process variables, such as temperature, pH, ionic strength, and cations, have a strong influence on the functional value of carrageenans as food additives. Upon heating and subsequent cooling, ι-carrageenan and κ-carrageenan form thermoreversible

TABLE 7.5

Potential Benefits of Alginate
as Dietary Fiber

Stimulates immune system
Reduces intestinal absorption
Increases satiety
Reduces glycemic index value
Modulates colonic microflora
Elevates colonic barrier function

gels in the presence of gel-promoting cations; therefore, to make carrageenan gel in water-based foods, salts must be added. Salts enhance the interaction effect in the following order: Na_2SO_4, NaCl, KCl, and NH_4Cl. K^+ salts must be added to the system before cooling below the gelling temperature.[78,79]

The ι-carrageenan form is particularly thixotropic; that is, it is a gel that is normally thick but can become less viscous and flow over time when shaken or agitated. ι-Carrageenan is often used in cold, filled, ready-to-eat desserts. Carrageenans are freeze–thaw stable. They are usually incorporated in foods at concentrations of 1 to 2%. To avoid agglomeration, the carrageenans are often premixed with high concentrations of other ingredients such as sugar, usually in a ratio of 1:10. If premixing is not possible, stirring with a high-speed

TABLE 7.6

Common Uses of Alginates in Food Products

Application	Remark
Foam stabilizer in beer	Propylene glycol alginate provides better foam retention and prevents foam-negative contaminants.
Texturized foods	Alginate gives food products thermostability and the desired consistency.
Bakery products	Alginate provides freeze–thaw stability and can reduce syneresis.
Fruit preserves	Alginate is commonly used as a thickening, gelling, and stabilizing agent in jams, marmalades, and fruit sauces. Alginate–pectin gels are heat reversible and give better gel strength than the individual components.
Ice cream	Alginate provides the ideal viscosity, prevents crystallization and shrinkage, and promotes homogeneous melting without whey separation; it is used in combination with other gums.
Other	Alginate is used in desserts, emulsions (e.g., low-fat mayonnaise), sauces, and extruded foods (noodles and pasta). Propylene glycol alginate is acid stable and resists loss of viscosity; it has unique suspension and foaming properties that make it useful in soft drinks, milk drinks, sorbet, ice cream, noodles, pasta, etc.

Source: Adapted from Brownlee, I.A. et al., *Crit. Rev. Food Sci. Nutr.*, 45, 497, 2005. With permission from Taylor & Francis, Ltd.

mixer together with the slow addition of carrageenan can prevent agglomeration. In instant preparations, carrageenan must be used as a powder to be mixed with cold water, when a thickening effect is caused by the swelling of the hydrocolloid. In solution, with a high content of soluble solids (>50%), the temperature is increased to a level favoring gelation of the polysaccharide.

A general method of preparing carrageenan-containing food products has been described in a patent application.[77] The compositions comprise approximately 55 to 85% by weight nutritive carbohydrate sweeteners, sufficient amounts of a gelling system to provide a gel strength of 1 to 8 kg/cm², and 10 to 20% moisture. The gelling system contains high levels of methoxyl pectin and κ-carrageenan and has low viscosity when maintained above 55°C. The method of production involves forming hot fluid slurry, shaping it into pieces by starch molding, and curing it to form a gelled product. In view of the large variations in functionality, technologists often use a mixture of carrageenans to derive the desired benefits. For most applications, λ-carrageenan and ι-carrageenan, extracted from *Kappaphycus alvarezii* (old name, *Eucheuma cottonii*) and *Eucheuma denticulatum* (old name, *Eucheuma spinosum*) are used.

7.5.1 Functional Benefits of Using Carrageenans in Food Products

Table 7.7 compares the various properties of carrageenans that are important in food systems.

TABLE 7.7

Comparison of Properties of Carrageenans

Medium	κ-Carrageenan	ι-Carrageenan	λ-Carrageenan
Hot water	Soluble at >60°C	Soluble at >60°C	Soluble
Cold water	Na salt soluble; K and Ca salts insoluble	Na salt soluble; K and Ca salts give thixotropic dispersion	Na salt soluble
Hot (80°C) milk	Soluble	Soluble	Soluble
Cold (20°C) milk	Na, Ca, K salts insoluble, but swell	Insoluble	Soluble, thicken
Gelation	Gels strongest with K⁺	Gels strongest with Ca²⁺	No gelation
Concentrated sugar solution	Soluble when hot	Soluble with difficulty	Soluble when hot
Concentrated salt solution	Insoluble	Soluble when hot	Soluble when hot
Stability			
Freeze–thaw	No	Yes	Yes
pH > 5	Stable	Stable	Stable
Syneresis	Yes	No	No
Salt tolerance	Poor	Good	Good

Source: Adapted from Rudolph, B., in *Marine & Freshwater Products Handbook*, Flick, G.G. and Martin, R.E., Eds., VCH Publishing, Lancaster, 2000, pp. 515–530. With permission from VCH Publishing.

7.5.1.1 Texture Modification

Carrageenans modify the textures of diverse food products through changes in water binding, emulsifying, and foaming properties. Textural modifications of food are influenced by the interactions of these polysaccharides with other food components, including proteins, other polysaccharides, and cations. The effects could be additive or sometimes synergistic. Mixed gels of locust bean gum (LBG) and carrageenan are brittle, slightly elastic gels, whereas xanthan gum and κ-carrageenan form soft cohesive gels. Starch has a notable influence on texture and cooking yield, increasing product hardness and resilience as the proportion of starch increases. Combinations of carrageenan with LBG and starch can be used to improve texture in such products as sausages. LBG and κ-carrageenan improve cooking yield and reduce expressible moisture in formulations containing higher proportions of potato starch.[78]

Ions and pH have an important influence on the functionality of carrageenans in food. Carrageenan is strongly negatively charged over the entire pH range encountered in food. As the pH value decreases below 5, carrageenan solutions become increasingly unstable when heated and a loss of viscosity results due to irreversible cleavage of the polymer chains. These factors need to be considered when developing foods containing carrageenans.[78,79]

7.5.1.2 Fat Reduction

Over the past few years, concerns about the high fat content of prepared foods have stimulated research into developing low-fat snacks for the benefit of health-conscious consumers. Fat replacers can be divided into three classes on the basis of their composition: protein-based, carbohydrate-based, and fat-based. Each has different functional properties that provide both advantages and limitations in specific applications. Currently, no single fat replacer contributes all of the desired sensory and functional qualities to all products. A combination of two or more wisely chosen fat replacers, coupled with formula and procedural changes, appears to be the best strategy today. The addition of limited amounts of carrageenan has been found to be a fat-free, economically viable solution to providing palatable, healthier, and convenient third-generation foods. Carrageenan at 0.5%, either alone or in combination with cellulose, functions as a gel-forming fat substitute.[78]

Numerous low-fat products have been developed using carrageenan. The hydrocolloid at 0.25 to 0.75% has been used in low-fat ground pork patties having less than 10% total fat. The low-fat product has a better cooking yield and higher moisture content, with sensory attributes similar to those of the high-fat control product. Incorporating carrageenan resulted in a reduction of total lipids and cholesterol by as much as 48 and 44%, respectively, and a reduction in calories of 31% as compared with the controls. The product was found to have good storage stability for 35 days at 4°C.[80]

Carrageenan was also a satisfactory fat substitute in emulsified meatballs. High acceptability scores were observed for low-fat (<10%) meatballs containing salt, polyphosphates, and κ-carrageenan up to 2%. The hydrocolloid (0.3 to 0.7%) alone or in combination with 20% pectin gave ideal physicochemical and textural properties to low-fat beef frankfurters and also resulted in significant reduction of cholesterol.[81] Due to the apparent effects of carrageenan on muscle, it may be necessary to modify processing conditions for the manufacture of low-fat, water-added, dark vs. white poultry meat products.[82] Aqueous dispersions of soluble hydrocolloids including carrageenan could replace lipids in process cheese spreads. About 40% and 50% fat reductions were obtained relative to a control cheese spread containing 25% fat by increasing moisture to 62% and 68%, respectively, and by eliminating a portion of fat from the formulation. Spreads with 2.2% λ-carrageenan had texture consistent with a high-fat cheese control. Above that level, cheese spread firmness increased and the melting point decreased.[83]

7.5.1.3 Salt Reduction

The consumption of high amounts of salt is known to have adverse effects on health; therefore, interest is growing in low-salt food products that can ward off high blood pressure and related ailments. Carrageenan has been shown to be useful in the reduction of dietary salt.[78] At a level of 0.5%, carrageenan was added to a sausage preparation to which KCl and/or $CaCl_2$ were added at the 0.5% (w/w) level. Cooking yield was increased by all the treatments, but expressible moisture levels were not significantly different, indicating that water was not entrapped by carrageenan under the ionic strength conditions employed. Myofibrillar proteins appeared to maintain good functionality under these conditions. It was concluded that the presence of carrageenan can significantly reduce the amount of NaCl added without detrimental effects on texture and sensory properties.[78]

7.5.1.4 Flavor Perception

Carrageenan can influence the flavor of processed products. At levels of 0.1 to 0.5% (w/w), λ-carrageenan suppressed the release of aroma compounds, including aldehydes, esters, ketones, and alcohols, in thickened viscous solutions containing 10% sucrose. The extent of suppression was dependent on the physicochemical characteristics of the aroma compounds, with the largest effect occurring for the most volatile compounds.[84,85] Release of a sweet flavor from the food to the human papillae is affected by the diffusion of the sweetener throughout the food. It was observed that κ-carrageenan enhanced mean diffusion constants for sucrose and aspartame in soft gels.[86] Carrageenan enhanced flavor in a formulation containing a mixture of spices, hydrolyzed vegetable protein, and salt that has been commercially adopted.[87]

7.5.1.5 Fiber Fortification

Despite the recognized benefits of dietary fiber (see Chapter 2), the intake of fiber around the world is far from adequate. Carrageenan at a concentration of 0.5 to 0.7% can be used to increase the fiber content of such low-fiber foods as fishery products, in addition to improving their viscosity and texture. The addition of carrageenan offers the seafood industry opportunities to manufacture a wide range of products such as fiber-rich salmon rolls, low-fat fish pâté, and fish burgers.[31]

7.5.1.6 Antioxidant Activity

Oligosaccharides from κ-carrageenan can exhibit significant antioxidant activities.[88] Nevertheless, for economic reasons, it is preferable to use the whole seaweed extracts to impart antioxidant benefits rather than the carrageenan oligosaccharides (see Section 7.2).

7.5.1.7 Antimicrobial Properties

Carrageenans (and other hydrocolloids such as pectins, xanthan gums, acacia gums, and agars) are anionic hydrocolloids. Complexes of these gums with cationic preservatives such as lauric acid have been reported to possess good antimicrobial properties. Such compounds are stable and can be stored under ambient temperature and humidity conditions for prolonged periods.[89] An interesting application of κ-carrageenan is for the control of pathogens in poultry and meat products. *Salmonella typhimurium* is a pathogen usually contaminating poultry. Similarly, *Escherichia coli* O157:H7 strain is another potent pathogen that bonds to the type I collagen of meat and poultry, and κ-carrageenan was found to almost completely prevent contamination of the poultry carcasses by the pathogen. This observation could aid the development of new strategies to prevent pathogen contamination.[90]

7.5.1.8 Antibrowning Activity

Carrageenans have been reported to exhibit antibrowning activity in apple juice and dried apples; they act synergistically with citric acid to inhibit browning. A combination of 0.1% of any of the carrageenans (κ, ι, or λ) and 0.5% citric acid was able to inhibit browning of unpasteurized apple juice containing 0.1% sodium benzoate for up to 3 months at 3°C.[91]

7.5.2 Applications of Carrageenans in Food Product Development

Different types of carrageenans have been incorporated in foods to benefit from one or more of their functional advantages. Their various applications for food product development are discussed below (see Table 7.8).

TABLE 7.8

Applications of Carrageenans in Food Product Development

Products	Action	Refs.
Bakery products	Carrageenan enhances loaf volume and water absorption and improves crumb grain score.	Kohajdová and Karovicová[49]; Ward and Andon[105]
Fishery products (e.g., surmi and other fish meat, novel fish products such as fish burgers and sausages)	Carrageenan and alginate enhance cooking yield, hardness, bind strength, texture, and fiber content.	García-García and Totosaus[78]; Karim and Rajiv[123]
Meat products (e.g., turkey, restructured beef products, low-fat meatballs, beef burgers)	Carrageenan increases yield; improves visual appearance, sliceability, and rigidity; decreases expressible juice; and enhances storage stability.	Bylaite et al.[84]; Trius and Sebranek[115]
Vegetable products	Carrageenan reduces or replaces pectin in jams and jellies and improves low-sugar products and texture.	Hamza-Chaftai[127]
Dairy products	Carrageenan (also agar and alginates) acts as a stabilizing, thickening, and gelling agent; κ-carrageenan–casein interactions stabilize ice cream; and carrageenan improves viscosity of goat's milk.	Hansen[104]
Flavored soy milk	ι-Carrageenan increases viscosity and sensory values.	Wang et al.[125]
Fruit juices	κ-, ι-, or λ-Carrageenan, alone or with citric acid, inhibits browning.	Tong and Hicks[91]; Hamza-Chaftai[127]
Wine and beer	Carrageenan and alginic acid clarify wine and provide colloidal stabilization in beer.	Cabello-Pasini et al.[129]
Reduced-sodium foods	Carrageenan maintains texture when sodium is replaced by potassium.	Tong and Hicks[91]
Parotta	Carrageenan improves textural characteristics.	Smitha et al.[124]
Novel food products involving carrageenan and food component interactions	κ-Carrageenan increases surface hydrophobicity and the oil-binding properties of proteins; κ-carrageenan–ovalbumin complexes have applications in food technology.	Chidanandaiah et al.[65]; Venugopal[68]

7.5.2.1 Dairy Products

Table 7.9 lists the typical dairy applications of carrageenans. In milk-based products, where gelation or structural viscosity is required, carrageenan is preferred over other gums for functional and economic reasons. Commercially available, ready-to-eat, milk-based desserts offer a wide variety of textures, flavors, and appearances. Hydrocolloids are used as food additives in many of these products to optimize their properties. Dairy desserts usually incorporate starch and carrageenan. Starch provides body and mouth feel to dairy desserts, and carrageenans provide the desired texture. In addition, carrageenans also function as emulsifiers and stabilizers in place of eggs, flour, and lecithin. Carrageenans impart smoothness and a sensation of richness to cheeses, ice cream, and eggless milk puddings. The polysaccharide also prevents separation of fat and syneresis and stabilizes casein against interactions with calcium ions. Carrageenans adhere to casein micelle surfaces and prevent bulk phase separation of the protein.

Carrageenan gels do not require refrigeration because they do not melt at room temperature. Usually, less than 0.5% (w/v) carrageenan is used. The thickening effect of κ-carrageenan in milk is 5 to 10 times greater than it is in water. The hydrocolloid helps maintain the structure of milk products after shearing. At a concentration of 0.025%, a weak thixotropic gel is formed in milk via interaction of κ-carrageenan with κ-casein micelles (i.e.,

TABLE 7.9

Typical Dairy Applications of Carrageenans

Product	Function	Product	Use Level (%)
Milk gels			
Cooked flans or custards	Gelation	K, K+I	0.20–0.30
Cooked prepared custards	Thickening	—	—
Cooked prepared custards with TSPP	Gelation	K, I, L	0.20–0.30
Pudding and pie fillings			
Dry mix cooked with milk	Level starch gelatinization	K	0.10–0.20
Ready-to-eat	Syneresis control, bodying agent	I	0.10–0.20
Whipped products	Stabilize overrun	L	0.05–0.15
Aerosol whipped cream	Stabilize overrun and emulsification	K	0.02–0.05
Cold prepared milks			
Instant breakfast mixes	Suspension, bodying agent	L	0.10–0.20
Shakes	Suspension, bodying agent, stabilize overrun	L	0.10–0.20

Abbreviations: K, κ-carrageenan; I, ι-carrageenan; L, λ-carrageenan; TSPP, tetrasodium pyrophosphate.

Source: Adapted from Rudolph, B., in *Marine & Freshwater Products Handbook*, Flick, G.G. and Martin, R.E., Eds., VCH Publishing, Lancaster, 2000, pp. 515–530. With permission from VCH Publishing.

the phenomenon of milk reactivity). Because it has higher gel strength in milk compared to other carrageenan types, κ-carrageenan is widely used in gelled milk products, such as ready-to-eat desserts and in powder preparations or puddings. λ-Carrageenan has the ability to disperse in milk at 5 to 10°C and thicken it without any salts; to incorporate this polysaccharide, the preparations are blended to form a variety of gels, which may be clear, turbid, heat stable, or thermally reversible.[92]

Chocolate milks are a well-established application of carrageenan. Stabilization of cocoa particles and fat suspensions in chocolate milk is obtained through the addition of 0.02 to 0.03% κ-carrageenan, which helps prevent fat separation in the milk. Carrageenans at this level can also be used to prevent the separation of whey from cream when ice cream thaws. Such stabilizing interactions are also important in producing evaporated milk, infant formulas, and whipped cream that must be stable to freeze–thaw cycles. Carrageenan gives evaporated skim milk the consistency of cream; the gel strength is influenced by the concentration of milk solids, which can vary from 2.5 to 20% (w/w); κ-carrageenan, from 0.1 to 0.4% (w/w); and cations.

λ-Carrageenan is nongelling and is used as a stabilizer and emulsifier in such products as whipped cream, ice cream, instant breakfast drinks, milk shakes, nondairy coffee creams, and dry-mix sauces.[95] Unlike κ-carrageenan and ι-carrageenan, λ-carrageenan is insensitive to the K^+ and Ca^{2+} present in milk. In these products, the carrageenans are usually blended with dextrose for uniform performance.[93,94]

Both κ-carrageenan and ι-carrageenan are commonly used to prepare dessert gels, whipped toppings, instant whipped desserts, and eggless custards and flavors. ι-Carrageenan is used as a functional ingredient for stabilization, thickening, and gelation in the preparation of products such as milk gels and ice cream. It has a significant reactivity with milk proteins and forms elastic, syneresis-free, thermally reversible gels in milk that are stable to repeated freeze–thaw cycles. In the presence of starches, however, ι-carrageenan exhibits syneresis. Gelation of the carrageenan can be delayed by stirring, even below the gelling temperature. The gel is formed in milk by decreasing the temperature from 60°C to 10°C. The gelation of ι-carrageenan is closely related to the helix–coil transition it undergoes at approximately 48°C in milk. The addition of κ-carrageenan, sodium alginate, and xanthan gum significantly reinforced the shear thinning behavior of ice cream, which was attributed to the gelation phenomenon. Sodium alginate had a better stabilizing effect, improving the textural quality and acceptance of the ice cream even after 16 weeks of storage, and κ-carrageenan contributed to this cryoprotection.[96]

Carrageenan with locust bean gum can give custards a smooth, consistent texture. The blend allows complete replacement of egg in instant custards such as flan. LBG and λ-carrageenan at 0.1% concentrations each reduced changes in the elastic properties of whipped dairy cream during freezing, suggesting a cryoprotective effect of the hydrocolloid in whipped cream.[97] Intensely heated desserts, particularly ultra-high-temperature (UHT)-treated desserts with a long shelf-life, dominate the European market of ready-to-eat dairy desserts. κ-Carrageenans are used to prepare these desserts, which usually contain skim milk powder, maize starch, sucrose, and water. A Tetra Pak indirect UHT pilot plant has been used to prepare dairy desserts containing κ-carrageenan, skimmed milk powder (SMP), acetyl-substituted waxy maize starch, sucrose, and water. The microstructure of dairy desserts required to provide specific textures is complex, involving various interactions between ι-carrageenan, starch, and milk proteins. The degree of shearing during production influences the starch granules and the interaction between carrageenan and casein micelles during subsequent cooling; also, it affects dessert rigidity, although gel strength is not significantly affected, as determined by a large deformation texture profile analysis. Application of the severe heat treatment associated with UHT influences gel structure due to extensive whey protein denaturation and subsequent complexation with casein micelles.[98,99]

Processed and imitation cheeses together comprise another major sector in the dairy industry. Depending on the type of processed cheese, manufacturing conditions vary in terms of temperature (75 to 140°C), duration, mechanical action (shear stress and rotation speed), and equipment (blades, bi-screw, scraped surface exchanger). The main challenges in cheese manufacture are achieving the correct viscosity and consistency during processing as well as a firm texture, thermostability, avoidance of fat exudation, and maximizing product yield when the block is grated or sliced. Incorporation of texturizers optimizes these properties. Also, texturizers regulate viscosity during processing, compensate for lost gel strength, improve sliceability, and optimize grating consistency. Melting behavior and spreadability can also be modified by using the correct additive or combination of texturants. The most suitable texturizers for cheese manufacture include carrageenans (and alginates) in addition to certain starches. ι-Carrageenan at concentrations up to 0.25% (w/w) was effective in preventing syneresis and increasing the rigidity of processed cheeses in comparison with κ-carrageenan. The effect of carrageenan was also influenced by the fat content of the cheese.[100,101]

The use of carrageenan along with other colloids such as LBG and carboxymethylcellulose (CMC) is very important in the stabilization of pasteurized, chilled, stable, and UHT-treated ice cream premixes. The effect of carrageenan is manifested in the freeze-concentrated aqueous phase of deep-frozen ice cream, resulting in firm cohesive gelation. The potential use of hybrid κ-/ι-carrageenans from the underexploited Portuguese seaweed *Mastocarpus stellatus* as a natural thickening and gelling agent for food applications has been recognized.[102] Multilayer emulsions containing carrageenan or other biopolymer provide better stability against droplet aggregation than single-layer emulsions under the same environmental conditions of pH, ionic concentrations, temperature, etc.[81,103,104] It may be mentioned that an ice structuring protein (ISP) developed recently has received EU approval, although it is expected that ISP will have little impact on the use of hydrocolloids, including carrageenan, in frozen dairy desserts.

TABLE 7.10

Advantages of Food Product Coatings

Crispy texture and appealing color and flavor
Enhanced nutritional quality
Moisture barrier during frozen storage and microwave reheating
Seal food and prevent the loss of natural juices
Structural reinforcement of the substrate
Increased bulk of the substrate, thus reduced finished product costs
Overall improvement of acceptability

7.5.2.2 Bakery Products

Incorporation of carrageenan improves batter quality and the properties of dough and pastes, including higher water absorption by doughs. A 10% replacement of wheat bran by carrageenan in wheat bran breads enhanced loaf volume and water absorption and improved crumb grain scores compared to these breads in the absence of the hydrocolloid. Also, the polysaccharide allows a greater amount of milk powder to be incorporated in bakery products. Breads made from such batters are free from shrinkage, which results when nonfat dry milk is added.[105] At the 1% level, carrageenan reduced the amount of freezable water and increased loaf volume, but there was a detrimental effect if the dough was frozen prior to baking. Breads prepared from this frozen dough had an inferior appearance and crumb hardness, and proof time was increased.[106]

Breading and battering have been extensively employed for poultry and fish products. Such battered, prefried, frozen products represent an extensive sector of the ready-to-eat market.[68] Table 7.10 lists the advantages of coated products, and Table 7.11 shows the various ingredients used in the development of coated products. The unit operations in the development of coated products are portioning/forming, predusting, battering, breading, flash frying, freezing, packaging, and storage, all of which are automated (see Figure 7.1). The commonly used ingredients for coatings are polysaccharides, proteins, fat/hydrogenated oils, seasonings, and water. Starch-based predust is generally used. The most important characteristic property of the batter is its viscosity, which affects the pickup and quality of the adhering batter, the handling properties of the battered product, its appearance, and final texture. Viscosity of the batter also determines its performance during frying and the quality of the finished products. Gums (usually guar or xanthan gum) are added to improve batter adhesion to the product through thermal gelation.

Batter containing carrageenan is excellent for coating chicken prior to frying. Other ingredients that have parallel or complementary effects are methylcellulose, hydroxypropylmethylcellulose, carboxymethylcellulose, and alginates. These compounds also provide improved viscosity, suspension

TABLE 7.11

Major Ingredients and Their Functions in Coated Products

Class of Ingredients	Components	Function in the Product
Polysaccharides	Wheat flour, corn flour, starch and modified starch, gums	Improve viscosity, emulsifying and foaming capacity, texture, and shelf life
Proteins	Milk powder, milk protein fraction, egg albumin, seed proteins, single cell proteins	Improve water absorption capacity of the flour and thus increase viscosity of the system
Fat/hydrogenated oils	Triglycerides, fatty acids	Add texture and flavor
Seasonings	Sugar, salt, spices	Enhance plasticizing effect and flavor and impart antioxidant and antibacterial properties
Leavening agents	Sodium bicarbonate, tartaric acid	Release carbon dioxide in tempura batters
Gums	Carrageenan, xanthan, gum arabic, etc.	Improve texture, viscosity, water-holding capacity, fiber content
Water	—	Gelatinize starch, hydrate proteins, improve batter viscosity

Source: Adapted from Fiszman, S.M. and Salvador, A., *Trends Food Sci. Technol.*, 14, 399, 2003. With permission from Elsevier/Rightslink.

characteristics, and emulsifying capacity, in addition to controlling the forms, texture, and shelf-life of the coated products through their interaction with proteins and lipids.[107,108] Regular consumption of fried products with significant fat levels, however, can have adverse effects on health.[108] Recent research has shown that the incorporation of a small amount of hydrocolloids (usually 1% of the formulated dry weight of the batter) can reduce oil absorption during frying due to the gelling ability of hydrocolloids and their hydrophilic nature.[107,109,110]

7.5.2.3 Meat Products

Meat products are characterized by their typical texture. The myofibrillar proteins of meat hold appreciable amounts of water when mixed with salt and polyphosphate and myosin, the principal muscle protein, is solubilized. This allows expansion of the myofibrillar lattice, thus improving water retention characteristics. Cooking loss is an important quality parameter of meat and meat products because it adversely affects the final weight of the product and its perceived juiciness and texture. Poultry processors are concerned about the loss of water during cooking and loss in texture and eating quality of the products. Carrageenan is used in meat products to improve firmness and color and to reduce cooking losses. Salt usually present in meat products is beneficial for the gelation of carrageenan prior to its interaction with

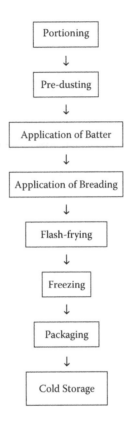

FIGURE 7.1
Process flowchart for the production of coated products.

meat proteins. There is a synergistic effect between meat proteins and carrageenan. Brine containing salt, phosphate, and carrageenan can be injected into the muscle of the meat. As the meat cooks, the carrageenan binds water within the poultry muscle, thereby reducing cooking losses with simultaneous improvement of texture and tenderness. In the poultry industry, carrageenan is also used to make restructured products from the meat trimmings, where the polysaccharide functions as an efficient binder. Such restructured products include turkey, rolls, chicken rolls, sausages, and cutlets. The addition of carrageenan markedly decreased purge loss and improved the texture of turkey breast rolls.[111–114]

The polysaccharide can also increase the gel strength and water-holding capacity of cooked ham-based products containing salt-soluble meat protein isolate and sodium and potassium chloride, at a pH value of 6.2. Both κ-carrageenan and ι-carrageenan could be effective in this respect.[29] The addition of κ-carrageenan and ι-carrageenan increased the water-holding ability, rigidity at 70°C, force to fracture, and true shear strain of meat batters.[115,116] The addition of these carrageenans at 1% (w/w) increased the water-holding ability, rigidity at 70°C, force to fracture, and true shear strain of structured lean beef rolls containing 4 to 5% fat, 33% water, 1 to 3% sodium chloride, and 0.35% sodium tripolyphosphate, cooked to 63°, 73°, and 83°C. Carrageenan also reduced purge of vacuum-packaged slices during refrigerated storage.[117] The addition of ι-carrageenan or starch increased the water binding of meat batters subjected to a combination of high pressure and temperature (200 MPa and 400 MPa, 70°C), although pressurizing caused a general decrease in color parameters. Both additives caused an increase in hardness and chewiness.[118]

7.5.2.4 Fishery Products

One of the popular gel products of fish muscle is surimi, a concentrate of fish myofibrillar proteins, which is used to develop restructured products having acceptable texture.[68] All three types of carrageenans (κ, ι, and λ), especially ι-carrageenan, improved the water-holding ability of the cooked gels of surimi prepared from Atlantic pollock (*Pollachius virens*) over three freeze–thaw cycles.[119] Gelation of washed blue whiting (*Micromesistius poutassou*) mince in the presence of either 0.5% κ-carrageenan or ι-carrageenan or sodium alginate and cations was evaluated. Mixtures were heat set (37°C for 30 minutes and 90°C for 50 minutes), cooled, and held for 24 hours at 4°C before characterization of the gel in terms of folding resistance, puncture properties, texture and stress relaxation, color, and water-holding capacity. NaCl mainly altered breaking strength in the gels made with ι-carrageenan. The salt also affected the yellowness value in the gels made with κ-carrageenan and adhesiveness of gels containing sodium alginate. KCl affected adhesiveness in the gels made with ι-carrageenan; it had no apparent effect on the gels made with sodium alginate. The combined action of the salts influenced the folding test value in the gels made with alginate and breaking strength in the gels made with ι-carrageenan.[120]

High pressure is known to affect the gelation of fish muscle proteins. Gelation of fish proteins with ι-carrageenan at atmospheric pressure produced gels that were more adhesive, less cohesive, lighter, and more yellowish than pressure-induced gels. For gels with κ-carrageenan, the heat-induced sample was particularly hard and adhesive, with low cohesiveness, more yellowness, and lower water-holding capacity than pressure-induced gels.[67]

Battering and breading techniques have contributed significantly to the development of coated seafood products (e.g., fish sticks). Some of these products are prepared from finfish such as cod, haddock, pollock, perch, and catfish; crustacean sources include shrimp, crab, and crawfish. Fish fillets are generally prepared onboard fishing vessels, where they are soaked in dilute brine to improve their color, taste, and texture. The fillets are placed in large fillet-shaped molds, compressed with a low-pressure ram, and kept frozen until the coating process.

A major disadvantage of coated fishery products is the adsorption of oil during flash frying—as much as 15 to 30% of their weight. Consumer concerns regarding the health hazards linked with the consumption of oil, such as obesity and heart disease, affects the marketability of coated products. Process modifications are therefore being examined to minimize oil absorption during flash frying of coated products. Incorporating carrageenan has been shown to reduce oil uptake during the frying of fishery products.[107] Traditional fish burgers are made mostly with added starch but generally fail to meet consumer expectations of eating quality characteristics. Promising functional ingredients to modify the organoleptic properties of such products include milk protein, citrus pectin, and bovine gelatin. The quality of fish burgers could be improved by introducing carrageenan (and other ingredients) to the formulation, such as soy protein, frozen vegetables, and dried fruits, to enhance such properties of the product as fiber content.[31,121]

7.5.2.4.1 Fish Gelatin

Fish gelatin is a possible alternative to mammalian gelatin; however, drawbacks of fish gelatin are its low gel strength as well as low gelling and melting temperatures. To overcome these shortcomings, the fish gelatin can be at least partially replaced with marine polysaccharides such as carrageenans, which can provide improved gel strength and gelling and melting temperatures. Combining carrageenan with LBG, konjac flour, and starch is another alternative that provides a variety of melting and nonmelting gels and gel textures to meet consumer requirements. Complexes of fish gelatin and κ-carrageenan at 4°C can give turbidity-free mixed systems. Compression measurements revealed a considerable increase in Young's modulus when the mixed solutions were allowed to gel at 4°C.[122] Long-life refrigerated mousse desserts, based on carrageenan and pectin rather than gelatin, are suitable for vegetarians.[123] Unlike in Asia, however, carrageenan water gel desserts are not popular in the United States or Europe.

7.5.2.5 Vegetable Products

Carrageenan has the potential to replace pectin in jams and jellies, particularly low-calorie (low-sugar) types, as pectin is not an effective gelling agent due to the low sugar content. Carrageenan and LBG may be used in fruit gels that do not melt or soften at room temperature. Relishes retain water better when carrageenan is used in the recipe. Gels made from ι-carrageenan have the disadvantage of a high melting temperature, so they are not as smooth to eat as gelatin gels. In tropical countries, however, this is an advantage, as the products do not melt and do not require refrigeration. A further advantage is that they do not toughen on storage. Salad dressings and sauces can be stabilized by carrageenan. Low-oil or no-oil salad dressings use ι-carrageenan or κ-carrageenan to help suspend herbs and other flavorings and to provide the mouth feel that is expected from a normal salad dressing.

The low oil content of reduced-oil mayonnaise normally gives a thin product; additives are needed to thicken it and to stabilize the oil-in-water emulsion. A combination of carrageenan and xanthan gum is effective for the purpose; however, care must be taken to prevent degradation of carrageenan under acidic conditions, such as during the processing of acidic sauces. The problem can be addressed by adding the acidic ingredients after heating. Addition of carrageenan to the packing syrup prior to freezing fruit results in better quality of the fruits upon thawing. Flavored gels are made by boiling strongly flavored fruit or fruit juice in water two to three times the volume of the juice. When the boiling is stopped, carrageenan powder is added for gelling. Conventional fruit jellies are based on pectin and a high sugar content to help set the jelly. In a low- or non-calorie jelly, the pectin must be replaced, and mixtures of κ-carrageenan and ι-carrageenan have proved to be suitable. Addition of λ-carrageenan adds body and provides a pleasant mouth feel.

Sorbet is a creamy alternative to ice cream with no fat; a combination of κ-carrageenan and ι-carrageenan together with LBG or pectin gives sorbet a smooth texture.[27] Demand for parotta, a popular wheat product in India, is increasing. The product is characterized by specific extensibility, a viscous nature, and water retention. Adding 0.5% (w/w) carrageenan increased water absorption and viscosity but decreased extensibility of the product. Among the hydrocolloids examined, guar gum brought about the greatest improvement in the quality of parotta. Other hydrocolloids in decreasing order were hydroxypropylmethylcellulose (HPMC), xanthan, carrageenan, and gum arabic.[124] The use of soy-based products as alternatives to dairy products has attracted much attention recently. In addition to being lactose free and lower in fat content, soy-based products are also a good source of protein. Hydrocolloids can be added to soy milk to improve viscosity and shear stress. Flavored soy milks containing chocolate flavoring and ι-carrageenan have improved sensory attributes compared to plain soy milk. The products are stable for 1 month under refrigerated conditions.[125]

A soy-based cream cheese has been developed employing food hydrocolloids that interact with the soy proteins to provide acceptable texture. Various amounts of blended tofu, oil, salt, carrageenan, pectin, and maltodextrin were used to develop the cheese product, which has texture properties similar to those of commercial dairy cream cheese. Carrageenan and salt were used to impart the desired firmness, while maltodextrin added body. Chemical analysis revealed that the soy cream cheese had lower fat and higher protein and fiber contents than the commercial dairy cream cheese. Rheological studies at 25°C and 4°C showed that the developed products mimicked the texture of the commercial sample, although they had less elasticity. The products were also stable against syneresis and oil separation during storage of 20 days.[126]

Carrageenan at 0.35% (w/l) had a profound effect on the apparent viscosity of banana gellified milk. The prepared product had good texture and acceptable taste.[127] Banana slices were subject to a 3-minute dip in a solution containing 1% (w/v) calcium chloride, 0.75% (w/v) ascorbic acid, and 0.75%

(w/v) cysteine, combined with a carrageenan coating and a controlled atmosphere (3% O_2 + 10% CO_2). Physicochemical and microbiological qualities were evaluated during 5 days of storage at 5°C. Dip treatment combined with controlled atmosphere treatment prevented weight loss of the product and increased polyphenol oxidase activity during the 5 days of storage. Color, firmness, pH, titratable acidity, and total soluble solids values and total phenolic content demonstrated the smallest changes.[128]

ι-Carrageenan influenced the gelation and retrogradation behavior of starch as observed in a system consisting of corn starch and carrageenan. During heat-induced gelatinization of the starch at 0.5% (w/v), the addition of 0.2% ι-carrageenan lowered the incipient swelling temperature of the composite system with decreased peak viscosity, due to thermodynamic incompatibility between these two substances. The behavior, however, was dependent on the presence of salts. The effect of the polysaccharide to depress gel-like characters was the greatest in the presence of LiCl and NaCl. Salts modified the coil–helix transition and subsequent interhelical associations of ι-carrageenan, affecting the behaviors of the composite system.[60]

7.5.2.6 Brewing

Carrageenan and alginic acid are good wine stabilizers. The interaction of carrageenan and protein can be used in the clarification of beer, with the complex precipitating from the wort. Protein flocculation and the precipitation capacities of carrageenan and alginic acid were two times greater than those of agar. Whereas alginic acid absorbed protein at a maximum concentration of <50 mg/mL, the maximum adsorption and precipitation capacity for carrageenan was observed at a protein concentration of >400 mg/mL.[129] Carrageenan was added to malt syrup at 0 to 40 ppm. The mixture was heated to boiling for 10 minutes and then rested for 1 to 2 hours. The lowest turbidity of the product was noted in samples containing 30 ppm carrageenan, while the control sample without carrageenan was most turbid. The clarification was due to precipitation of proteins by the hydrocolloid.[130]

7.5.2.7 Miscellaneous Food-Related Applications

The ability of κ-carrageenan gel to sequester paralytic shellfish poison (PSP) is dependent on the gel surface area, interaction time, and polysaccharide concentration. The interaction was also found to be affected by high concentrations of monovalent cations, suggesting that the polysaccharide gel can be utilized as an agent to alleviate PSP intoxication.[131]

Extracts of selected seaweed such as *Ascophyllum*, *Sargassum*, and *Laminaria* are used as fertilizer for organic and nonpolluting farming. Super-Grow organic fertilizer (kelp extract) is made from these nutrient-rich seaweed species and is suitable for all field crops, vegetable and flower gardens, orchards, and turf grass. The extract promotes balanced growth of crops, boosts immunity and resistance, improves crop quality, and increases yield. The market for seaweed extracts as fertilizer is growing because of wider recognition of the usefulness of these products and the increasing popularity of organic farming, where they are especially effective.[132]

Seaweed extracts rich in λ-carrageenan have been reported to elicit an array of plant defense responses, possibly because of the high sulfate content. Their effect on the tobacco parasite *Phytophthora parasitica* var. *nicotianae* was studied. Carrageenans efficiently induced signaling and defense gene expression in the tobacco leaves. Defense genes encoding sesquiterpene cylase, chitinase, and proteinase inhibitor were induced locally, and the signaling pathways mediated by ethylene and salicylic acid were triggered. Some effects lasted for at least a week. The result suggests that the seaweed extract has the potential to protect against certain plant diseases.[133]

7.6 Furcellaran

Furcellaran forms thermally reversible aqueous gels by a mechanism involving double-helix formation, similar to κ-carrageenan. The gelling is influenced by the cations present; K^+, NH_4^+, and Cs^+ form very stable gels, whereas Na^+ prevents gel formation. The addition of sugar affects the gel texture, which changes from a brittle to an elastic texture. The gels retain good stability against food-grade acids. Furcellaran with milk provides good gels and therefore is used in puddings. It is also suitable for cake fillings and icings. Furcellaran has the advantage over pectin in marmalades, as it allows stable gels at sugar concentrations even below 50 to 60%. The required concentration of furcellaran is 0.2 to 0.5%, depending on sugar content and required gel strength. Furcellaran is also used in processed meat products, such as spreadable meats, pastes, and pastry fillings. It facilitates protein precipitation and hence clarifies the drink.[134]

7.7 Fucoidan and Laminarin

Fucoidans and laminarin are particularly important for their health benefits rather than their food applications. Fucoidan is a natural antioxidant and has great potential for preventing free-radical-mediated diseases. Because of their antioxidant properties, fucoidan and laminarin have been found useful for controlling lipid oxidation in muscle foods such as pork patties. Gamma irradiation is known to enhance lipid oxidation of muscle foods, the extent of oxidation being dependent on the irradiation dose. Fucoidan can control radiation-induced lipid oxidation in meat products. The Hunter ($L^*a^*b^*$) color values of pork patty with fucoidan added improved significantly with an increase in irradiation dose. The hardness profiles of patties containing fucoidan and laminarin decreased, and the amount of water in the patties increased. Further, the polysaccharides in combination were also effective in controlling microbial growth in the product.[135] The antiinflammatory, anti-angiogenic, anticoagulant, and antiadhesive properties of fucoidans have also been well recognized. Also, fucoidans are reported to have antitumor, antimutagenic, anticomplementary, antiviral, immunomodulating, hypogly-cemic, and hypolipidemic properties (see Chapter 11).

7.8 Ulvan

Ulvan is a polysaccharide from *Ulva*, commonly referred as sea lettuce. This seaweed, one of the most commonly consumed macroalgae, contains 16.5% water-soluble and 13.3% insoluble dietary fibers, consisting mainly of ulvan. It is not fermented by colonic bacteria, because of its particular chemical structure; consequently, dietary fibers from sea lettuce could be expected to act mainly as bulking agents with little effect on nutrient metabolism.[136]

7.9 Floridean Starch from Red Seaweed

Floridean starch, isolated from red seaweed, exhibits low gelatinization temperature, low viscosity, high clarity, and little or no retrogradation upon repetitive freeze–thaw cycles. The low gelatinization and pasting tempera-tures and high clarity of Floridean starch can be attributed to the absence of amylose and to the relatively short average chain length of 18 glucose units and higher degree of branching frequency. The rheological properties of Floridean starch make it suitable for various applications, such as in instant noodles and deep-frozen food.[137]

7.10 Microalgae

Marine microalgae are being hailed as a new "super food" because they are an almost unlimited, natural source of novel functional food ingredients, as well as bioactive compounds. The advantages of microalgae are their huge diversity and being able to cultivate them under controlled conditions.[138]

Marine microalgae are rich in many bioavailable nutrients. Some of the important well-studied microalgae include the species of *Chlorella*, *Spirulina*, and *Dunaliella*. *Chlorella* is a genus of single-celled green algae belonging to the phylum Chlorophyta. The chloroplasts of this alga contain the green photosynthetic pigments chlorophyll *a* and chlorophyll *b*. *Chlorella* multiplies rapidly through photosynthesis, requiring only CO_2, water, sunlight, and minerals for growth. Its high photosynthetic activity and level of proteins make *Chlorella* an attractive potential food. *Spirulina* is a microscopic, unicellular, photosynthetic, blue–green alga. It is a thallophyte (i.e., it has no clear distinction between leaf, stem, and root). This alga occurs naturally in warm, alkaline, salty, brackish lakes. Its color is derived from the green pigment of chlorophyll and from the blue color of a protein called phycocyanin. The types of *Spirulina* used as human and animal food supplements are the two species of cyanobacteria: *S. maxima* and *S. platensis*. The genus *Dunaliella* includes halotolerant, unicellular, motile green algae with exceptional morphological and physiological properties belonging to the family Chlorophyceae. These algae grow in salt concentrations as high as 1.5 *M*. They are devoid of rigid cell walls and have a single, large, cup-shaped chlorophast.[139]

Another marine alga, *Nannochloropsis*, contains 37.6% (w/w) available carbohydrates, 28.8% crude protein, and 18.4% total lipids. The mineral contents of 100 g of dry biomass were Ca, 972 mg; K, 533 mg; Na, 659 mg; Mg, 316 mg; Zn, 103 mg; Fe, 136 mg; Mn, 3.4 mg; Cu, 35.0 mg; Ni, 0.22 mg; and Co, <0.1 mg. It also contained significant amounts of unsaturated fatty acids. Under cultivation conditions, the nutrient composition of the biomass was highly influenced by residence time in the photobioreactor. The biomass harvested for short residence times was richer in protein and eicosapentaenoic acid than biomass harvested for high residence time.[140]

7.10.1 Microalgal Polysaccharides

Cyanobacterial extracellular polysaccharides have found some food applications. The red unicellular rhodophyte microalga *Porphyridium* produces a polysaccharide with potential as a thickening agent and food additive because of its high viscosity over a wide range of pH, temperature, and salinity. The polysaccharide also exhibits a variety of biological activities with potential for medical and cosmetic uses.[141,142]

7.11 Commercial Aspects

The global seaweed industry has an estimated total annual production of about 8 mt of wet algae (harvested and cultivated), valued at approximately US$6 billion The food and pharmaceutical products isolated from the algae are worth about US$5 billion.[143,144] China is the largest producer of edible seaweed, harvesting about 5 mt, mostly for kombu, which is produced from the brown seaweed *Laminaria japonica*. The Republic of Korea grows about 1 mt of three different species, about half of which is used to make wakame, which is produced from a different brown seaweed, *Undaria pinnatifida*, The most valuable seaweed, perhaps, is nori, around 600,000 t of which are cultivated in the coastal waters of Japan at a value of over US$2 billion. Japan exports seaweed to countries including Taiwan, Netherlands, Vietnam, China, and the United States.[29]

Alginate, carrageenan, and agar are the most important seaweed polysaccharides used in many industries, including food, pharmaceuticals, and biotechnology. Production of these hydrocolloids has shown significant growth in the last decade: from US$644 million in 1999 to US$1020 million in 2009.[145] In 2009, production totals (on a dry weight basis) of agar, alginate, and carrageenan were 72,000, 95,000, and 169,000 t, respectively.[145] Annual food uses of agar, alginate, and carrageenan were approximately 10,000, 14,000, and 44,000 t, valued at US$220, 300, and 550 million, respectively (D. Seisun, IMR Int., pers. comm.). The main producers of agar are Japan, United States, Thailand, Indonesia, India, and Russia, with more than 50% of the production being used for food and pharmaceutical purposes. Agar is produced primarily from two types of red seaweed, *Gelidium* (particularly *G. chilenis*) and *Gracilaria* (*G. gracilis*).[27,146]

Currently, alginate is produced from such seaweed as *Ascophyllum*, *Durvillaea*, *Laminaria*, *Lessonia*, *Sargassum*, and *Macrocystis*, but primarily *M. pyrifera*, *Laminaria hyperborea*, and *Laminaria japonica*.[147] The abundant supply of *M. pyrifera* along the California coastline is harvested for alginate production. About 30% of total production is used by the food industry, the rest being used in industrial, pharmaceutical, and dental applications. Sodium and calcium alginates are used as additives for food and medicine, textiles, printing, dyeing, and papermaking. The fiber content of alginate is of significant interest for the development of fiber-rich food products. At present, the only derivative of alginate that has commercial application is propylene glycol alginate.[149]

Commercial carrageenans have been designated as either refined or natural grade. Refined carrageenan is the original carrageenan, and until the early 1980s it was simply referred to as carrageenan. It is now sometimes called filtered carrageenan. The refined material is made by dissolving the carrageenan in alkali before purifying it. Natural carrageenans are undissolved, with the impurities extracted. Apart from *Chondrus crispus* (Irish moss), carrageenan is also produced from two species, *Kappaphycus alvarezii* and *Eucheuma denticulatum*, cultivated in the Philippines.[27] Numerous products incorporating carrageenan are commercially available.[143,147] Carrageenan as an ingredient is identified as a "natural food stabilizer" by some food companies. A recent survey of supermarkets, health-food stores, and food manufacturers identified 198 carrageenan-containing food products. A demographic and food frequency data analysis revealed that most consumers of these products were over 30 years old, college educated, and female.[148] The carrageenan market has been somewhat unstable recently, due to the introduction of less expensive, lower-refined grades such as processed eucheuma seaweed (PES), which can compete efficiently with the traditional purified grades of carrageenan in applications where gel clarity is not important.[149] Fucoidans are now being marketed as nutraceuticals ("miracle drugs") and as food supplements because of its significant biological functions.[147]

Textural modification of food products is one of the most important uses of seaweed polysaccharides. It has been found that, in general, the potential of various seaweed polysaccharides to modify gel texture from firm and brittle to soft vary: agar > κ-carrageenan > high-G alginate > high-M alginate > ι-carrageenan.[29] A commercial food-grade stabilizer developed by Gum Technology Corporation (Tucson, AZ) consists of carrageenan, xanthan, and gum arabic and is claimed to reduce ice crystal growth, to control syneresis, to emulsify oil in a number of food products. The product has been approved by the U.S. Food and Drug Administration. A blend consisting of locust bean gum, carrageenan, and potassium citrate is being used as a stabilizer for clear dessert gels but it can also be used in prepared fruit applications. Table 7.12 shows functional claims made by agar, alginate, and carrageenan in commercial food products, and Table 7.13 gives recent food applications for

TABLE 7.12

Functional Claims by Commercial Seaweed Polysaccharide Preparations in Various Food Products

Polysaccharide	Functional Claim
Agar	Forms gel
	Adds texture
	Reduces sugar bloom
Alginate	Forms gel
	Fat replacer
	Adds smooth texture
	Creates creamy mouth feel
	Dairy fat replacer
Carrageenan	Forms heat stable gel
	Controls syneresis
	Adds to mouth feel
	Adds viscosity
	Stabilization
	Moisture retention
	Enhances texture
	Forms creamy and smooth gel

Note: Product categories include baked goods, beverages, confectionery, dairy, desserts, dressings and dips, fried foods, frozen foods, meat analogs, meat products, pasta, restructured products, sauces and gravies, snack foods, soups.

TABLE 7.13

Uses of Agar, Alginate, and Carrageenan in 2009

Market Segment	Agar (t)	Alginate (t)	Carrageenans (t)
Confectionery/water gels	2800	—	8500
Baking	2300	—	—
Retail (gel powder)	1200	—	—
Meat products	200	—	18,500
Other (e.g., dairy)	300	—	14,000
Bacto/pharma/agarose	700	—	—
Technical grades	—	11,000	—
Animal feed/pet food	—	4000	5000
Propylene glycol alginate (PGA)	—	2000	—
Food/pharmaceuticals	—	8000	—
Toothpaste	—	—	2000
Other	—	—	2000
Grand total	7500	25,000	50,000

Source: Adapted from Bixler, H.J. and Porse, H., *J. Appl. Phycol.*, 2010 (DOI 10.1007/s10811-010-9529-3).

agar, alginate, and carrageenans. The increasing applications of seaweed and seaweed products in food products and rising international trade call for regularization and harmonization of standards of these products.[149–151]

References

1. Tressler, D. K. and Lemon, J. M. W., The red algae of economic importance, in *Marine Products of Commerce*, Tressler, D. K. and Lemon, J. M. W., Eds., Reinhold, New York, 1951, chap. 5.
2. Chapman, V. J. and Chapman, D. J., Sea vegetables: algae as food for man, in *Seaweed and Their Uses*, 3rd ed., Chapman & Hall, London, 1980, pp. 62–97.
3. Teas, J. et al., Algae: a poor man's HAART, *Med. Hypotheses*, 62, 507, 2004.
4. Mabeau, S. and Fleurence, J., Seaweed in food products: biochemical and nutritional aspects, *Trends Food Sci. Technol.*, 4, 103, 1993.
5. Ito, K. and Hori, K., Seaweed: chemical composition and potential food uses, *Food Rev. Int.*, 5, 101, 1989.
6. Kumar, C. S. et al., Seaweed as a source of nutritionally beneficial compounds: a review, *J. Food Sci. Technol.*, 45, 1, 2008.
7. Lahaye, M., Marine algae as sources of fibres: determination of soluble and insoluble dietary fibre contents in some sea vegetables, *J. Sci. Food Agric.*, 54, 587, 1991.
8. Jimenez-Escrig, A. and Muniz, F.J., Dietary fiber from edible seaweed: chemical structure, physicochemical properties, and effects on cholesterol metabolism, *Nutr. Res.*, 20, 585, 2000.

9. Jefferson, A. and Cowbrough, K., Carbohydrates and fibre: a review of functionality in health and wellbeing, in *Food Science and Technology Bulletin: Functional Foods*, Vol. 2, IFIS Publishing, Reading, U.K., 2006, pp. 31–37.

10. Venugopal, V., *Marine Products for Healthcare*, CRC Press, Boca Raton, 2008.

11. Athukorala, Y. et al., Potential antioxidant activity of some marine red alga, *Grateloupia filicina*, extracts, *J. Food Lipids*, 10, 251, 2003.

12. Heo, S.-J. et al., Antioxidant activities of enzymatic extracts from brown seaweed *Biores. Technol.*, 96, 1613, 2005.

13. Lim, S. N. et al., Evaluation of antioxidative activity of extracts from a brown seaweed, *Sargassum siliquastrum*, *J. Agric. Food Chem.*, 50, 3862, 2002.

14. Huang, H.-L. and Wang, B.-G., Antioxidant capacity and lipophilic content of seaweed collected from the Qingdao coastline, *J. Agric. Food Chem.*, 52, 4993, 2004.

15. Rajagopal, S. V., Antioxidants and free radical scavenging activity of brown algae of Visakhapatnam coast, *Asian J. Chem.*, 20, 5347, 2008.

16. Kuda, T. et al., Antioxidant properties of dried product of 'haba-nori,' an edible brown alga, *Petalonia binghamiae*, *Food Chem.*, 98, 545, 2006.

17. Yan, X. et al., Prevention of fish oil rancidity by phlorotannins from *Sargassum kjellmanianum*, *J. Appl. Phycol.*, 8, 201, 1996.

18. Nagar, T. and Yukimoto, T., Preparation and functional properties of beverages made from sea algae, *Food Chem.*, 81, 327, 2003.

19. Wang, P. et al., *In vitro* antioxidative activities of three marine oligosaccharides, *Nat. Prod. Res.*, 21, 646, 2007.

20. Kamenarska, Z. et al., Antibacterial, antiviral, and cytotoxic activities of some red and brown seaweed from the Black Sea, *Botanica Marina*, 52, 80, 2009.

21. Mandal, A. et al., Biopotentials of seaweed collected from southwest coast of India, *J. Mar. Sci. Technol.*, 17, 67, 2009.

22. Stirk, W. A. et al., Seasonal variation in antifungal, antibacterial and acetylcholinesterase activity in seven South African seaweeds, *J. Appl. Phycol.*, 19(3), 1573, 2007.

23. Melany, P. S. E., Jensen, P. P. R., and Fenical, W., Antimicrobial activities of extracts from tropical Atlantic marine plants against marine pathogens and saprophytes, *Mar. Biol.*, 149, 991, 2006.

24. Darcyvrillon, B., Nutritional aspects of developing uses of marine macroalgae for the human food industry, *Int. J. Food Sci. Nutr.*, 44, S23, 1993.

25. Noël Dierick, B., Ovyn, A., and De Smet, S., Effect of feeding intact brown seaweed *A. nodosum* on some digestive parameters and on iodine content in edible tissues in pigs, *J. Sci. Food Agric.*, 89, 584, 2009.

26. Carillo, S. et al., Potential use of seaweed in the laying hen ration to improve the quality of n-3 fatty acid enriched eggs, *J. Appl. Phycol.*, 20, 721, 2008.

27. McHugh, D.J., *A Guide to the Seaweed Industry*, FAO Fisheries Technical Paper. No. 441, Food and Agriculture Organization, Rome, 2003.

28. Anon., *Seaweed Benefits for Health and Slimming*, http://www.greenfootsteps.com/seaweed-benefits.html.

29. Rudolph, B., Red algae of economic significance, in *Marine & Freshwater Products Handbook*, Flick, G. G. and Martin, R. E., Eds., VCH Publishing, Lancaster, 2000, pp. 515–530.

30. Nussinovitch, A., *Hydrocolloid Applications: Gum Technology in the Food and Other Industries*, Blackie Academic, New York, 1997.

31. Borderías, A. J., Sánchez-Alonso, I., and Pérez-Mateos, M., New applications of fibers in foods: addition to fishery products, *Trends Food Sci. Technol.*, 16(10), 458, 2006.

32. Senthil, A. et al., Effect of using seaweed (*eucheuma*) powder on the quality of fish cutlet, *Int. J. Food Sci. Nutr.*, 56, 327, 2005.

33. Cofrades, S. et al., Influence of different types and proportions of added edible seaweed on characteristics of low-salt gel/emulsion meat systems, *Meat Sci.*, 79, 767, 2008.

34. Daniells, S., Simple, low-cost microparticles could replace fat, texturise, FoodNaviator.com, August 29, 2008.

35. Irish Seaweed, http://www.irishseaweed.eu/onlinestore.html.

36. Glicksman, M., Utilization of seaweed hydrocolloids in the food industry, *Hydrobiologia*, 151/152, 31, 1987.

37. Embuscado, M. E. and BeMiller, J. N., Characterization of processed *Eucheuma* seaweed [abstract], in *IFT Annual Meeting Book of Abstracts*, Institute of Food Technologists, Chicago, IL, 1996.

38. Chang, H.-C. and Wu, L.-C., Texture and quality properties of Chinese fresh egg noodles formulated with green seaweed (*Monostroma nitidum*) powder, *J. Food Sci.*, 73, S398, 2008.

39. Modifilan Seaweed Extract, *Modifilan: Pure Brown Seaweed Extract* (http://www.modifilan-seaweed-extract.com).

40. Wu-Liu, X. et al., Diet Food Formula for Overweight People and Diabetics, U.S. Patent 6472002B2, 2002.

41. Taylor, C., *The Seaweed Jelly-Diet Cookbook Guide*, Lulu.com, 2008 (http://seaweedjellydiet.com).

42. Ghosh, P. K., Central Salt and Marine Chemicals Research Institute, Bhavanagar, India, personal communication, May 21, 2007.

43. Armisen, R., Agar, in *Thickening and Gelling Agents for Food*, 2nd ed., Imeson, A., Ed., Blackie Academic, London, 1999, pp. 1–21.

44. Sikora, M., Juszczak, L., and Sady, M., Hydrocolloids in forming properties of cocoa syrups, *Int. J. Food Prop.*, 6, 213, 2003.

45. Mandala, I., Karabela, D., and Kostaropoulos, A., Physical properties of breads containing hydrocolloids stored at low temperature. I. Effect of chilling, *Food Hydrocoll.*, 21, 1397, 2007.

46. Gray, J. A. and BeMiller, J. N., Bread staling: molecular basis and control, *Comp. Rev. Food Sci. Food Safety*, 2, 1, 2003.

47. Rosell, C. M., Rojas, J. A., and De Barber, B. C., Influence of hydrocolloids on dough rheology and bread quality, *Food Hydrocoll.*, 15, 75, 2001.

48. Azizi, M. H. and Rao, G. V., Influence of selected surfactant gels and gums on dough rheological characteristics and quality. *J. Food Qual.*, 27, 320, 2004.

49. Kohajdová, Z. and Karovicová, J., Application of hydrocolloids as baking improvers, *Chem. Papers*, 63(1), 26, 2009.

50. Shi, X. and BeMiller, J. N., Effects of food gums on viscosities of starch suspensions during pasting, *Carb. Polym.*, 50, 7, 2007.

51. Anton, A. A. and Artfield, S. R., Hydrocolloids in gluten-free breads: a review, *Int. J. Food Sci. Nutr.*, 59, 11, 2008.

52. Gallagher, E., Gormley, T. R., and Arendt, E. K., Recent advances in the formulation of gluten-free cereal-based products, *Trends Food Sci. Technol.*, 15, 143, 2004.

53. Karim, A. A., Norziah, M. H., and Seow, C. C., Methods for the study of starch retrogradation, *Food Chem.*, 71, 9, 2000.
54. Maga, A. and Fapojuwo, O. O., Extrusion of corn grits containing various levels of hydrocolloids, *Int. J. Food Sci. Technol.*, 21, 61, 1986.
55. Gurkin, S., Hydrocolloids: ingredients that add flexibility to tortilla processing, *Cereal Food World*, 47, 41, 2003.
56. Maeda, H. et al., Effects of agar (kanten) diet on obese patients with impaired glucose tolerance and type 2 diabetes, *Diab. Obes. Metab.*, 7, 40, 2005.
57. Ross, K. A., Pyrak-Nolate, L. J., and Campanella, O. H., The effect of mixing conditions on the material properties of an agar gel: microstructural and macrostructural considerations, *Food Hydrocoll.*, 20, 79, 2006.
58. Meena, R., Prasad, K., and Siddhanta, A. K., Development of a stable hydrogel network based on agar–kappa-carrageenan blend cross-linked with genipin, *Food Hydrocoll.*, 23, 497, 2009.
59. Onsøyen, E., Alginates, in *Thickening and Gelling Agents for Food*, Imeson, A., Ed., Blackie Academic, London, 1997, pp. 22–44.
60. Guarda, A. et al., Different hydrocolloids as bread improvers and anti-staling agents, *Food Hydrocoll.*, 18, 241, 2004.
61. Gómez, M. et al., Functionality of different hydrocolloids on the quality and shelf-life of yellow layer cakes, *Food Hydrocoll.*, 21, 167, 2007.
62. Lee, S. et al., Physicochemical, textural and noodle-making properties of wheat dough containing alginate, *J. Text. Stud.*, 39, 393, 2009.
63. Kohajdová, Z. et al., Marine biotechnology for production of food ingredients, *Adv. Food Nutr. Res.*, 52, 237, 2007.
64. Weilin, K. and Keeton, J. T., Textural and physicochemical properties of low-fat, precooked ground beef patties containing carrageenan and sodium alginate, *J. Food Sci.*, 63, 571, 1998.
65. Chidanandaiah, K., Leshri, R. C., and Sanyal, M. K., Effect of sodium alginate coating with preservatives on the quality of meat patties during refrigerated storage, *J. Muscle Foods*, 3, 275, 2009.
66. Ponte, D. J. B., Roozen, J. P., and Pilnik, W., Effects of additions on the stability of frozen stored minced fillets of whiting. I. Various anionic hydrocolloids, *J. Food Qual.*, 8, 51, 1985.
67. Perez-Mateos, M. et al., Carrageenans and alginate effects on properties of combined pressure and temperature in fish mince gels, *Foods Hydrocoll.*, 16, 225, 2002.
68. Venugopal, V., *Seafood Processing: Adding Value Through Quick Freezing, Retortable Packaging and Cook-Chilling*, CRC Press, Boca Raton, FL, 2006.
69. Lian, P. Z., Lee, C. M., and Hufnagel, L., Physicochemical properties of frozen red hake (*Urophycis chuss*) mince as affected by cryoprotective ingredients, *J. Food Sci.*, 65, 1117, 2000.
70. Hershko, V. and Nussinovitch, A., Physical properties of alginate-coated onion (*Allium cepa*) skin, *Food Hydrocoll.*, 12, 115, 1998.
71. Terada, A., Hara H., and Mitsuoka, T., Effect of dietary alginate on the faecal microbiota and faecal metabolic activity in humans, *Microb. Ecol. Health Dis.*, 8, 259, 1995.
72. Draget, K. I. and Skjåk-Bræk, G., Alginates in relation to human health: from commodity to high cost products?, in *Gums and Stabilizers for the Food Industry 11*, Williams, P. A. and Phillips, G. O., Eds., Royal Society of Chemistry, Cambridge, U.K., 2002, pp. 356–364.

73. Brownlee, I. A. et al., Alginates as a source of dietary fiber, *Crit. Rev. Food Sci. Nutr.*, 45, 497, 2005.
74. Reifen, R., Edris, M., and Nussinovitch A., A novel, vitamin A-fortified, edible hydrocolloid sponge for children, *Food Hydrocoll.*, 12, 111, 1998.
75. Sugimoto, S. et al., Improvement of organoleptic quality of fermented soybean beverage by additions of propylene glycol alginate and calcium lactate, *J. Food Process. Pres.*, 5, 83, 1982.
76. Mollah, M. Z. I. et al., Effect of gamma irradiated sodium alginate on red amaranth (*Amaranthus cruentus* L.) as growth promoter, *Rad. Phys. Chem.*, 78, 61, 2009.
77. Soumya, R. and Ryan, A. L., Method of Preparing Food Products with Carrageenan, U.S. Patent 6663910B2, 2003.
78. García-García, E. and Totosaus, A., Low-fat sodium-reduced sausages: effect of the interaction between locust bean gum, potato starch and κ-carrageenan by a mixture design approach, *Meat Sci.*, 78, 406, 2008.
79. Ray, D. K. and Labuza, T. P., Characterization of the effect of solutes on the water-binding and gel strength properties of carrageenan, *J. Food Sci.*, 46, 786, 1981.
80. Sharma, B. D., The storage stability and textural, physico-chemical and sensory quality of low-fat ground pork patties with carrageenan as fat replacer, *Int. J. Food Sci. Technol.*, 39, 31, 2004.
81. Candogan, K. and Kolsarici, N., The effects of carrageenan and pectin on some quality characteristics of low-fat beef frankfurters, *Meat Sci.*, 64, 199, 2003.
82. Amako, D. E. N. and Xiong, Y. L., Effects of carrageenan on thermal stability of proteins from chicken thigh and breast muscles, *Food Res. Int.*, 34, 247, 2001.
83. Brummel, S. E. and Lee, K., Soluble hydrocolloids enable fat reduction in process cheese spreads, *J. Food Sci.*, 55, 1290, 1990.
84. Bylaite, E. et al., Influence of κ-carrageenan on the release of systematic series of volatile flavor compounds from viscous food model systems, *J. Agric. Food Chem.*, 52, 3542, 2004.
85. Juteau, A. et al., Flavor release from polysaccharide gels: different approaches for the determination of kinetic parameters, *Trends Food Sci. Technol.*, 15, 394, 2004.
86. Bayarri, S. et al., Diffusion of sucrose and aspartame in kappa-carrageenan and gellan gum gels, *Food Hydrocoll.*, 15, 67, 2001.
87. Mahungu, S. M., Hansen, S. L., and Artz, W. E., Fat substitutes and replacers, in *Food Additives*, Branen, A. L., Davidson, P. M., Salminen, S., and Thornage III, J. H., Eds., Marcel Dekker, New York, 2002, pp. 311–337.
88. Humano, Y. et al., Preparation and *in vitro* antioxidant activity of κ-carrageenan oligosaccharides and their oversulfated, acetylated, and phosphorylated derivatives, *Carbohydr. Res.*, 340, 685, 2005.
89. Seguer Bonaventura, J. et al., New Preservatives and Protective Systems, U.S. Patent No. 7662417 B2, 2010.
90. Medina, M. B., Binding interaction studies of the immobilized *Salmonella typhimurium* with extracellular matrix and muscle proteins, and polysaccharides, *Int. J. Food Microbiol.*, 93, 63, 2004.
91. Tong, C. B. S. and Hicks, K. B., Sulfated polysaccharides inhibit browning of apple juice and diced apples, *J. Agric. Food Chem.*, 39, 1719, 1991.
92. Spagnuolo, P. A. et al., Kappa-carrageenan interactions in systems containing casein micelles and polysaccharide stabilizers, *Food Hydrocoll.*, 19, 371, 2005.

93. Puvanenthiran, A. et al., Milk-based gels made with κ-carrageenan, *J. Food Sci.*, 63, 137, 2003.
94. Anon., *Crystallizing the Importance of Stabilizers in Ice Cream*, IFIS Publishing, Reading, U.K., 2003 (http://www.foodsciencecentral.com/fsc/ixid12673).
95. Nussinovitch, A., Hydrocolloids for coatings and adhesives, in *Handbook of Food Hydrocolloids*, 2nd ed., Phillips, G. O. and Williams, P. A., Eds., CRC Press, Boca Raton, FL, 2010, pp. 347–366.
96. Soukoulis, C., Chandrinos, I., and Constantina Tzia, C., Study of the functionality of selected hydrocolloids and their blends with κ-carrageenan on storage quality of vanilla ice cream, *LWT Food Sci. Technol.*, 41(10), 1816, 2008.
97. Camacho, M. M., Martinez, N. N., and Chiralt, A., Stability of whipped dairy creams containing locust bean gum/λ-carrageenan mixtures during freezing–thawing processes, *Food Res. Int.*, 34(10), 887, 2001.
98. Depypere, E. et al., Rheological properties of dairy desserts prepared in an indirect UHT pilot plant, *J. Food Eng.*, 91, 140, 2009.
99. Verbeken, D. et al., Textural properties of gelled dairy desserts containing κ-carrageenan and starch, *Food Hydrocoll.*, 18, 817, 2004.
100. Cernikova, M. et al., Effect of carrageenan type on viscoelastic properties of processed cheese, *Food Hydrocoll.*, 22, 1054, 2008.
101. Rassameesangpetch, P., Texturising Cheese, paper presented at FHM 2009, Kuala Lumpur, Malaysia, August 11–14, 2009.
102. Hilliou, L., Larontonoda, F. D. S., Serreno, A. M., and Goncalves, M. P., Thermal and viscoelastic properties of κ/ι-hybrid carrageenan gels obtained from the Portuguese seaweed *Mastocarpus stellatus*, *J. Agric. Food Chem.*, 54, 7870, 2006.
103. Gu, Y. S., Decker, A. F., and McClements, D. J., Production and characterization of oil-in water emulsions containing droplets stabilized by multilayer membranes consisting of beta-lactoglobulin, ι-carrageenan and gelatin. *Langmuir*, 21, 5752, 2005.
104. Hansen, P. M. T., Food hydrocolloids in the dairy industry, in *Food Hydrocolloids*, Nishinari, K. and Doi, E., Eds., New York, Plenum Press, 1994, pp. 211–224.
105. Ward, F. M. and Andon, S. A., Hydrocolloids as film formers, adhesives, and gelling agents for bakery and cereal products, *Cereal Food World*, 47, 52, 2002.
106. Selomulyo, V. O. and Zhou, W., Frozen bread dough: effects of freezing storage and dough improvers. *J. Cereal Sci.*, 45, 1, 2007.
107. Fiszman, S. M. and Salvador, A., Recent developments in coating batters, *Trends Food Sci. Technol.*, 14, 399, 2003.
108. Shih, F. and Daigle, K., Oil uptake properties of fried batters from rice flour, *J. Agric. Food Chem.*, 47, 1611, 1999.
109. Annapure, U. S., Singhal, R. S., Kulkarni, P. R., Screening of hydrocolloids for reduction in oil uptake of a model deep fat fried product, *Lipid*, 101, 217, 1999.
110. Mellema, M., Mechanism and reduction of fat uptake in deep-fat fried foods, *Trends Food Sci. Technol.*, 14, 364, 2003.
111. Daigle, S. P. et al., PSE-like turkey breast enhancement through adjunct incorporation in a chunked and formed deli roll, *Meat Sci.*, 69, 319, 2005.
112. Cierach, M. et al., The influence of carrageenan on the properties of low-fat frankfurters, *Meat Sci.*, 82, 295, 2009.
113. Verbeken, D. et al., Influence of κ-carrageenan on the thermal gelation of salt-soluble meat proteins, *Meat Sci.*, 70, 161, 2005.

114. Day, L. et al., Incorporation of functional ingredients into foods [review], *Trends Food Sci. Technol.*, 11, 1, 2008.
115. Trius, A. and Sebranek, J. G., Carrageenans and their use in meat product, *Crit. Rev. Food Sci. Nutr.*, 36, 69, 1996.
116. Xiong, Y. L., Noel, D. C., and Moody, W. G., Textural and sensory properties of low-fat beef sausages with added water and polysaccharides as affected by pH and salt, *J. Food Sci.*, 64, 550, 1999.
117. Shand, P. J., Sofos, J. N., and Schmidt, G. R., Kappa-carrageenan, sodium chloride and temperature affect yield and texture of structured beef rolls, *J. Food Sci.*, 59, 282, 1994.
118. Fernandez, P. et al., High pressure-cooking of chicken meat batters with starch, egg white, and iota carrageenan, *J. Food Sci.*, 63, 267, 1998.
119. Llanto, M. G. et al., Effects of carrageenan on gelling potential of surimi prepared from Atlantic pollock, in *Advances in Fisheries Technology and Biotechnology for Increased Profitability*, Voigt, M. N. and Botta, J. R., Eds., Technomic, Lancaster, PA, 1990, pp. 305–312.
120. Montero, P. and Perez-Mateos, M., Effects of Na$^+$, K$^+$ and Ca^{2+} on gels formed from fish mince containing a carrageenan or alginate, *Food Hydrocoll.*, 16, 375, 2002.
121. Kasapis, S. et al., Scientific and technological aspects of fish product development. Part I. Handshaking instrumental texture with consumer preference in burgers, *Int. J. Food Prop.*, 7, 449, 2003.
122. Haug, I. J. et al., Physical behaviour of fish gelatin–κ-carrageenan mixtures, *Carb. Polym.*, 56, 11, 2004.
123. Karim, A. A. and Rajiv, B., Gelatin alternatives for the food industry: recent developments, challenges and prospects, *Trends Food Sci. Technol.*, 19(12), 644, 2008.
124. Smitha, S. et al., Effect of hydrocolloids on rheological, microstructural and quality characteristics of parotta: an unleavened Indian flat bread, *J. Text. Stud.*, 39, 267, 2008.
125. Wang, B., Xiong, Y. L., and Wang, C., Physicochemical and sensory characteristics of flavored soymilk during refrigeration storage, *J. Food Qual.*, 24, 513, 2001.
126. Zulkurnain, M. et al., Development of a soy-based cream cheese, *J. Tex. Stud.*, 39, 635, 2003.
127. Hamza-Chaftai, A., Effect of manufacturing conditions on rheology of banana gellified milk: optimization of the technology, *J. Food Sci.*, 55, 1630, 1990.
128. Bico, S. L. S. et al., Combined effects of chemical dip and/or carrageenan coating and/or controlled atmosphere on quality of fresh-cut banana, *Food Control*, 20, 508, 2009.
129. Cabello-Pasini, A. et al., Clarification of wines using polysaccharides extracted from seaweed, *Am. J. Enol. Viticuli.*, 56, 52, 2005.
130. Wang, J. C., Application of carrageenan to wine brewing, *Food Sci. China*, 20, 37, 1999.
131. Canete, S. J. P. and Montano, M. N. E., Kappa-carrageenan gel as agent to sequester paralytic shellfish poison, *Mar. Biotechnol.*, 4, 565, 2002.
132. East Coast Seaweed, Inc., Tamil Nadu, India (www.seaweedindia.com/seaweed-extracts.html).

133. Mercier, L. et al., The algal polysaccharide carrageenans can act as an elicitor of plant defence, *New Phytol.*, 149, 43, 2001.
134. Belitz, H. D., Grosch, W., and Schieberle, P., *Food Chemistry*, 3rd ed., Springer-Verlag, Heidelberg, 2004, chap. 4.
135. Kim, H.-J. et al., Effects of combined treatment of gamma irradiation and addition of fucoidan/laminarin on ready-to-eat pork patty, *Korean J. Food Sci. Anim. Res.*, 29, 34, 2009.
136. Dubigeon, C. B., Lahaye, M., and Barry, J.-L., Human colonic bacterial degradability of dietary fibres from sea lettuce (*Ulva* sp.), *J. Sci. Food Agric.*, 73, 149, 1997.
137. Yu, S. et al., Physico-chemical characterization of Floridean starch of red algae, *Starch/Stärke*, 54, 66, 2002.
138. Plaza, M. et al., Innovative natural functional ingredients from microalgae, *J. Agric. Food Chem.*, 57, 7159, 2009.
139. Matsunaga, T. et al., Marine microalgae, *Adv. Biochem. Eng. Biotechnol.*, 96, 965, 2005.
140. Rebolloso-Fuentes, M. M. et al., Biomass nutrient profiles of the microalga *Nannochloropsis*, *J. Agric. Food Chem.*, 49, 2966, 2001.
141. Phillippis, R. and Vincenzini, M., Exocellular polysaccharides from cyanobacteria and their possible application, *FEMS Microbiol. Rev.*, 22, 151, 1998.
142. Ginzberg, A., Korin, E., and Arad, S., Effect of drying on the biological activities of a red microalgal polysaccharide, *Biotechnol. Bioeng.*, 99, 411, 2008.
143. Williams, P. A. and Phillips, G. O., Eds., *Gums and Stabilizers for the Food Industry 11*, Royal Society of Chemistry, Cambridge, U.K., 2002.
144. Raasmussen, R. S. and Morrissey, M. T., Marine biotechnology for production of food ingredients, *Adv. Food Nutr. Res.*, 52, 237, 2007.
145. Bixler, H. J. and Porse, H., A decade of change in the seaweed hydrocolloids industry, *J. Appl. Phycol.*, 2010 (DOI 10.1007/s10811-010-9529-3).
146. West, J. and Miller, K. A., Agarophytes and carrageenophytes, in *California's Living Marine Resource: A Status Report*, Leet, W. S., Dewees, C. M., Klingbeil, R., and Larson, E. J., Eds., California Department of Fish and Game, 2001, pp. 286–287.
147. Thomas, D. N., *Seaweeds*, Natural History Museum, London, 2002.
148. Shah, Z. C. and Huffman, F. G., Current availability and consumption of carrageenan-containing foods, *Ecol. Food Nutr.*, 42, 357, 2003.
149. Seisun, D., President, IMR International, San Diego, CA, October 21, 2009.
150. Matsunaga, T., Takeyama, H., and Takano, H., Algal culture, in *The Encyclopedia of Bioprocess Technology*. Vol. 1. *Fermentation, Biocatalysis, and Bioseparation*, Flickinger, M. C. and Drew, S. W., Eds., John Wiley & Sons, New York, 1999, p. 69.
151. Phillips, G. O. and Williams, P. A., Eds., *Handbook of Hydrocolloids*, 2nd ed., Woodhead Publishing, Cambridge, U.K., 2009.

8

Extracellular Polysaccharides from Non-Marine and Marine Microorganisms: Food Applications

8.1 Introduction

Extracellular polysaccharides (EPSs) of microorganisms from non-marine habitats have been used as additives in the food, biotechnology, and pharmaceutical fields for several years. Some of these compounds are xanthan, pullulan, levan, and curdlan. The advantages of these compounds include easy production by controlled fermentation of the microorganisms, potential improvement of their functionalities by chemical modifications, biodegradability, nontoxicity, and comparatively low prices. Because of these advantages, microbial EPSs, apart from their use in foods, have found applications in other areas, including biotechnology, textiles, medicine, drug delivery, oil recovery, water purification, and metal removal in mining and industrial waste treatments. Most of these applications take advantage of their rheological properties, ability to form hydrogels, and stability at high temperatures and variable pH conditions.[1–3] This chapter discusses the potential applications of both non-marine and marine EPSs in food product development. It is presumed that some discussion of non-marine polysaccharides will be beneficial to understanding and exploring the variety of ways marine exopolysaccharides can be utilized (see Table 8.1).

8.2 Functional Properties of Exopolysaccharides Influencing Their Uses in Food

The salient properties of EPSs for their use in food products include their rheological properties, interactions with other food components, stability, and metal binding capabilities. The chemical composition of EPSs is important,

TABLE 8.1

Commercial Microbial Exopolysaccharides and Their Food Applications

Organism	Polysaccharide	Applications
Acinetobacter calcoaceticus	Emulsan	Emulsifying agent
Sphingomonas paucimobilis	Gellan	Texture modifier; solidifying culture media, especially for studying marine microorganisms
Xanthomonas spp.	Xanthan	Thickener and stabilizer in foods, often used in combination with guar
Acetobacter spp.	Cellulose	Artificial skin to aid in the healing of burns or surgical wounds; natural nondigestible fibers; hollow fibers or membranes for specific separation technologies
Rhizobium meliloti, *Agrobacterium radiobacter*	Curdlan	Gelling, stabilizing, and thickening agent; dietetic foods
Lactic acid bacteria	Dextran and other exopolysaccharides	Texturization
Pseudomonas and *Azotobacter* spp.	Bacterial alginate	Emulsifier; texture modifier; nutrient encapsulation; immobilization matrix; coating for roots of seedlings and plants to prevent desiccation; microencapsulation matrix for nutrients; hypoallergenic wound-healing tissue
Biopolymyxa spp., *Aerobacter levanicum,* *Erwinia amylovora,* *Rahnella aquatilis,* *Lactobacillus reuieri,* *Serratia levanicum*	Levan	Viscosity modifier; stabilizer, emulsifier, and gelling agents; encapsulating agent for nutraceuticals
Aureobasidium pullulans	Pullulan	Stabilizer; binder; dietetic foods; prebiotic

as modifying the composition offers opportunities to enhance their functionality (see Chapter 5). Behavior of a particular EPS is also governed by its interactions with food components, including other polysaccharides in the system, and by their sensitivity to physical parameters, including pH, temperature, and ionic strength. Interactions between negatively charged EPSs and milk proteins (casein and whey proteins) at pH between 5.4 and 6.1 have been reported to influence the texture of fermented milk. The solubility of the caseins was only slightly affected by the EPSs, but the solubility of the whey protein isolates depended strongly on the added EPSs. Charge density, molecular weight, and chain stiffness of EPSs were other factors determining these interactions.[4]

Rheological properties primarily determine the applicability of a particular EPS, as these properties influence texture, flow, mouth feel, flavor binding, and other sensory properties of the food in which the EPS is incorporated. The relevant aspects related to rheological behavior, gelation kinetics, and thermal scanning rheology (TSR) of polysaccharides of diverse origin, including those from microorganisms, have been discussed in greater detail.[5-7] The rheological properties of some commercially important microbial polysaccharides such as xanthan, gellan, pullulan, and bacterial alginate from non-marine microorganisms are briefly highlighted here, with a view to comparing these compounds with EPSs from marine microorganisms.

The rheological properties of xanthan are characterized by its ability to undergo yield stress and shear thinning behavior. At a very low shear rate, xanthan solutions may display Newtonian behavior; however, at a certain critical value, there is a transition from Newtonian to viscoelastic behavior. The critical shear rate required depends on the concentration: the higher the concentration, the lower the shear rate required. At high shear, the molecules align in the direction of applied force, and the xanthan solution flows easily. The yield value and the shear thinning behavior of xanthan are more pronounced than those of other gums.[8,9]

Another commercially important EPS, pullulan has relatively low viscosity in solution, resembling gum arabic.[10] Gellan, the EPS produced by *Pseudomonas elodea*, has excellent thermal and acid stability as well as gel elasticity. Gellan exhibits the least shear thinning. The physical properties of gellan vary considerably depending upon the concentrations used, temperature, aqueous environment, and presence of cations. A very low concentration of gellan produces a "fluid gel." Chemical deacylation of gellan in the native form produces a change from soft, weak elastic thermoreversible gels to harder, firmer, stronger brittle gels under optimal gelling conditions.[11]

Aqueous suspensions of curdlan can be thermally induced to produce high-set gels that will not return to a liquid state upon heating. The gel strength increases with temperature. Increased junction zones are involved in rigid cross-links in higher concentration gels. As the gel concentration increases, apparent viscosity also increases, reducing the mobility of the polymer chains. Curdlan gels have been reported to be intermediary between the highly elastic gel of gelatin and the brittleness of agar gel, and they are stable against freezing and thawing. The pseudoplastic flow behavior of curdlan solutions allows the compounds to be used as thickeners and stabilizers in liquid foods such as salad dressings and spreads.[5,12,13]

Exopolysaccharides from lactic acid bacteria have several food applications. *Leuconostoc* is commercially exploited to produce glucan, a homopolymer containing glucose as a single carbohydrate unit. These EPSs are used as texturizing, stabilizing, emulsifying, sweetening, gelling, or water-binding agents in food as well as non-food products. Solution viscosity decreases with an increase in shear stress and exhibits typical non-Newtonian pseudoplastic behavior.[14]

Alginate has been produced from *Azotobacter vinelandii* for its favorable rheological properties.[15] Gelrite®, a commercial EPS produced from the bacterium *Pseudomonas elodea* (now designated as *Sphingomonas paucimobilis*), is of greater potential value, as it functions as an effective gelling agent. Deacetylation of certain bacterial alginates significantly increases their ion binding capacities, making them more similar in their properties to algal alginates.[16] An extracellular microbial polysaccharide named XM-6 shows unusual gelation properties. Its sol–gel transition is unusually sharp. In the sol state, the polysaccharide shows the shear rate and temperature viscosity dependence typical of a disordered (random coil) polymer solution. Gelation of the EPS involves interchain association through ordered junction zones, with specific incorporation of cations within the ordered structures. Optimum gelation has been observed in the presence of Na^+ and Ca^{2+}. Both gel strength and melting temperature increase with increasing salt concentration. By suitable adjustment of the salt concentration, gelation can be made to occur just below body temperature (e.g., 30 to 35°C), which has obvious implications for biomedical or food applications. Gels having the strength required for normal industrial or food applications may be obtained at EPS concentrations of 0.3% (w/v) in the presence of 1% NaCl (w/v).[17]

Findings such as these have led to the general conclusion that EPSs from non-marine organisms have interesting and potentially useful rheological properties, such as non-Newtonian behavior with pseudoplasticity. These properties give these polysaccharides significant potential to serve as additives that enhance the sensory properties of food products, as discussed below and in recent articles.[3,18–21]

8.3 Food Applications for Non-Marine Exopolysaccharides

A number of microbial bacterial polysaccharides (e.g., xanthan, curdlan, pullulan) have found commercial applications in food processing, replacing some of the traditionally used plant gums. The following brief discussion is intended to summarize the uses of EPSs from non-marine sources as food additives and to highlight the possible uses of EPSs from marine sources in food products.

8.3.1 Xanthan

Xanthan is an anionic polysaccharide, whose properties are briefly discussed in Chapter 5. In most foods, xanthan is used at 0.5%. Xanthan gum helps suspend solid particles, such as spices, and in frozen foods such as ice creams it contributes to the pleasant texture (along with guar gum and locust bean gum). In beverages, xanthan provides body and a good mouth feel; it also stabilizes pulp, especially in combination with other polysaccharide such as carboxymethylcellulose. Its pseudoplasticity makes it useful for salad dressings, and its heat stability and stabilizing action are useful for canned foods. Xanthan also has valuable applications in bakery products, where it improves the viscoelastic properties of dough; for example, it prevents lump formation during kneading to improve dough homogeneity, enhances gas absorption in the dough, and improves pumping performance during production. It also improves the quality of bread in terms of specific volume, crumb hardness, sensory properties, and overall acceptability. The polysaccharide also aids in the suspension of large solids particles such as fruits and nuts in baked products, enhances texture and moisture retention in cake batters, and controls starch retrogradation in cereal products. Xanthan gum is used in gluten-free baking to give the dough or batter the "stickiness" that would otherwise be achieved with the gluten. Xanthan in combination with locust bean gum produces a melt-in-the mouth gelling system for foods, adds emulsion stability and viscosity to dairy products (such as milk shakes, ice cream, and whipped desserts), and prevents crystal formation in these products.

Xanthan gum can prevent oil separation by stabilizing the emulsion; hence, it is used in low-fat meat products and spreads and in low-fat biscuits. In coated products such as onion rings or fish sticks, the polysaccharide helps the batter coating adhere better and controls syneresis. Xanthan is added to food sauces, gravies, and dry mixes for improved viscosity and thermal stability. The intestinal digestibility of xanthan is low, so it is suitable for use in low-calorie foods and functions as a dietary fiber, although at low pH values it interacts with proteins. As noted, in starch products, xanthan controls the tendency of starch to retrograde. It improved the texture of chapati, a popular flat unleavened Indian bread made of whole-wheat flour, by retaining its extensibility during storage at room temperature.[22–27] Several applications of xanthan are summarized in Table 8.2.

TABLE 8.2

Different Properties of Xanthan Used in Foods

Function	Applications
Adhesion	Icings and glazes
Binding agent	Pet foods
Coating	Confectionery
Emulsifying agent	Salad dressings
Encapsulation	Powdered flavors
Film formation	Protection coatings, sausage casings
Foam stabilizer	Beer
Stabilizer	Ice cream, salad dressing
Swelling agent	Processed meat products
Thickening agent	Jam, sauces, syrups, pie fillings

8.3.2 Levan

Levan is a biopolymer that is naturally produced by microorganisms. Levan has potential applications in the food industry as a stabilizing agent, thickener, emulsifying agent, flavor modifier, and encapsulating agent for nutraceuticals and drugs. Levan may act as a prebiotic to change the intestinal microflora in a beneficial way, and it has been observed to lower cholesterol and triglyceride levels, making it useful in dietetic foods. It is also used as a sugar substitute offering low calories and cariogenicity. A number of Japanese companies use microbial levans as additives in dairy products containing *Lactobacillus* spp.[28–30]

8.3.3 Curdlan

This polysaccharide is useful as a gelling material to improve the textural quality, water-holding capacity, and thermal stability of various foods. The rheological properties of curdlan, including its flow behavior, suggest a role as a thickener and stabilizer in foods such as salad dressings and spreads. In the production process, the polymer can be added to foods before heating, either as a powder or as a suspension in water or aqueous alcohol at concentrations less than 1%. Curdlan gels are known to efficiently absorb high concentrations of sugars from syrups and are relatively resistant to syneresis, making them useful as binders and texturizers in sweet jellies and similar foods. The polysaccharide has been well accepted as an effective ingredient to replace fat in meat products such as sausages.[31]

Because it is stable against extreme conditions, curdlan is used in products requiring resistance against freezing or retorting. The thermal stability of curdlan gels has also been found to reduce oil uptake and moisture loss in food products during deep-fat frying.[32] At concentrations as low as 0.5% (w/w), curdlan gels prevented oil uptake and moisture loss in deep-fat-fried doughnuts and were superior in this respect to many cellulose derivatives.[33] Because curdlan is not readily degraded by human digestive enzymes, it can be used in low-calorie foods. Curdlan is also incorporated into fish feed to improve immune activity.[34] Several Japanese patents have been approved for a variety of foods employing curdlan as an additive. These foods include soybean curd (tofu), sweet bean paste, boiled fish paste, noodles, sausages, jellies, and jams marketed in Japan. Curdlan is also used to produce edible coatings and films for foods.[35,36]

8.3.4 Gellan

Physical properties such as high viscosity and high resistance to heat make gellan interesting for commercial food applications. Gellan is used to improve the texture of food products, physical stability of liquid nutritional products, and water-holding capacity during cooking and storage. Food products that typically incorporate gellan include dessert gels, icings and glazes, sauces, puddings, and microwavable foods. Gellan can also form films and coatings that can be used in breading and batters. Films offer several advantages, particularly their ability to provide an effective barrier against oil absorption. Spraying a cold solution of gellan onto the surface of food products such as nuts, crisps, and pretzels produces an instant thin layer of gel that facilitates the adhesion of salt on the food surface. The addition of 0.25% gellan reduced oil uptake in legume-based noodles during deep-fat frying.[37] Food-grade gellans used as gelling agents in foods and personal care applications are commercially available in three forms: no, low, and high acyl contents, with the respective names of Gelrite®, Kelcogel® F, and Kelcogel®.[38] In spite of the benefits it offers, gellan is the most expensive food gum. Its low yield and requirement for expensive downstream processing have hindered the economic viability of its microbial production.[25,39,40]

8.3.5 Pullulan

Pullulan offers a wide range of commercial and industrial applications in the food technology and healthcare fields. Because of its resistance to mammalian amylases, it provides few calories and can be used in diet foods. Dietary pullulan functions as a prebiotic, promoting the growth of beneficial bifidobacteria. Pullulan can serve as a partial replacement for starch in pasta or bakery goods and as a stabilizer or binder. It can also be used to bind nuts in cookies. The viscosity of a pullulan solution is stable to heating, changes in pH, and most metal ions. It can be used as a low-viscosity component of beverages and sauces and to stabilize the quality and texture of mayonnaise.[10,41]

8.3.6 Dextran

Dextrans produced by lactic acid bacteria increase viscosity, bind hydration water, and reduce syneresis in food products. Because of their remarkable thickening and shear thinning properties, these polysaccharides play a major role in the manufacture of fermented dairy products such as yogurt, cheese, fermented cream, and milk-based desserts, as they contribute to the texture, mouth feel, taste perception, and stability of the final products. Their functionality, however, depends on their charge density; consequently, different strains of lactic acid bacteria that produce either charged or neutral EPSs can be selected to fulfill design textural requirements.[42] The viscosity of foods can be improved by adding *Leuconostoc mesenteroides* to the food, as it secretes a dextran that simultaneously functions as a thickening and gelling agent in the food.[43] The chemically cross-linked product Sephadex®, used for gel filtration, represents a major technical use of dextran. Dextran of low molecular weight (75,000 ± 25,000) is used in the pharmaceutical industry as a plasma substitute.

8.3.7 Others

Guar and locust bean gum (LBG) are commonly used in ice cream to improve texture and reduce ice cream meltdown. LBG is also used extensively in cream cheeses, fruit preparations, and salad dressings. Guar gum and locust bean gum provide an optimal improvement in consistency when added at a maximum level of 1% (w/w) to formulated ketchups.[44] The EPSs produced by certain fungi such as *Sclerotium rolfsii* are capable of minimizing syneresis in cooked corn starch pastes during refrigeration for a maximum of 20 days. The EPSs increased viscosity with a non-Newtonian, pseudoplastic behavior. The consistency coefficients were higher and the flow behavior indexes were lower. EPS at a 1% concentration was capable of minimizing liquid separation (syneresis) in cooked corn starch.[45] The exopolysaccharide from the yeast *Cryptococcus laurentii* was more stable than xanthan gum and other exopolysaccharides under the most vigorous of denaturing conditions.[46]

8.4 Microbial Emulsifiers

Emulsifiers are used in the areas of food processing, agriculture, cosmetics, pharmaceuticals, detergents, personal care products, textile manufacturing, laundry supplies, metal treatment, pulp and paper processing, and paints. These emulsifiers are surface-active compounds of microbial origin capable of reducing surface tension at the interfaces between liquids, solids, and gases, thereby allowing them to mix or disperse readily as emulsions in water or other liquids. The enormous market demand for surfactants is currently met by numerous synthetic, mainly petroleum-based, chemical surfactants, many of which are usually toxic to the environment and are nonbiodegradable. Tightening environmental regulations and increasing awareness for the need to protect the ecosystem have effectively resulted in an increasing interest in bioemulsifiers as possible alternatives to chemical surfactants. They have advantages over their chemical counterparts in biodegradability and effectiveness at extreme temperature or pH and in having lower toxicity. The problems of frequent oil spillage and associated environmental pollution on the high seas encouraged research on microbial solutions for oil removal, which has led to the identification of a number of marine microbes offering superior crude oil emulsifying capacities.[47–51] Emulsan, produced by *Acinetobacter calcoaceticus*, is perhaps the first promising commercial emulsifier of bacterial origin.[51]

In food processing, microbial emulsifiers could be an interesting proposition because of their unique functional properties.[52] Gum arabic was the original plant gum used in dilute emulsion systems. Later, other gums such as galactomannans, xanthan, and pectin, among others, were found to improve the stability of oil-in-water emulsions.[53] The polysaccharide galactomannan is a component of the cell wall of the mold *Aspergillus* and consists of a mannose backbone with galactose side groups. Other emulsifiers isolated from non-marine organisms include EPSs from *Bacillus coagulans* that showed significant emulsifying activities in different vegetable oils/hydrocarbon substrates, as well as EPSs from *Candida* spp.[54,55]

In addition to polysaccharides, an oligosaccharide having thermal stability and emulsification capacity has been isolated from nonpathogenic *Gluconacetobacter hansenii*. The compound is a single-sugar, α-linked, glucuronic-acid-based, water-soluble oligosaccharide. At a concentration of 0.15%, the oligosaccharide demonstrated maximum emulsification capacity, with the emulsion exhibiting moderate stability. The oligosaccharide has potential applications in food and pharmaceutical preparations and as a therapeutic agent in biomedical fields.[20,56] Further details on applications of microbial polysaccharides can be found in a number of recent articles.[3,18–20]

8.5 Exopolysaccharides from Marine Organisms

8.5.1 Rheological Properties

A number of marine microorganisms have been shown to produce EPSs, as discussed in Chapter 5. The rheological properties of many of these EPSs have been determined with a view to identifying potential uses for these EPSs. Some salient findings of these studies are pointed out in this section. EPSs released by two marine bacteria isolated from an intertidal zone exhibit high specific viscosity, pseudoplasticity, and stability over a wide range of pH values in the presence of a variety of salts. The viscosity of one EPS was relatively insensitive to increasing temperature, whereas that of the other showed an irreversible drop on heating. These properties were comparable to those of xanthan, suggesting their potential to replace xanthan.[57] The rheological properties of EPSs from a few marine organisms belonging to *Pseudomonas* spp. have been reported. Two bacterial strains belonging to *P. aeruginosa* from marine sources produce commercially important biopolymers. One strain produced a highly viscous, water-absorbing extracellular acidic polysaccharide when it was grown aerobically in minimal medium containing glucose, fructose, or sucrose as the sole source of carbon. The biopolymer has the ability to absorb water 400 times its dry weight. This property was superior to that of currently used nondegradable synthetic water absorbents.[58]

When incorporated in food, EPSs isolated from another *Pseudomonas* sp. displayed higher viscosity-enhancing properties than both carboxymethylcellulose and alginate. The exopolysaccharide also demonstrated a high flocculating capacity (82.6%), similar to that of xanthan gum, guar gum, and carboxymethylcellulose and much higher than that of sodium alginate. It also had significant emulsifying activity against *n*-hexadecane, analogous to the value observed for xanthan gum.[59]

The copious amounts of EPSs produced by the marine bacterium *Pseudoalteromonas ruthenica* were found to be pseudoplastic in nature, to have good shearing properties, and to be stable at high pH values, thus demonstrating a potential for use in the food, oil, and textile industries.[60] The exopolysaccharides of *Pseudomonas oleovorans* in aqueous solutions have viscoelastic properties similar to those of guar gum. They offer good flocculating and emulsifying properties and film-forming capacity. These properties make the polymers good alternatives to more expensive polysaccharides, such as guar gum, in the food, pharmaceuticals, cosmetics, textile, paper, and petroleum industries.[59,60] An EPS produced by a *Pseudomonas* sp. forms a brittle, firm, and optically clear gel that is superior to agar.[61]

The EPS secreted by the marine bacterium *Halomonas chejuensis* in a 0.25 to 1% (w/w) aqueous dispersion showed pseudoplastic character associated with a marked shear thinning property. At 1% concentration, the consistency index and flow behavior index were 1410 and 0.73, respectively. The polysaccharide had an average molecular weight of 2.2×10^3 kDa and was stable against pH and salts. The rheological behavior of the EPS dispersion indicated formation of a structure intermediate between that of a random-coil polysaccharide and a weak gel.[62] An EPS from *Halomonas eurihalina* strain H96, isolated from hypersaline habitats and cultivated in defined medium, has been characterized rheologically. When the pH of the polymer solution was decreased to 3.0, a gel with a viscosity of 30,000 cP formed.[63] *H. eurihalina* produces an EPS having high viscosity. Sodium, potassium, magnesium, and calcium increased the viscosity of the organism in the order of KCl > NaCl > $MgCl_2$ > $CaCl_2$. The highest viscosity value was measured in acidic 10^{-4}-*M* KCl. A loss of sulfate content seemed to correlate with the increase in viscosity. The addition of hydrophobic substrates to the culture medium produced changes in the chemical composition and emulsifying activity of the EPS.[64]

An acidic polysaccharide from haloalkalophilic *Bacillus* sp. I-450 exhibited pseudoplastic behavior with a shear thinning effect, while the compression test indicated that the polymer had high gel strength.[65] The extremely halophilic archebacterium *Haloferax mediterranei* produces a heteropolysaccharide that exhibits a pseudoplastic behavior and a high apparent viscosity at relatively low concentrations. Viscosity is remarkably resistant to extremes of pH, temperature, or salinity. These characteristics make this polymer interesting for enhanced oil recovery and other applications for which a very resistant thickening agent is required.[79]

The marine bacterium *Alteromonas* sp. strain 1644 produced two types of EPSs, one secreted to the medium and another bound to the cells. The chemical and rheological characteristics of the bound polysaccharide were different from those of the one secreted by the organism. At low ionic concentration, irrespective of the nature of the ions, solutions of the polysaccharide had very low viscosities; however, above 0.03 M, it formed a gel even at low polymer concentrations. The viscosity of the gel was maintained even at high temperatures. This behavior was dependent on the nature of the ions, with an order of $NH^{4+} > Mg^{2+} > Na^+ > Li^+ > K^+$, and $Br^- > NO_3^- > SO_4^{2-} > Cl^- > I$, for cations and anions, respectively.[66] The acidic EPS released by *Alteromonas* has potential for use as a thickening agent.[67]

Enterobacter cloacae isolated from marine sediment collected from the Gujarat coast in India produced copious amounts of EPSs that had excellent rheological properties under extreme conditions. It had good viscosity even at temperatures as high as 90°C and at acidic and alkaline pH values in the presence of monovalent and divalent cations. It also exhibited viscosity at low temperatures, which can be of importance to food dressings that must be stored at 4°C in a viscous-flowing form.[68] Rheological studies of two EPSs produced by *Vibrio alginolyticus* showed that they possessed good shearing properties; however, one was unstable at high temperatures and high pH.[69,70] The exopolysaccharide from the yeast *Cryptococcus laurentii* was more stable than xanthan gum and other EPSs under the most vigorous of denaturing conditions.[46] Two exopolysaccharides, EPS I and EPS II, produced by the fungus *Sclerotium rolfsii* after cultivation for 48 and 72 hours, respectively, were able to minimize syneresis in cooked corn starch pastes during refrigeration for up to 20 days. EPS II was able to prevent syneresis without affecting pH, gelling properties, hardness, or color, suggesting that the compound has potential for use as food stabilizer. Corn starch/EPS blends exhibited an increase in viscosity, higher consistency coefficients, and lower flow behavior indices.[71]

8.5.2 Other Food-Related Functional Properties of Marine Exopolysaccharides

An EPS designated as PE12, which exhibits high emulsifying activities against a range of oil substrates, has been isolated and characterized. The high-molecular-mass (>2000 kDa) polymer was produced from a *Pseudoalteromonas* strain. Xylose was found to be the major sugar component of the EPS at an unusually high level of 27.7%. The polymer was also shown to absorb various metal species from marine sediment and therefore has potential use as a metal-chelating agent.[72]

The metal-binding properties of four bacterial exopolysaccharides originating from deep-sea hydrothermal vents have been reported. The compounds could remove toxic elements such as lead, cadmium, and zinc. The maximum uptake capacities reached 316, 154, and 77 mg/g for lead, cadmium, and zinc, respectively. The strong chelating properties of these EPSs for lead, cadmium, and zinc make them a possible alternative to other physical or chemical biosorbents in wastewater treatment.[73] The free-radical-

scavenging and antioxidant activities of a polysaccharide from the myce-
lium of a marine filamentous fungus, *Phoma herbarum*, have been reported.
Sulfation of its hydroxyl groups enhanced antioxidant activity of the EPS,
measured in terms of superoxide and hydroxyl radicals scavenging activities, metal chelating action, lipid peroxidation, and inhibition of linoleic acid
oxidation. Furthermore, sulfated polysaccharide also protected erythrocytes
against oxidative damage.[74]

An exopolysaccharide designated as EPS 71a was produced by the moderately halophilic bacterium *Enterobacter cloacae* and emulsified several hydrocarbons and vegetable oils. Emulsions with hexane and groundnut oil were
stable for up to 10 days between pH 2 and 10 in the presence of traces of
sodium chloride at 35°C. The emulsion stabilizing property of EPS 71a was
comparable to that of many commercial gums, including xanthan. The polysaccharide can be an interesting additive for the food industry as a viscosity-
enhancing agent, especially in products containing edible acids such as citric
acid and ascorbic acid.[75] Another EPS produced by *E. cloacae* emulsified oils
of cottonseed, coconut, jojoba, castor, groundnut, and sunflower. It formed
stable emulsions with groundnut oil at an optimal concentration of 1 mg/
mL, and it could emulsify paraffin oil, hexane, benzene, xylene, and kero-
sene. Emulsions with groundnut oil and hexane were stable up to 10 days
between pH 2 and 10 and in the presence of sodium chloride in the range of 5
to 50 mg/mL at 35°C. The exopolysaccharide had good viscosity even at high
temperature, which makes it a probable candidate for microbial enhanced
oil recovery.[76,77]

Exopolymers synthesized from halophilic bacteria could find use as raw
material for biodegradable plastics, in oil exploration efforts, or in other
commercial processes.[78] An exopolysaccharide produced by the marine bacterium *Halomonas eurihalina* increased the viscosity of acidic solutions and
served as an emulsifier.[76] Table 8.3 summarizes the rheological and other
functional properties of EPSs from some marine microorganisms that are
useful in food applications.

8.6 Comparison of Marine Exopolysaccharides and Commercial Polysaccharides

The previous section discussed the rheological properties of some marine
extracellular polysaccharides, and it is interesting to compare these properties with those of EPSs from a few non-marine organisms. An EPS produced
by *Pseudomonas elodea* (now designated as *Sphingomonas paucimobilis*) is comparable to commercial gellan.[21] The exopolysaccharide has five times greater
viscosity than that of xanthan, which also emulsified various hydrocarbons

TABLE 8.3

Rheological and Other Functional Properties of EPSs from Marine Microorganisms Useful in Food Applications

Microorganism	Rheological and Related Properties	Refs.
Alcaligenes spp.	Least shear thinning polysaccharide	Toledo et al.[50]
Nova Scotian marine bacteria	Two EPSs exhibited high specific viscosity, pseudoplasticity, and stability over a wide range of pH and salts; the viscosity of one was insensitive to increasing temperature, and the other to heating	Boyle and Reade[57]
Pseudoalteromonas ruthenica	Good shearing property, pseudoplastic in nature, and stable at higher pH levels	Saravanan and Jayachandran[60]
Halomonas spp.	Superior capacity to emulsify a wide range of food oils under neutral and acidic pH conditions; good stability	Yim, J. H. et al.[62]; Bejar et al.[63]
Enterobacter cloacae	Emulsifying properties	Iyer et al.[68,75,77]
Haloalkalophilic *Bacillus* spp.	Pseudoplastic behavior, good gel strength	Kumar et al.[65]
Alteromonas spp.	Viscosity at low concentration and at high temperatures	Samain et al.[66]
Vibrio spp.	Water-soluble polysaccharide having potential texturizing property	Jean et al.[70]
Pseudoalteromonas sp. strain TG12	Emulsifying and metal-chelating agent	Gutierrez[72]
Phoma herbarum (fungus)	Antioxidant properties	Yang et al.[74]
Cryptococcus laurentii (yeast)	More stable than xanthan under the most vigorous of denaturing conditions	Peterson et al.[82]
Halomonas spp.	High-viscosity polymer; exhibited pseudoplastic behavior	Bouchotroch et al.[86]
Antarctic sea ice bacterium	Potential cryoprotective role	Nichols et al.[93]
Alteromonas macleodii subsp. *fijiensis*	Polymer having viscosity comparable to xanthan	Raguenes et al.[94]

and oils. EPSs secreted by a marine *Bacillus* and a thermophilic yeast are comparable to commercially important xanthan in their rheological properties.[80–82] The marine bacterium *Antarctobacter* produced a compound designated as AE22 that compared well with xanthan and gum arabic in its oil-emulsifying properties. AE22, although not a pure polysaccharide, also exhibited stabilizing properties due to its viscosity properties in solution, attributed to the presence of certain chemical groups or protein moieties found on the polymer. The compound shows particular promise as an emulsion–stabilizing agent in biotechnological applications.[83]

An EPS secreted by the marine bacterium *Hahella chejuensis*, designated as EPS-R, at a concentration of 0.5% (w/w) demonstrated an oil emulsification capacity as high as 89%, which was higher than that of such commercial polysaccharides as xanthan (67.8%), gellan (2.01%), and sodium alginate (1.02%).[62] Bioemulsifiers produced by two new marine *Halomonas* spp., TG39 and TG67, have been isolated and their emulsifying activities compared with those of commercial emulsifiers. Both of the EPSs effectively emulsified a wide range of food oils under both neutral and acidic pH conditions and could withstand acid and high temperature. Their emulsification capacities were better than those of some commercial emulsifiers such as xanthan gum, gum arabic, and lecithin.[84]

8.7 Food Applications of Marine Exopolysaccharides

As indicated in the previous discussion, most EPSs from marine microorganisms have remarkable rheological properties that are generally comparable with those of commercial EPSs of non-marine origin, suggesting that marine EPSs are capable of at least partially replacing commercial EPSs as food additives. Achieving this goal, however, depends on large-scale isolation of marine EPSs at economic level and elucidation of their functional properties. At present, commercial-scale biotechnological and food applications of extracellular polysaccharides from marine organisms are still in their infancy, essentially due to difficulties in obtaining these compounds in sufficient quantities at affordable prices. A few authors have noted the potential uses of marine EPSs as thickener, gelling, emulsifying, and stabilizing agents in food products. Some of them can also be useful for medical applications,[1,85] including EPSs from halotolerant microorganisms that have played an essential role in food biotechnology for the production of fermented food and food supplements and are reported to have good oil emulsification properties.[86]

Similarly, the EPS of a *Pseudomonas* strain has good flocculating and emulsifying properties and film-forming capacity, which make the polymer a potentially good alternative to more expensive natural polysaccharides, such as guar gum, for applications in the food, pharmaceutical, cosmetics, textiles, paper, and petroleum industries. Despite these findings, practical uses have not been so far explored.[59] As noted earlier, an extracellular microbial polysaccharide designated as XM-6 has been isolated from cultures of an *Enterobacter* sp. that shows unusual gelation properties of potential technological significance.[67] Rougeaux's group examined EPSs from numerous microbial isolates from deep-sea hydrothermal vents for their interesting properties, such as gelation or high viscosity in aqueous solution. One EPS, from an *Alteromonas* marine organism, had a viscosity comparable with that of xanthan and has potential for use as a thickening agent.[87,88] Table 8.4 lists some marine sources of exopolysaccharides and their potential applications.

TABLE 8.4

Some Marine Sources of Exopolysaccharides and Their Potential Applications

Microorganism	Source	Function	Refs.
Cyanobacteria, Desulfovibrio, Desulfobacter	Deep-sea hydrothermal vents, Polynesian mats	Potential as texturizer due to good shearing properties	Rougeaux et al.[87,88]
Vibrio alginolyticus	Marine fouling material	Adhesives and binders	Jayaraman and Seetharaman[89]
Enterobacter cloaceae	Arabian Sea	Viscosity-enhancing agent	Iyer et al.[68,75,77]
Alteromonas spp.	East Pacific Rise, Antarctic ice, North Fiji Basin	Texturizer due to pseudoplastic nature; provides stability and flow; tolerant to salt	Loaëc et al.[73]; Raguenes et al.[91]
Pseudoalteromonas ruthenica	Marine	Potential use as texturizer due to good shearing property	Saravanan and Jayachandran[60]
Pseudoalteromonas spp.	Sea ice in Southern Ocean	Texturizer	Nichols[92]
Geobacillus spp.	Ischin Island	Two extracellular polysaccharides from extremophiles	Nicolaus[80]
Zooglea spp.	Hypersaline habitats	Metal-binding activity	Anton[79]
Hahella chejuensis	—	Emulsifier	Yim et al.[62]
Halomonas eurihalina	Hypersaline habitats	Emulsifier	Calvo et al.[64]
Marine archibacteria			
Thermococcus spp.	Mediterranean Sea	Potent emulsifiers and adhesives	Weiner[48]; Zhenming and Yan[90]
Haloferax mediterranei	Mediterranean Sea	Texturizer due to pseudoplastic nature and high apparent viscosity at relatively low concentrations	Anton et al.[79]

8.8 Commercial Status

During the last 50 years a number of bacterial polysaccharides have been introduced, yet few of them remain commercially available as major products. Polysaccharides such as xanthan and pullulan have been well-accepted food additives for many years and present many novel properties that are not offered by plant gums or synthetic polymers. The isolation of xanthan is probably the only commercially viable alternative currently; the production of xanthan exceeds 20,000 t per year from several commercial sources. Curdlan from lactic acid bacteria has also emerged as an important class of food additives. The major obstacle to commercialization of these polysaccharides lies in identifying novel or superior properties compared to existing products, and it is very unlikely that many will survive the rigors of extensive evaluation and market research to reach production. Recent data indicate that novel environments such as deep-sea hydrothermal vents are yielding new bacterial strains and polysaccharides. A few of them are claimed to have interesting properties; however, none is yet used commercially or marketed in spite of their functional potential as texturizing, stabilizing, and emulsifying agents. Complicating their practical application are such factors as the economic viability of large-scale cultivation of marine organisms, isolation of sufficient amounts, competition from traditional and well-recognized polysaccharides, and safety evaluation and regulatory approval of the new compounds. The future of utilizing polysaccharides from marine organisms depends on a number of factors, including: (1) successful cultivation of marine microorganisms in sufficient amounts to recover adequate amounts of polysaccharides, (2) identification of properties of these compounds and comparing them with those from non-marine sources, (3) development of specific need-based applications and new markets, and (4) economics of production.

References

1. Sutherland, I. W., Novel and established applications of microbial polysaccharides, *Trends Biotechnol.*, 16, 41, 1998.
2. Flemming, F. C. and Wingender, J., Relevance of microbial extracellular polymeric substances (EPSs). Part II. Technical aspects, *Water Sci. Technol.*, 43, 9, 2001.
3. Khan, T., Park, J. K., and Kwon, J. H., Functional biopolymers produced by biochemical technology considering applications in food engineering, *Korean J. Chem. Eng.*, 24, 816, 2007.
4. Girard, M. and Schaffer-Lequart, C., Attractive interactions between selected anionic exopolysaccharides and milk proteins, *Food Hydrocoll.*, 22, 1425, 2006.

5. Lapasin, R. and Pricl, S., *Rheology of Industrial Polysaccharide: Theory and Applications*, Chapman & Hall, New York, 1995, p. 595.
6. Nishinan, K., Rheological and related studies on industrially important polysaccharides and proteins, *J. Central South Univ. Technol.*, 14(Suppl.), 498, 2007.
7. Nishinan, K., Rheological and DSC study of sol–gel transition in aqueous dispersions of industrially important polymers and colloids, *Colloid Polym. Sci.*, 275, 1093, 1997.
8. Stokke, B. J. et al., Macromolecular properties of xanthan, in *Polysaccharides: Structural Diversity and Functional Versatility*, Dumitriu, S., Ed., Marcel Dekker, New York, 1998, pp. 433–472.
9. Milas, M. et al., Flow and viscoelasatic properties of xanthan gum solutions, *Macromolecules*, 23, 2506, 1990.
10. Leathers, T. D., Pullulan, in *Polysaccharides and Polyamides in the Food Industry*, Vol. 1, Steinbüchel, A. and Rhee, S. K., Eds., John Wiley & Sons, New York, 2005, pp. 387–422.
11. Rinaudo, M. and Milas, M., Gellan gum, a bacterial gelling polymer, in *Novel Macromolecules in Food Systems*, Doxastakis, G. and Kiosseoglou, V., Eds., Elsevier, Amsterdam, 2000, pp. 239–264.
12. Lo, Y. M. et al., Viscoelastic effects on the diffusion properties of curdlan gels, *J. Food Sci.*, 68, 2057, 2003.
13. Zhang, H. et al., A molecular description of the gelation mechanism of curdlan, *Int. J. Biol. Macromol.*, 30, 7, 2002.
14. Majumder, A. and Goyal, A., Rheological and gelling properties of a novel glucan from *Leuconostoc dextranicum* NRRL B-1146, *Food Res. Int.*, 45, 525, 2009.
15. Remminghorst, U. and Rehm, R. H. A., Bacterial alginates: from biosynthesis to applications, *Biotech. Lett.*, 28, 1701, 2006.
16. Sutherland, I. W., Structure–function relationships in microbial exopolysaccharides, *Biotechnol. Adv.*, 12, 393, 2004.
17. Nishet, B. A. et al., XM-6: a new gel-forming bacterial polysaccharide, *Carb. Polym.*, 4, 377, 1984.
18. Vuyst, L. De and Vaningelgem, F., Developing new polysaccharides, in *Texture in Food*. Vol. 1. *Semi-Solid Foods*, McKenna, B. M., Ed., Aspen, Gaithersburg, MD, 2003, pp. 275–320.
19. Walkenström, P., The creation of new food structures and textures by processing, in *Texture in Food*. Vol. 1. *Semi-Solid Foods*, McKenna, B. M., Ed., Aspen, Gaithersburg, MD, 2003, pp. 201–215.
20. Kumar, A. S., Mody, K., and, Jha, B., Bacterial exopolysaccharides: a perception, *J. Basic Microbiol.*, 47, 103, 2007.
21. Kwon, B. D., Foss, P. A., and Rha, C., Rheological characteristics of high viscosity polysaccharides, in *Industrial Polysaccharides: Genetic Engineering Structure/ Property Relations and Applications*, Yalpani, M., Ed., Elsevier, Amsterdam, 1987, pp. 253–266.
22. Muadklay, J. and Charoenrein, S., Effects of hydrocolloids and freezing rates on freeze–thaw stability of tapioca starch gels, *Food Hydrocoll.*, 22, 1268, 2008.
23. Shalini, K. G. and Laxmi, A., Influence of additives on rheological characteristics of whole-wheat dough and quality of chapatti (Indian unleavened flat bread). Part I. Hydrocolloids, *Food Hydrocoll.*, 21, 110, 2007.

24. Gujral, H. S., Haros, M., and Rosell, C. M., Improving the texture and delaying staling in rice flour chapati with hydrocolloids and α-amylase, *J. Food Eng.*, 65, 89, 2004.
25. Kohajdová, C. and Karovicová, J., Application of hydrocolloids as baking improvers, *Chem. Papers*, 63, 26, 2009.
26. Guarda, A. et al., Different hydrocolloids as bread improvers and anti-staling agents, *Food Hydrocoll.*, 18, 241, 2004.
27. Turabi, E., Sumnu, G., and Sahin, S., Rheological properties and quality of rice cakes formulated with different gums and an emulsifier blend, *Food Hydrocoll.*, 22, 305, 2008.
28. van Kranerber, R., Genetics and engineering of microbial exopolysaccharides for food: approaches for the production of existing and novel polysaccharides, *Curr. Opin. Biotechnol.*, 19, 498, 1999.
29. Suzuki, M. and Chatterton, N. J., *Science and Technology of Fructans*, CRC Press, Boca Raton, FL, 1993.
30. Kang, S. A. et al., Levan: applications and perspectives, in *Microbial Production of Biopolymers and Polymer Precursors*, Rehm, B. H. A., Ed., Caister Academic Press, Norfolk, U.K., 2009.
31. Funami, T., Yada, H., and Nakao, Y., Curdlan properties for application in fat mimetics for meat products, *J. Food Sci.*, 63, 283, 1998.
32. Patrick, W. D. et al., Texture stability of hydrogel complex containing curdlan gum over multiple freeze–thaw cycles, *J. Food Proc. Pres.*, 33, 126, 2009.
33. Funami, T. et al., Decreasing oil uptake of doughnuts during deep-fat frying using curdlan, *J. Food Sci.*, 64., 889, 1999.
34. Mashiro, K. et al., Food Composition for Fish Kind Cultivation, Japanese Patent 10313794, 1998.
35. Lee, J.-Y., Curdlan, in *Polysaccharides and Polyamides in the Food Industry*, Vol. 1, Steinbüchel, A. and Rhee, S. K., Eds., John Wiley & Sons, New York, 2005, pp. 209–232.
36. Nakao, Y. et al., Curdlan: properties and application to foods, *J. Food Sci.*, 56, 769, 1991.
37. Bajaj, I. and Singhal, R., Gellan gum for reducing oil uptake in sev, a legume based product, during deep-fat frying, *Food Chem.*, 04, 1472, 2007.
38. Giavasis, I., Harvey, L. M., and McNeil, B., Gellan gum, *Crit. Rev. Biotechnol.*, 20, 177, 2000.
39. Banik, R. M., Kanari, B., and Upadhyay, S. N., Exopolysaccharide of the gellan family: prospects and potential, *World J. Microbiol. Biotechnol.*, 16, 407, 2000.
40. Fialho, A. M. et al., Occurrence, production, and applications of gellan: current state and perspectives, *Appl. Microbiol. Biotechnol.*, 79, 889, 2008.
41. Tsujisaka, Y. and Mitsuhashi, M., Pullulan, in *Industrial Gums: Polysaccharide and Their Derivatives*, 3rd ed., Whistler, R. L. and BeMiller, J. N., Eds., Academic Press, London, 1993, pp. 447–460.
42. Duboc, P. and Mollet, B., Applications of exopolysaccharides in the dairy industry, *Int. Dairy J.*, 11, 759, 2001.
43. Purama, R. K. and Goyal, A., Dextransucrase production by *Leuconostoc mesenteroides*, *Ind. J. Microbiol.*, 2, 89, 2005.
44. Sahin, H. and Ozdemir, F., Effect of some hydrocolloids on the rheological properties of different formulated ketchups, *Food Hydrocoll.*, 18, 1015, 2004.

45. Vinarta, S. C. et al., A further insight into the practical applications of exopolysaccharides from *Sclerotium rolfsii*, *Food Hydrocoll.*, 20, 5619, 2006.
46. Peterson, G. R. et al., Rheologically interesting polysaccharides from yeasts, *Appl. Biochem. Biotechnol.*, 20, 845, 1987.
47. Banat, J. M. et al., Potential commercial applications of microbial surfactants, *Appl. Microbiol. Biotechnol.*, 53, 495, 2000.
48. Weiner, R. M., Biopolymers from marine prokaryotes, *Trends Biotechnol.*, 15, 390, 1997.
49. Ashtaputre, A. A. and Shah, A. K., Emulsifying property of a viscous exopolysaccharide from *Sphingomonas paucimobilis*, *World J. Microbiol. Biotechnol.*, 11, 219, 1995.
50. Toledo, F. I. et al., Production of bioemulsifier by *Bacillus subtilis*, *Alcaligenes faecalis* and *Enterobacter* species in liquid culture, *Biores. Technol.*, 98, 8470, 2008.
51. Rosenberg, E. and Ron, E. Z., Bioemulsans: microbial polymeric emulsifiers, *Curr. Opin. Biotechnol.*, 8, 313, 1997.
52. Faergemand, M. and Krog, N., Using emulsifiers to improve food texture, in *Texture in Food*. Vol. 1. *Semi-Solid Foods*, McKenna, B. M., Ed., CRC Press, Boca Raton, FL, pp. 216–250.
53. Garti, N., Hydrocolloids as emulsifying agents for oil-in-water emulsions, *J. Disp. Sci. Technol.*, 20, 327, 1999.
54. Kodali, V. P., Das, S., and Sen, R., An exopolysaccharide from a probiotic: biosynthesis dynamics, composition and emulsifying activity, *Food Res. Int.*, 42, 695, 2009.
55. Cirigliano, M. C. and Carman, G. M., Purification and characterization of liposan, a bioemulsifier from *Candida lipolytica*, *Appl. Environ. Microbiol.*, 50, 846, 1985.
56. Khan, T. et al., Physical properties of a single sugar α-linked glucuronic acid-based oligosaccharide produced by a *Gluconacetobacter hansenii* strain, *Proc. Biochem.*, 42, 252, 2007.
57. Boyle, C. D. and Reade, A. E., Characterization of two extracellular polysaccharides from marine bacteria, *Appl. Environ. Microbiol.*, 46, 392, 1983.
58. Jamil, N. and Ahmed, N., Production of biopolymers by *Pseudomonas aeruginosa* isolated from marine source, *Brazilian Arch. Biol. Technol.*, 51, 457, 2008.
59. Freitas, F. et al., Characterization of an extracellular polysaccharide produced by a *Pseudomonas* strain grown on glycerol, *Bioresource Technol.*, 100, 859, 2009.
60. Saravanan, S. and Jayachandran, P., Preliminary characterization of exopolysaccharides produced by a marine biofilm-forming bacterium *Pseudoalteromonas ruthenica* (SBT 033), *Lett. Appl. Microbiol.*, 46, 1, 2008.
61. Kang, K.S. et al., Agar-like polysaccharide produced by a *Pseudomonas* spp.: production and basic properties. *Appl. Environ. Microbiol.*, 43, 1086, 1982.
62. Yim, J. H. et al., Physicochemical and rheological properties of a novel emulsifier, EPS-R, produced by the marine bacterium *Hahella chejuensis*, *Biotechnol. Bioprocess Eng.*, 9, 405, 2004.
63. Bejar, V. et al., Characterization of exopolysaccharides produced by 19 halophilic strains of the species *Halomonas eurihalina*, *J. Biotechnol.*, 61, 135, 1998.
64. Calvo, C. et al., Effect of cations, pH and sulfate content on the viscosity and emulsifying activity of the *Halomonas eurihalina* exopolysaccharide, *J. Ind. Microbiol. Biotechnol.*, 20, 205, 1998.

65. Kumar, C. G. et al., Purification and characterization of an extracellular polysaccharide from haloalkalophilic *Bacillus* sp. I-450, *Enz. Microb. Technol.*, 34, 673, 2004.

66. Samain, E. et al., Simultaneous production of two different gel-forming exopolysaccharides by an *Alteromonas* strain originating from deep sea hydrothermal vents, *Carb. Polym.*, 34, 235, 1998.

67. Talmont, F. et al., Structural investigation of an acidic EPS from a deep-sea hydrothermal vent marine bacteria, *Food Hydrocoll.*, 5, 171, 1991.

68. Iyer, A., Mody, K., and Jha, B., Rheological properties of an exopolysaccharide produced by a marine *Enterobacter cloacae*, *Nat. Acad. Sci. Lett.*, 28, 119, 2005.

69. Muralidharan, J. and Jayachandran, S., Physicochemical analyses of the EPSs produced by a marine biofouling bacterium, *Vibrio alginolyticus*, *Proc. Biochem.*, 38, 841, 2003.

70. Jean, G., et al., MarineBacterial Strain of the Genus *Vibrio*, Water-Soluble Polysaccharides Produced by Said Strain and Their Uses, European patent, EP0975791 (A1), 2000.

71. Viñarta, S. C. et al., A further insight into the practical applications of exopolysaccharides from *Sclerotium rolfsii*, *Food Hydrocoll.*, 20, 619, 2006.

72. Gutierrez, T., Emulsifying and metal ion binding activity of a glycoprotein exopolymer produced by *Pseudoalteromonas* sp. strain TG12, *Appl. Environ. Microbiol.*, 74, 4867, 2008.

73. Loaëc, M., Olier, R., and Guezennec, J., Chelating properties of bacterial exopolysaccharides from deep-sea hydrothermal vents, *Carb. Polym.*, 35, 65, 1998.

74. Yang, X. B., Gao, X. D., and Han, R. X., Sulfation of a polysaccharide produced by a marine filamentous fungus *Phoma herbarum* YS4108 alters its antioxidant properties *in vitro*, *Biochim. Biophys. Acta (Gen. Sub.)*, 1725, 120, 2005.

75. Iyer, A., Mody, K., and Jha, B., Emulsifying properties of a marine bacterial EPS, *Enz. Microb. Technol.*, 38, 220, 2006.

76. Martínez-Checa, F. et al., Yield production, chemical composition, and functional properties of emulsifier H28 synthesized by *Halomonas eurihalina* strain H-28 in media containing various hydrocarbons, *Appl. Environ. Microbiol.*, 58, 358, 2002.

77. Iyer, A., Mody, K., and Jha, B., Characterization of an exopolysaccharide produced by a marine *Enterobacter cloacae*, *Indian J. Exp. Biol.*, 43, 467, 2005.

78. Hough, D. W. and Danson, M. J., Extremozymes, *Curr. Opin. Chem. Biol.*, 3, 39, 1999.

79. Anton, J. et al., Production of an extracellular polysaccharide by *Haloferax mediterranei*, *Appl. Environ. Microbiol.*, 54, 2381, 1988.

80. Nicolaus, B. et al., Polysaccharides from extremophilic microorganisms, *Origins Life Evol. Biosphere*, 34, 159, 2004.

81. Manca, M. C. et al., Chemical composition of two exopolysaccharides from *Bacillus thermoantarcticus*, *Appl. Environ. Microbiol.*, 62, 3265, 1996.

82. Peterson, G. R. et al., Yeasts producing EPSs with drag-reducing activity, *Enz. Microbiol. Technol.*, 12, 255, 1990.

83. Gutierrez, T. et al., Partial purification and chemical characterization of a glycoprotein (putative hydrocolloid) emulsifier produced by a marine bacterium, *Antarctobacter*, *Appl. Microbiol. Biotechnol.*, 76, 1017, 2007.

84. Gutuirrez, T. et al., Glycoprotein emulsifiers from two marine *Halomonas* species: chemical and physical characterization, *J. Appl. Microbiol.*, 103, 1716, 2007.

85. Raguénès, G. et al., A novel, highly viscous polysaccharide excreted by an *Alteromonas* isolated from a deep-sea hydrothermal vent shrimp, *Curr. Microbiol.*, 46, 448, 2003.

86. Bouchotroch, S. et al., Bacterial exopolysaccharides produced by newly discovered bacteria belonging to the genus *Halomonas*, isolated from hypersaline habitats in Morocco, *J. Ind. Microbiol. Biotechnol.*, 24, 374, 2000.

87. Rougeaux, H. et al., Novel bacterial exopolysaccharide from deep-sea hydrothermal vents, *Carbohydr. Polym.*, 31, 237, 1996.

88. Rougeaux, H. et al., Microbial communities and exopolysaccharides from Polynesian mats, *Mar. Biotechnol.*, 3, 181, 2001.

89. Jayaraman, M. and Seetharaman, J., Physiochemical analysis of the EPSs produced by a marine biofouling bacterium, *Vibrio alginolyticus*, *Proc. Biochem.*, 38, 841, 2003.

90. Zhenming, C. and Yan, F., EPSs from marine bacteria, *Ocean Univ. China*, 30, 67, 2005.

91. Raguenes, G. et al., *Alteromonas infernos* spp. nov., a new polysaccharide producing bacterium isolated from deep sea hydrothermal vent, *J. Appl. Bacteriol.*, 82, 422, 1997.

92. Nichols, C. M., Bacterial EPSs from extreme marine environments with special consideration of the southern ocean, sea ice, and deep-sea hydrothermal vents: a review, *Mar. Biotechnol.*, 7, 253, 2005.

93. Nichols, C. M., Bowman, J. P., and Guezennec, J., Effects of incubation temperature on growth and production of EPSs by an Antarctic sea ice bacterium grown in batch culture, *Appl. Environ. Microbiol.*, 71, 3519, 2005.

94. Raguenes, G. et al., Description of a new polymer-secreting bacterium from a deep-sea hydrothermal vent, *Alteromonas macleodii* subsp. *fijiensis*, and preliminary characterization of the polymer, *Appl. Environ. Microbiol.*, 62, 67, 1996.

9

Edible Films and Carrier Matrices from Marine Polysaccharides

9.1 Introduction

Packaging is important to the preservation and marketing of food products. The basic functions of packaging are to protect the food, provide easy handling and convenience for consumers, and display process information (see Table 9.1). In developing food packaging, many factors must be considered, including gas and water vapor permeability; mechanical properties; sealing capability; thermoforming properties; resistance to water, grease, acid, and ultraviolet light; transparency; antifogging capacity; printability; availability; and cost. For several decades, petroleum-based food packaging has been utilized—single or multicomponent laminates that satisfy the above requirements. Recently, however, a major drawback of synthetic packaging materials has been recognized—namely, the serious environmental consequences of their being nonbiodegradable. For this reason, efforts are being focused on developing biodegradable and edible films from natural, renewable resources. The classes of compounds generally used for such films include polysaccharides (e.g., starch, cellulose derivatives, pectin, alginate, carragenan, chitosan, pullulan), proteins (e.g., casein, whey protein, collagen, gelatin, fish myofibrillar protein, soy protein, wheat gluten, corn zein), and lipids (e.g., neutral lids, fatty acids, wax). Table 9.2 lists advantages of these biopolymer-based packaging materials. Challenges in the development of these films include optimization of required permeability and mechanical strength in order to obtain ideal storage life for the packaged food and to resist stress of handling and marketing.[1,2] Table 9.3 summarizes the various techniques necessary to optimize edible packaging for optimal food quality. This chapter discusses the potential for marine polysaccharides to be used in the development of edible packaging, for active packaging, and for the encapsulation of nutrients, as well as the use of nanotechnology to improve their functionality. Following is a brief discussion on the advantages of polysaccharides as packaging materials and new developments in the field.

9.2 Advantages of Polysaccharides as Packaging Material

Polysaccharides offer several advantages as packaging material. They are nontoxic, are amenable to biodegradation by enzymes, and do not produce environmentally harmful byproducts, as already mentioned in Chapter 1. Polysaccharides can easily interact among themselves or with other polymers under reasonably mild and environmentally benign conditions to form gels that can be suitably cast into films. Being of natural origin and generally edible, their migration into the packaged food does not pose any health

TABLE 9.1

Requirements for Food Packaging

Packaging should maintain or enhance food quality in terms of sensory properties.

Packaging should maintain the necessary microbiological standards (should not support growth of unwanted microorganisms).

Packaging should be amenable to pasteurization or sterilization, if necessary.

Manufacturing process should be simple and economic.

Packaging should have adequate mechanical properties.

Packaging should be sealable.

Packaging should facilitate distribution.

Packaging should conform to industry requirements (e.g., size, palletization).

Packaging should carry the required codes (e.g., bar code, sell-by).

Packaging should have an aesthetically pleasing appearance.

Packaging should reflect culture-specific consumer preferences.

TABLE 9.2

Advantages of Biopolymer-Based Packaging Materials

Edible

Biodegradable

Low-cost and abundant renewable raw material

Can be applied directly on the food surface

Free of toxic materials and safe for health

Good adhesives properties

Retain nutritional value of foods

Control microbial spoilage and hazards

Retain sensory characteristics of food

Function as carriers for antimicrobial and antioxidant agents

Control over intercomponent migration of moisture, gases, lipids, solutes

Individual packaging of small particulate foods

Microencapsulation and controlled release of flavor-bearing compounds

Potential to develop multicomponent food packaging

Reduced packaging volume, weight, and waste

hazard, unlike the case of synthetic packaging materials. As a packaging material, polysaccharides impart structural cohesion. They are generally less permeable to gases but are poor barriers to water vapor; nevertheless, their permeability characteristics can be modified. Polysaccharide-based films can function as active packaging designed for microbial control, lipid oxidation reduction, and flavor retention.[2,3]

9.3 Some Recent Concepts and Techniques

9.3.1 Hurdle Technology

The microbial stability and safety of most foods depend on a system of several barriers (hurdles) against harmful microorganisms. In a stable product, these hurdles control microbial spoilage and prevent food poisoning. The application of this concept is referred to as *hurdle technology, combined processes, combination preservation*, or *barrier technology*. This approach has proved to be very effective, as an intelligent combination of hurdles maintains microbial stability and safety as well as the sensory, nutritive, and economic properties of a food. About 40 hurdles have been identified, including temperature, water activity, acidity (pH), redox potential (Eh), competitive microorganisms and preservatives, oxygen tension, smoking, modified atmosphere, pressure, radiation and other physical processes, microstructure of the food, and packaging (e.g., selective permeable films, advanced edible coatings).[4] The

TABLE 9.3

Efforts Required in Optimization of Edible Packaging for Food Packaging

Parameter	Quality Changes to Control or Improve	Actions Required
Chemical	Fat degradation and rancidity development, flavor changes, browning, protein degradation, loss of vitamins	Control of oxygen permeability, active packaging containing oxygen scavengers, multicomponent packaging, nanotechnology
Microbiological	Growth of spoilage-causing microorganisms, control of microbial hazards	Improved oxygen barrier, inclusion of oxygen absorbers, improved moisture barrier, multicomponent packaging, nanotechnology, antimicrobial activity of film, active packaging containing antimicrobial agents in film
Physical handling	Texture of packaging material	Improved casting/film-making techniques, chemical modification of raw material, multicomponent packaging, nanotechnology

hurdles in stable and safe foods differ in quality and intensity, depending on the particular product. These hurdles must keep the normal population of microorganisms in the food under control and not allow them to overcome the hurdles that are in place. In foods preserved by hurdle technology, the various hurdles in a food might have not only an additive effect on stability but also a synergistic one.

9.3.2 Modified Atmosphere Packaging

Modification of the gas atmosphere inside a package has been shown to enhance the shelf life of food products contained inside the pouch. Such modifications increase the shelf life of a product by slowing down degradative processes, particularly microbial spoilage reactions. Another advantage is control of lipid oxidation and the associated development of rancidity and loss of sensory quality of the food by reducing the level of oxygen in the package. The technique involves flushing and filling the food packages with nitrogen or a combination of different gases such as carbon dioxide and nitrogen. During storage, the gas composition may change as a consequence of varying permeabilities of packaging material and the chemical and biological activities of the packaged food. Storage temperature also influences the gas concentration. A decrease in carbon dioxide level in packages has been reported in shellfish stored in a modified atmosphere due to dissolution of the gas in the muscle tissue. This technique can significantly extend the shelf life of perishable commodities, including fishery products; reduce economic losses; and allow distribution of products over longer distances with less frequent deliveries at the points of purchase. The potential exists to replace synthetic packaging, at least partially, with edible films, many of which may possess good barrier properties against gases and moisture and have antimicrobial and antioxidant activities.[5]

9.3.3 Active Packaging

In active packaging, sometimes referred to as *smart packaging*, the packaging and the environment interact during food preparation and storage, resulting either in improved product quality, safety, and an extended shelf life or in the attainment of some product characteristics that cannot be obtained by any other means. The role of active packaging, in conjunction with other food processing and packaging, is to enhance the preservation of contained food and beverage products. Active packaging senses and responds to internal or external environmental changes in a package with a view to maintaining the quality of the product. The primary active packaging systems currently in use involve oxygen scavenging, moisture absorption and control, carbon dioxide and ethanol generation, and antimicrobial systems. The most developed active packaging technology is perhaps oxygen scavenging.

The reduction of oxygen in a package can inhibit oxidative reactions as well as the growth of aerobic microorganisms; however, a reduction in oxygen concentration to a very low level may encourage the growth of anaerobic pathogenic microorganisms such as *Clostridium botulinum* in the package.

Active packaging containing antioxidants can be developed to control lipid oxidation in foods. Antioxidants incorporated into films control oxidation of lipids in foods coated with the films. Coatings containing antioxidants can slow the browning reactions and loss of vitamin C. The technology has attracted much attention from the food industry because of increased consumer demand for minimally processed, preservative-free products.

Biopack Environmental Solutions has developed a new concept in food packaging that involves the use of oxygen scavengers, which are preservatives encapsulated in cyclodextrins incorporated into polylactic acid and poly 3-hydroxybutyrate as active protective agents. This system was designed to improve the quality and safety of cheeses in Europe. The Biopack principles can be applied to the packaging of other foods.[6,7]

Antimicrobial packaging is a promising form of active packaging aimed at enhancing the shelf life of food products through the control of microbial growth. The compounds used may be of animal (lactoperoxidase, lysozyme, chitosan), plant (essential oils, aldehydes, esters, herbs, spices), or microbial (nisin) origin. Most of the new antimicrobial packaging materials being developed exploit natural agents to control common foodborne microorganisms. The bacteriocins nisin and lysozyme are the antimicrobials most commonly incorporated into films. Lysozyme (1,4-β-N-acetylmuramidase C, EC.3.2.1.17) is a 14,600-Da enzyme present in avian eggs, milk, tears, and fish. It is active against Gram-positive bacteria. Nisin is a 34-amino-acid peptide secreted by *Lactococcus lactis* that inhibits Gram-positive bacteria; it has a molecular weight of 3500 Da and usually exists as a dimer. Other materials include food-grade acids and salts, chitosan (due to its antimicrobial activities), plant extracts, and the enzyme lactoperoxidase.

Antimicrobial films function based on the diffusivity of the preservatives incorporated into the films. Only low levels of these compounds should come into contact with the food. The antimicrobial substances in the packaging materials function by extending the lag period of microbes in the food. Care needs to be taken in developing antimicrobial films so that the process of incorporation of the compounds does not adversely affect their bioactivity. In addition, the films must be safe for food use, and an emphasis on safety features associated with the addition of antimicrobial agents is important in the development of antimicrobial food packaging.[8–11] The use of antimicrobial film is still a challenge for researchers due to regulatory restrictions and the concentrations necessary to ensure the safety of the food products. The U.S. Food and Drug Administration requires a reduction of microorganisms of up to 5 log CFU/g with respect to the most resistant pathogenic microorganism.[12]

Methods to determine the penetration rate of microorganisms from the outside environment through packaging materials to the contained food product or *vice versa* include the closed-bottle biological oxygen demand (BOD) test, modified Sturm test, and semi-closed-bottle test. In the BOD method, filled food packages are immersed in a tank of bacteria-inoculated water and incubated for several weeks. Microbial permeation can be observed when the immersed food spoils before packaged food not exposed to the tank water due to enhanced microbial growth rate, pH changes, gas production, etc. Results are expressed as the difference in the $\log_{10}$ CFU of a test organism exposed to a control film and the $\log_{10}$ CFU of the organism exposed to the antimicrobial film. For antimicrobials such as nisin or antimicrobial acids with a long history of studying their incorporation into antimicrobial films, the majority of results have centered around a reduction of 2 log cycles.[8–11]

9.3.4 Encapsulation and Delivery of Nutraceuticals

Encapsulation of nutraceuticals using a continuous film or coating protects them against the surrounding conditions. Microencapsulation is encapsulation around microscopic particles having sizes on the order of microns. The active materials form the core, which is surrounded by a protective sheath that is edible.[13,14] Preferably it should be insoluble in aqueous solution, solid at oral cavity temperatures, and degradable by stomach and intestinal enzymes. Depending on the size of the particle, encapsulated particles are considered to be microencapsules when their size ranges between 0.2 and 5000 μm, macrocapsules when the range is larger than 5000 μm, and nanocapsules when they are smaller than 0.2 μm. Encapsulation protects sensitive materials such as vitamins, minerals, oils, aromas, sweeteners, colorants, preservatives, antioxidants, amino acids, nutrients, and probiotic bacteria against heat, moisture, light, and redox changes. Encapsulation prolongs the shelf life of the product, provides better delivery, masks flavors or odors, prevents ingredient interactions and keeps incompatible substances separate, and controls the release of core material through the surrounding shell.[1] Bioactive molecules and beneficial microorganisms may be protected by encapsulation during transit in the digestive system to their absorption sites.

The use of encapsulation and microencapsulation of bioactive substances, nutrients, and probiotics is growing in the food and pharmaceutical industries. The major bioactive lipids and other nutraceuticals that are delivered within the food industry include omega-3 fatty acids and phytosterols, among others. The use of microcapsules containing aroma compounds, vitamins, or additives in hydrocolloid-based edible films has been reported Techniques for encapsulated delivery of these compounds include emulsion-based technologies and hydrogel particles. Emulsion technology is particularly suited for the design and fabrication of delivery systems for encapsulating bioactive lipids. Oil-in-water (O/W) emulsions are currently the most widely used encapsulated delivery systems and carriers for such major nutraceuticals as omega-3 fatty acids; conjugated linoleic acids; carotenoids such as astaxanthine, lycopene, and β-carotene; and phytosterols in the food, medical, and pharmaceutical industries. O/W emulsions stabilized by polysaccharides have been used as delivery systems for omega-3 fatty acids to incorporate them into various food products, including milk, yogurts, ice cream, and meat patties.

Hydrophilic components (e.g., minerals, vitamins, enzymes, proteins, bioactive peptides, and fibers) can be trapped within the internal water phase, and emulsion-based delivery systems can be specifically designed to prevent oxidation of polyunsaturated fats within the lipid droplets. Encapsulated forms of ingredients such as fish oils are useful as supplements in various products.[14,15] Hydrogels are also useful as delivery systems. Polyelectrolyte hydrogels formed by alginate and chitosan can be used for the encapsulation and controlled release of food ingredients, cells, enzymes, and therapeutic agents.

It is important to note that the addition of active molecules may sometimes cause interactions with the film or encapsulating matrix. Physicochemical properties of the aroma compounds, such as molecular weight, structure, hydrophobicity, chemical function, solubility, volatility, partial vapor pressure, and polarity, have an effect on the affinity for the polymer matrix and on the transfer kinetics. These interactions, which determine the film stability and properties, must be monitored.[14] The application of nanotechnology to develop nanoparticles is a promising tool for drug and nutraceutical delivery (see Section 9.13). Table 9.4 show some major lipophilic nutraceutical components that are delivered in foods, and Table 9.5 lists the features and requirements for a typical delivery system.

TABLE 9.4

Major Lipophilic Nutraceutical Components Delivered into Foods

Class	Components	Potential Nutritional Benefits
Fatty acids	Omega-3 fatty acids, conjugated linoleic acid	Prevention of coronary heart disease, stroke, and cancer; bone health; immune response disorders; weight gain; improves bone health; vision
Carotenoids	β-Carotene, lycophene, lutein, and zeaxanthin	Prevention of cancer, coronary heart disease, macular degeneration, and cataracts
Antioxidants	Tocopherols, flavonoids, polyphenols	Prevention of cancer and coronary heart disease
Phytosterols	Stigmasterol, β-sitosterol, and campesterol	Prevention of coronary heart disease

9.4 Edible Films

An edible film or coating has been defined as a packaging, film, coating, or thin protective layer that is an integral part of the food and can be eaten with it.[15,16] These films are variably referred to as biodegradable, biocompatible, environmentally friendly, renewable, biopolymer, edible, or green. Edible coatings have been in use for some time; for example, soy films have traditionally been employed in the Orient to wrap and shape ground meats or vegetables. Similarly, gelatin has been used as sausage casing. Current interest in this area is directed toward replacing, at least partially, synthetic materials that have been associated with environmental problems. Edible films, like conventional synthetic films, can preserve food quality by controlling moisture transfer, oxygen uptake, or the loss of volatile aroma compounds. Being edible, these materials can also be used as direct coatings on the food surface to create a modified atmosphere within the food.

TABLE 9.5

Features and Requirements for a Typical Delivery System

Features	Characteristics
Food grade	The delivery system must be fabricated entirely from permitted food ingredients using processing operations that have regulatory approval.
Economic production	The delivery system should be capable of being economically manufactured from inexpensive ingredients; it should have improved shelf life, enhanced marketability, novel functionality, better bioavailability, etc. to outweigh the additional costs associated with encapsulation.
Protection against chemical degradation	The delivery system should protect an encapsulated bioactive lipid against some forms of chemical degradation (e.g., oxidation and hydrolysis).
Loading capacity and retention	The delivery system should be capable of encapsulating a relatively large amount of bioactive lipid per unit mass of carrier material and should efficiently retain the encapsulated component until delivery.
Delivery mechanism	The delivery system should release the bioactive lipid at a particular site of action, at a controlled rate or in response to a specific environmental stimulus such as pH or temperature.
Food matrix compatibility	The delivery system should be compatible with the surrounding food matrix without adversely affecting the appearance, texture, flavor, or stability of the final product.
Bioavailability/bioactivity	The delivery system should enhance (or at least not adversely affect) the bioavailability/bioactivity of the encapsulated component.

Source: Adapted from McClements, D.J. et al., *Crit. Rev. Food Sci. Nutr.*, 49, 577, 2009. With permission from Taylor & Francis, Ltd.

Raw materials for such environmentally friendly and edible packaging are available from agriculture and include various types of polysaccharides and proteins, such as starch and cellulose, pectin (agrowastes), zein (corn protein), butylhydroxybutyrate (microbial product), polylactic acid and whey (milk protein), and collagen and gelatin (animal skin, tendon, and connective tissue). These materials are biocompatible and replenishable and can be recovered from the processing discards of ever-growing agriculture wastes. Biodegradability is a characteristic property of these films, as they are amenable to degradation to carbon dioxide within a reasonable period of time.[3,17–19]

9.4.1 Casting of Edible Films

Various methods for casting biodegradable, edible films have been reported (e.g., wet casting, spraying, dipping, enrobing). Generally, before casting, the polysaccharides or proteins in aqueous solutions are allowed to gel under appropriate conditions—for example, in the presence of certain cations. Plasticizers are generally incorporated in the film to improve its properties. For wet casting, the dilute aqueous solution at a concentration of about 1% (w/v) is deaerated and poured in a thin layer on a suitable base material and then subjected to drying, generally at room temperature, for 2 to 3 days. The film is then easily peeled off without any tearing or wrinkling. For emulsion-based films, high-melting-point fat is dispersed in the film-making solution to enhance the moisture barrier properties of the film. After drying under standardized conditions of temperature (e.g., 30°C) and relative humidity (40%), emulsified films were found to contain varying contents of fat (30 to 90% dry basis); submitting them to heat treatment (100 to 200°C) for 1 to 8 minutes led to an apparent bilayer structure.[20] Extrusion technology, which is employed in the food industry for the manufacture of a wide range of food products, can be used to prepare these films. In using extrusion technology, apart from the type of raw material, a number of parameters must be optimized, such as the temperature of extrusion, feeding rate of material, screw speed, and concentration of plasticizer such as glycerol.[21]

Edible coatings can also be directly applied to the surfaces of food products by dipping the food in solutions of these compounds, instead of using their films to contain the food. Coating vegetative materials with gelling agents enhances the storage life of these products. The gelled film collapses and adheres to the vegetative tissue during storage. Critical surface tensions of the solid object to be coated, wettability of the surfaces by the gum (polysaccharide) solution, the composition and polarity of the films, and the surface tension of gum coating solutions are among the critical properties that influence performance of a coating. Better compatibility between the coated object and the coating films can be achieved by incorporating surface-active agents within the coating gum solution. Several edible coatings for preserving fruits such as oranges, apples, and grapefruits have been successfully applied, with their success being dependent on the control of internal gas composition. Guidelines for selecting coatings based on their gas permeation properties relative to controlling internal gas composition have been developed to maximize the quality and shelf life of fresh fruits and vegetables.[22,23]

9.4.2 Functional Properties of Edible Films

The most important functional property of a food packaging film, whether synthetic or edible, is its water vapor permeability (WVP), which influences moisture migration from the packaged food. Control of moisture migration in composite food products or between a food product and its environment remains a major challenge in food preservation. WVP involves the sorption, diffusion, and adsorption of water molecules and is largely governed by interactions between the polymer and the water. Water permeation through a film usually occurs through the hydrophilic part of the film; therefore, the ratio of the hydrophilic and hydrophobic portions is important to the WVP of a film. Polymers with high hydrogen bonding produce films that are susceptible to moisture, whereas polymers with hydrophobic groups make excellent barriers to moisture. WVP is also dependent on the pore size of the film. WVP tends to increase with polarity, degree of unsaturation, and degree of ramification of the lipids used (if any). The biopackaging compounds should have low water solubility so their films can efficiently protect foodstuffs having high or intermediate water activity, without dissolving into the moisture. Water activity (a_w) is important in determining the growth of microorganisms in the food. Microorganisms require certain critical a_w values for proliferation. Generally, the growth of bacteria, yeast, and mold is inhibited below a_w values of 0.85, 0.70, and 0.60, respectively. Deteriorative chemical and enzymatic reactions are also strongly influenced by a_w. WVP gives a relative indication of the barrier efficacy in controlling deteriorative processes. For an ideal film, the water vapor permeability should be as low as possible because the film is expected to retard moisture transfer between the food and the environment, or between two components of a heterogeneous food product.[24,25]

In addition to control of moisture transfer, most foods require specific gas compositions to sustain their freshness and overall quality during storage. The packaging material must be able to protect against the transfer of oxygen, carbon dioxide, and odor compounds. The oxygen permeability of films influences lipid oxidation and associated sensory changes in the food. Films with low oxygen permeability retard lipid oxidation. Low gas permeability also prevents the loss of odor-bearing compounds from the food matrix. In most packaging applications, the gas mixture inside the package is modified by introducing varying combinations of carbon dioxide, oxygen, and nitrogen (see Section 9.3.2). The packaging film should be able to maintain the gas composition inside the package.[26]

Relative humidity also affects gas permeabilities. The oxygen and carbon dioxide permeabilities for several biodegradable and synthetic packaging films have been determined: pectin, 259 and 4132 (87% RH); chitosan, 91.4 and 1553 (96% RH); pullulan, 3.3 and 14 (93% RH); and the synthetic packaging film PET, 169 and 2156 (77% RH), respectively.[27] Apart from permeability characteristics, specific mechanical properties of films including tensile strength (TS) and elongation at break (%E) are also important. These properties determine their resistance to handling during transportation and storage.

Edible coatings have proved to be suitable as carriers of preservatives such as antimicrobials and antioxidants through active packaging (see earlier discussion), thus further contributing to maintaining the quality, appearance, structure, and stability of foods. An added advantage of such active packaging is that very small amounts of preservatives are required because the compound is concentrated at the product surface. In addition, the films are capable of carrying various nutrients, thus offering the potential for food fortification; these films can serve as excellent encapsulation systems for the release and delivery of various drugs and nutraceuticals within the food and pharmaceutical industries.

Recently, a multiscale approach was used to investigate the mass transfer properties of edible films, including wetting and water absorption. The studies showed that a critical water content threshold was required to induce an increase of the molecular mobility. Nuclear magnetic resonance (NMR) studies contributed to a better understanding and identification of the interactions between the polymer and the diffusant and also the diffusion coefficient of the liquid water in the film.[28]

Edible films can be used to protect fresh agricultural products. Low gas barrier properties maintain the fresh appearance of fruits and vegetable by controlling oxygen-induced pigment oxidation. A low WVP protects against remoistening or drying of cereal-based products. When the edible coatings are incorporated with active compounds such as antimicrobials or antioxidants, such packaging and coatings may very efficiently protect food against microbial spoilage.[16,19,29]

9.4.3 Modification of Film Properties

The functional and mechanical properties of edible films are greatly dependent on the chemical nature of the macromolecules used to prepare the film. It is possible to modify the physical properties of films through physical and chemical treatments. Plasticizers such as glycerol modify the WVP, transparency, and mechanical properties such as tensile strength and elongation at break. In general, tensile strength and elastic modulus are more sensitive to changes in glycerol content and a_w. The oxygen permeability of films can be modified by incorporating additives such as stearic acid (in the case of methylcellulose film); inclusion of ascorbic acid has an opposite effect.[26] Physical treatments such as ohmic treatment can also modify the properties of films. Ohmic heating of film significantly affects water vapor, oxygen, and carbon dioxide permeability and the strength of chitosan film, in addition to giving the film a more uniform appearance, presumably due to rearrangement of the chitosan molecules in the gel as a result of applying the electric field.[30] Chemical modifications such as sulfation, oxidation, and amidation, which decrease or increase the hydrophilic character of the polymer, can have a significant effect on the film properties.[31] The barrier properties of edible films can also be optimized through development of multiple-layer packaging (see Section 9.12).

9.4.4 Challenges in Developing Bio-Based Packaging

Bio-based packaging, like conventional packaging, should minimize deteriorative changes in packaged food products. The bio-based material must remain stable, maintaining mechanical and barrier properties throughout the stipulated storage period of the food. Because the material is biodegradable, environmental conditions conducive to biodegradation must be avoided during storage of the food product, whereas optimal conditions for biodegradation must exist after discarding. This is a major challenge because many of the factors that influence biodegradation (e.g., water activity, presence of microorganisms, temperature, composition of bio-material) also affect the rate of deterioration of the packaged food. In the case of edible films, they may be required to operate as localized packages providing barriers to moisture or gases while the food is stored, yet they must become part of the food at the point of consumption. Bio-based packaging, through its barrier properties, must have the capacity to control one or more of the deteriorative changes, such as microbial decay or lipid oxidation, that occur in foods. The functionality and life of the packaging material should match the durability of the product shelf-life. Further, like conventional packaging, producers of bio-based packaging may need to supply consumers with mandatory product information as well as optional information such as cooking directions, recipes, etc.

These requirements pose challenges in the development of bio-based packaging. It is unlikely that one polymer will meet all of the required properties (very low gas permeability or high water resistance) for a particular product. Ways to address these challenges include the use of multiple materials in a composite, a laminate or co-extruded material, and cross-linking of the components to make copolymers to improve their stability, compatibility, and barrier properties. These techniques are also beneficial to improve the applications of polysaccharide films as supplementary coatings to conventional packaging materials.[16]

9.5 Edible, Biodegradable Films from Marine Polysaccharides

Marine polysaccharides such as chitosan, alginate, carrageenan, agar, and fucoidan are ideal raw materials for edible, biodegradable films because of their ability to form gels. It is also possible to develop films from the microbial polysaccharides secreted during cultivation of many marine microorganisms. These polysaccharides, under appropriate conditions of temperature, pH, and the presence of cations, form gels involving the intermolecular association or cross-linking of polymer chains resulting in semi-rigid, three-dimensional matrices that immobilize large volumes of water. These gels could be converted into films for packaging a variety of food products. Polysaccharide coatings, however, are generally less permeable to gases and more permeable to water vapor. The high water vapor permeability is due to the presence of significant amounts of hydrophilic groups in their chemical structures, particularly hydroxyl groups. As noted earlier, these properties can be modified by employing appropriate techniques.[19]

9.6 Chitosan

Chitin and its deacetylated form, chitosan, are inexpensive, nontoxic, biodegradable, and biocompatible. Chitosan is more versatile as compared to its precursor, chitin, because of its structural features and potential to develop films having different barrier and functional properties. Chitosan films can be cast from acetic, formic, or dilute hydrochloric acid solutions. Glycerin-plasticized films were cast from a 3% (w/w) solution of chitosan in 1% acetic acid. In addition to films, chitosan can also be formed into fibers, films, gels, sponges, beads, or even nanoparticles. Microcrystalline chitosan has superior film-forming properties.[32–34]

9.6.1 Barrier Properties

Chitosan films are highly impermeable to gases such as oxygen and CO_2; however, they have relatively high water vapor permeability due to the hydrophilic nature of chitosan. Barrier and mechanical properties depend on a number of factors, such as molecular weight, acetyl groups, plasticizer used, pH, and storage period, among others.[35] In a recent study, WVP, tensile strength (TS), and elongation (E) were found to be significantly affected by acid type, pH, and degree of deacetylation (DA). Low DA chitosan films had lower WVP and higher TS compared with high DA chitosan films. The E values were not affected by DA. As pH increased, the WVP of chitosan films tended to increase while TS decreased significantly. Chitosan films with acetic and propionic acid solvents had low WVP and high TS, while films containing lactic acid had high E and the lowest TS. Also, deacetylated chitosan may have lower crystallinity and maximum swelling.[36,37]

The water vapor permeability of films increases with increasing relative humidity (which determines water vapor pressure). In one study, the mean WVP ranged from 6.7 to 1146 (g/m^2/day) over the range of 11 to 84% RH. For chitosan film plasticized with glycerin, the permeability decreased during the first 2 weeks of storage. After that, the mean oxygen permeability (4 × 10^{-5} cc/m·atm·day) and mean ethylene permeability (2.3 × 10^{-4} cc/m·atm·day) remained constant while mean water permeability (2.2 × 10^{-1} g/m·day) decreased with respect to storage time. Tensile strength values (15 to 30 MPa) decreased and percentage elongation (%E) values (25 to 45%) increased during storage of edible chitosan films.[32] The films of acetate of chitosan from crawfish maintained lower moisture contents at any relative humidity level compared with chitosan formate films. The molecular weight of chitosan significantly influenced the sorption isotherm of chitosan formate films but not chitosan acetate films. The apparent viscosities of the coatings were dependent on the extent of acetyl groups.[34] In another study, the oxygen, carbon dioxide, nitrogen, and water vapor permeabilities of the chitosan film were reported to be 5.34 × 10^{-2} (cm^3/m·atm·day), 0.17 (cm^3/m·atm·day), 0.03 (cm^3/m·atm·day), and 2.92 × 10^{-10} (g water m/m^2·s·Pa), respectively. The glass transition temperature (T_g) of the film was 80°C.[38]

The addition of glycerol and other polyols as plasticizers influences the barrier and mechanical properties of chitosan film. In a recent study, the film properties of chitosan film were reported to be as follows: WVP, 3.3 to 9.6 × 10^{-11} g/m·s·Pa; tensile strength, 16.8 to 51.1 MPa; and %E, 1.3 to 10.7%. Both WVP and E% decreased with an increase in both glycerol concentration and pH of the film-forming solution, while tensile strength increased.[39] In another report, the tensile strength of chitosan films decreased with the addition of glycerol and other polyols (sorbitol and polyethylene glycol) and fatty acids (stearic and palmitic acids) in the film-making solution, whereas the percent elongation was increased in films containing polyols. Films containing fatty acids showed no significant differences in the properties. Glycerol blend films showed a decrease, whereas sorbitol and polyethylene glycol (PEG) blend films showed an increase in the water vapor permeability values. No considerable differences in WVP were observed in fatty acid blend films. The equilibrium moisture content of all of the films was low at lower water activity (a_w), but increased at higher a_w.[40,41] Infrared drying of wet-cast chitosan film was better than conventional oven drying at 80° to 100°C. No significant differences were observed in elongation values. Water vapor and oxygen transmission rate values were slightly reduced in oven- and infrared-dried films compared to those of 27°C dried films.[42] A patent has been granted that relates to the development of chitosan film containing plasticizers and/or emulsifiers and/or waxes for biomedical and food uses; the additives contain one or more of various acids—namely, acetic, formic, lactic, citric, oxalic, tartaric, maleic, and adipic.[43]

The mechanical and barrier properties of chitosan can be modified by chemical derivatization of chitosan, such as alkylation. The alkyl moiety may reduce the inter- and intrachain hydrogen bonds, introducing plastic characteristics to the derivative, as well as improving its WVP. Also, the alkyl moiety can play an important role in increasing the antimicrobial activity of chitosan film by conferring a cationic polyelectrolyte character to the polysaccharide. This modification can be accomplished by quaternization of nitrogen atoms in the amino groups via extensive methylation and alkylation processes. Cross-linked chitosan films offer greater strength and resistance for handling in many of the above-mentioned applications.[18] In general, the alkyl–chitosan derivatives appear to be more plastic than chitosan films but less resistant. Thus, while butyl chitosan exhibited a maximum elongation strain of 13.1%, tensile strength of 13.4 MPa, and Young's modulus of 171 MPa, the quaternary salt of dodecyl chitosan exhibited a maximum strain of 2.6%. However, the latter had a tensile strength of 38.3 MPa and Young's modulus of 1792 MPa (Table 9.6). By dynamic mechanical analysis, all chitosan films and their derivatives revealed similar nonlinear viscoelastic behavior, exhibiting a stress–strain curve typical of brittle material characterized by a decrease in the percentage of elongation at break[44] (see Figure 9.1). The water-soluble *N,O*-carboxymethyl chitosan (NOCC), produced by reacting chitin with chloroacetic acid under alkaline conditions, can be a candidate for edible coating. A commercial NOCC coating reportedly had some success as a selectively permeable postharvest coating for fresh fruit.[16] Table 9.7 compares the permeabilities of chitosan and some synthetic packaging materials.

TABLE 9.6

Young's Modulus, Tensile Strength, and Maximum Strain Values for Chitosan Films and Derivatives

Sample	Young's Modulus	Tensile Strength	Maximum Strain
Chitosan	2383	>44	2.8
Chitosan quaternary salt	1529	38.8	3.7
Butyl chitosan	171	13.4	13.1
Butyl chitosan quaternary salt	1875	45.8	3.1
Octyl chitosan	642	24.3	5.6
Octyl chitosan quaternary salt	1516	40.5	3.5
Dodecyl chitosan	511	16.9	4.0
Dodecyl chitosan quaternary salt	1792	38.3	2.6

Source: de Britto, D. and de Assis, O.B.G., *Int. J. Biol. Macromol.*, 41, 198, 2007. With permission from Rightslink.

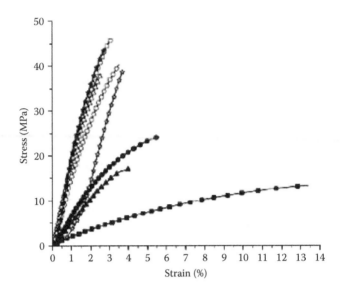

FIGURE 9.1

Stress–strain relationship for films of chitosan (■), butyl chitosan (▦), octyl chitosan (●), and dodecyl chitosan (▲) at room temperature. The open symbols represent films of corresponding quaternary salts. (From de Britto, D. and de Assis, O.B.G., *Int. J. Biol. Macromol.*, 41, 198, 2007. With permission from Rightslink.)

9.6.2 Antimicrobial Activities

The antimicrobial activity of chitosan was discussed in Chapter 5. Chitosan films exhibit antimicrobial activities against a variety of microorganisms. The activity is also observed when food portions are dipped or sprayed with film-forming solutions of chitosan.[45] Diffusion of acetic or propionic

TABLE 9.7

Comparison of Chitosan and Synthetic Packaging Material Permeabilities

Film	O Permeability $(m^3 \cdot m/m^2 \cdot s \cdot Pa)$	CO Permeability $(m^3 \cdot m/m^2 \cdot s \cdot Pa)$	Water Vapor Permeability $(g \cdot m/m^2 \cdot s \cdot Pa)$
Polyester	2.69×10^{-19}	2.61×10^{-17}	3.6×10^{-13}
Polypropylene (PP)	5.5×10^{-17}	—	6.5×10^{-13}
Low-density polyethylene (LDPE)	2.25×10^{-17}	—	8.1×10^{-13}
High-density polyethylene (HDPE)	5.02×10^{-18}	—	2.52×10^{-13}
Polyvinyl chloride (PVC)	5.15×10^{-19}	$1.35 \times 10^{-18} – 2.7 \times 10^{-17}$	2.16×10^{-11}
Chitosan	1.4×10^{-19}	—	$.9 \times 10^{-10}$

Note: Values measured at $25 \pm 2°C$ and 50 to 70% relative humidity.

Source: Adapted from Lin, D. and Zhao, Y., *Comp. Rev. Food Sci. Food Safety*, 6, 60, 2007. With permission from Taylor & Francis, Ltd.

acids from the chitosan films in which they are incorporated can also be responsible for antimicrobial activity, suggesting the possible use of chitosan films for antimicrobial packaging.[46]

Chitosan acetate films maintained lower moisture contents at any relative humidity level compared with chitosan formate films. The type of chitosan significantly influenced the sorption isotherms of chitosan formate films but not chitosan acetate films.[47]

The antimicrobial activity of a chitosan coating increases the lag phase of microorganisms, including *Staphylococcus aureus*, *Propionobacterium propionicum*, *Listeria monocytogenes*, *Pseudomonas aeruginosa*, Enterobacteriaceae, and *Serratia liquefaciens*.[48] *Aspergillus niger* was almost completed inhibited when either a film-forming solution or chitosan film at a concentration of 0.1% (w/v) was used. These properties can have potential for the preservation of various food products, including dairy items.[49,50]

The method of production can influence the antimicrobial activities of chitosan films. Films made by both the heat-press method and casting inhibited microorganisms; cast films with a biopolymer coating exhibited higher antimicrobial activity than heat-pressed films, due to its ability to release more nisin.[51,52] Chitosan acetate films cast at 37°C and 80°C demonstrated a significant inhibitory effect on *S. aureus* and *Salmonella*, The antimicrobial capacity diminished to a large extent during storage at 23°C and 75% RH for 2 months or alternatively when stored at 37°C and 0% RH over the same period of time.[53]

The antimicrobial properties of chitosan could be enhanced by active packaging through the inclusion of antimicrobials such as nisin, acetic acid, and propionic acid. When acetic acid was incorporated in the film, diffusion of the acid from film 44 to 54 μm thick was found to be unaffected by pH in the range of 5.7 to 7.0, but was reduced with a decrease in temperature from 24°C to 4°C. Mixed films consisting of chitosan and lauric acid have lower water permeability, providing improved moisture barriers. Incorporation of lauric acid or essential oils (cinnamaldehyde or eugenol) into the chitosan film at the time of preparation produced a subsequent reduction in the diffusion of acetic or propionic acid. Maximum inhibitory effects were obtained with lauric acid and cinnamaldehyde incorporated to final concentrations of 1.0% and 0.5% (w/w), respectively.[54]

Antimicrobial film was made by incorporating garlic oil (GO) as a natural antimicrobial agent. Incorporation of GO up to 0.4% resulted in a decrease in tensile strength and elongation at break, and a slight increase in water vapor permeability. The film containing 0.1% GO was shown to have an inhibitory effect on *Staphylococcus aureus* and *Listeria monocytogenes*.[55]

Antimicrobial and physicochemical properties of chitosan films can be obtained by enriching the films with essential oils such as anise, basil, coriander, and oregano oil. The intensity of antimicrobial efficacy was in the following order: oregano > coriander > basil > anise. The antibacterial effects of the oils against *L. monocytogenes* and *Escherichia coli* O157:H7 were similar when applied alone or incorporated in the films. The chitosan films and chitosan–oregano films were applied on inoculated bologna samples and stored 5 days at 10°C. Pure chitosan films reduced the numbers of *L. monocytogenes* by 2 logs, whereas films with 1% and 2% oregano essential oil decreased the numbers of *L. monocytogenes* by 3.6 to 4 logs and *E. coli* by 3 logs. Inclusion of 1% and 2% oregano essential oil increased the thickness of films from 89 μm to 220 and 318 μm, respectively. The essential oil addition also decreased the water vapor permeability and puncture and tensile strength but increased the elasticity of the films. When applied to bologna, the films absorbed moisture, resulting in final thicknesses of 143, 242, and 333 μm, respectively.[56]

Incorporation of lysozyme also enhanced the antimicrobial activity of chitosan. A 10% lysozyme solution was incorporated into a 2% chitosan film-forming solution that was cast by solvent evaporation. The films were able to release lysozyme proportionally with increasing initial concentration in the film matrix in a natural log relationship with time. The films with 60% lysozyme enhanced the inhibition efficacy of the films against both *Streptococcus faecalis* and *Escherichia coli*, whereas a 3.8 log cycle reduction in *S. faecalis* and a 2.7 log cycle reduction in *E. coli* were achieved. The WVP of the chitosan films was not affected by lysozyme, but the tensile strength and percent elongation values decreased with an increase in lysozyme concentration.[57] The solutions of chitosan and lysozyme could be stored for up to 6 months without losing their antimicrobial activities against *E. coli* and *Listeria monocytogenes*. Increased water solubility, lysozyme release, and decreased tensile strength and elongation were observed in films made from solutions stored at 37°C, while WVP was not significantly affected. The results indicated that premade chitosan–lysozyme solutions may be distributed as a commercial product for coating or film applications within 6 months of preparation.[58]

9.6.3 Antioxidant Activity

Films made of chitosan and its derivatives also possess antioxidant activity, which can be attributed to the chelating action of chitosan on metal ions and on its ability to complex with lipids. The poor oxygen permeability of chitosan films also controls lipid oxidation. Chitosan coatings enriched with rosemary and olive oleoresins have been reported to exhibit antioxidant activity due to the inhibition of peroxidase and polyphenoloxidase and the control of browning reactions, which typically result in a loss of quality in fruits and vegetables.[60]

9.6.4 Other Benefits

Chitosan reduces undesirable changes in the emulsifying properties and fat-binding, water-holding, and gel-forming capacities caused by denaturation of myofibrillar proteins during frozen storage. A novel natural polymer chitosan–cellulose blend prepared by a spray-drying process was found to have potential applications for odor treatment as well as metal ion adsorption.[61] Chitosan coatings could also be used to coat paper used for packaging.[62] Chitosan possesses barrier properties that could be modified to suit a wide range of food applications.

9.6.5 Chitosan Film Food Applications

9.6.5.1 Agricultural Produce

Fruits and vegetables undergo many physiological changes during post-harvest storage, including tissue softening, increased sugar levels, decreased organic acids, degradation of chlorophyll, production and loss of volatile flavor compounds, decreased phenolic and amino acid contents, and breakdown of cell materials. Controlling respiration can significantly improve the storability and shelf life of fresh produce, as a certain level of respiration activity is required to prevent plant tissues from senescing and dying. In minimally processed agricultural produce, the most important quality attributes contributing to marketability are appearance, color, texture, flavor, nutritional content, and microbial quality. The marketability of agricultural produce, therefore, demands efficient control of one or more of these quality changes.[25]

Due to its barrier properties, chitosan film can prevent moisture loss and drip formation, reduce lipid oxidation, improve flavor attributes, retain color, and improve microbial stability, thus extending the shelf life of a variety of fruits and vegetables. The film controls gas exchange and reduces water loss and browning in fruits. Rather than packaging produce within a chitosan film, dipping the produce in a dilute solution of chitosan and dilute acetic acid can be performed. When fruits, tomatoes, and cucumbers are immersed in a 1% polysaccharide solution containing 0.1% $CaCl_2$, a film forms over the produce that helps control sensory changes during refrigerated storage. The technique also allows the incorporation of calcium, vitamin E, and potassium to enhance the nutritional value of fruits.[63]

Manually sliced strawberries were treated with a solution of 1% chitosan, packaged in a modified atmosphere with high (80%) and low (5%) oxygen, and stored at 4 to 15°C. The chitosan coating inhibited the growth of microorganisms and enhanced the storage stability of the products, in addition to maintaining color.[64] A chitosan coating also substantially reduced vapor loss in bell peppers and cucumbers[65] and reduced the respiration rate of peaches, pears, and kiwifruit. The coating also reduced the ethylene production of peaches markedly and increased the internal CO_2 and decreased the internal O_2 levels of pears. Chitosan significantly inhibited the growth of *Botrytis cinerea* and fungi isolated from decaying pears.[66] As noted earlier, chitosan coatings enriched with rosemary and olive oleoresins exhibited antioxidant activity when applied to butternut squash by inhibiting the enzyme peroxidase for up to 5 days of refrigerated storage.[60] Coating papaya fruit with medium-molecular-weight chitosan at 2% (w/v) suppressed the mesophilic plate count, growth of molds and yeast, and activity of the enzymes polygalacturonase and pectin methylesterase, thereby reducing deteriorative processes during refrigerated storage of the fruit.[67] Mangoes ripen rapidly after harvest and are easily infected by fungal organisms, leading to considerable post-harvest losses. Use of synthetic fungicides can lead to development of fungicide resistance and adverse environmental effects. A 2% chitosan coating can delay ripening and decay of mangoes during storage at 15°C and 85 to 90% RH. Chitosan coatings also inhibit respiration rate, loss of firmness, and color change in mangoes; furthermore, declines in titratable acidity, ascorbic acid, and fruit weight are also effectively inhibited.[68,69] In processed products such as frozen and freeze-dried fruits and vegetables, edible coatings can also improve structural integrity and prevent moisture absorption and oxidation.[25] Chitosan coatings can also control enzymatic browning in fruits.[70]

Edible coatings based on high-molecular-weight chitosan, pure or combined with methylcellulose or oleic acid, were applied to fresh-cut carrots by simple immersion and by applying a vacuum pulse (5 kPa for 4 min). Coatings improved sample appearance by controlling water vapor transmission and preserved the sample color.[71] The addition of plasticizers such as glycerol and sorbitol or Tween® 80 at 0.1% in 1.5% (w/v) chitosan film helped retain the wettability of surfaces of carrot and tomato.[72] Fresh-cut sweet potatoes coated with 1% chitosan having a molecular weight of 470 kDa retained their color. No significant changes in lightness values of 470-kDa-chitosan-coated fresh-cut sweet potatoes were observed during 17 days of storage. The coating had no effect on texture or weight loss of the product during storage.[73]

Current recommendations to extend the quality and shelf life of green asparagus suggest a storage temperature between 2 and 4°C and modified atmosphere packaging (MAP). MAP storage, however, has not gained wide acceptance in the industry. Alternatively, coating with chitosan–beeswax significantly reduced the fresh weight loss of the product compared to untreated spears when stored at 4°C for 14 days. The coating markedly reduced foul odors, thus suggesting antimicrobial activity by chitosan; however, the chitosan coating caused spears to become abnormally stiff or inflexible and yellow in color, with an astringent taste. The suitability of these coatings for use on asparagus will depend on improvements in formulations and application techniques.[74] A patented method to mask the unpleasant flavor involves the addition of catechin from tea at about a 0.5% level to the chitosan-containing food or drink.[75]

Litchi pulp is very perishable and thus has a short shelf life due to post-harvest pericarp browning. Browning occurs within 6 hours of exposure to ambient temperature when fruit are removed from storage at 2°C. Coating litchi fruit with 2% (w/v) chitosan followed by storage for 20 days at 2°C and 90 to 95% RH enhanced shelf life, delayed browning, decreased antho-cyanin content, increased polyphenol oxidase activity, reduced the decrease in concentrations of total soluble solids and titrable acidity, and partially inhibited microbial decay.[76] In another study, manually peeled litchi fruits were treated with aqueous solutions of 1 to 3% chitosan, placed into trays wrapped with plastic film, and stored at –1°C. The coating retarded weight loss, slowed declines in sensory quality, and suppressed enzymatic brown-ing; higher contents of total soluble solids, titrable acid, and ascorbic acid were noted.[77] Table 9.8 provides examples of edible coatings of marine poly-saccharides and modified atmosphere packaging for agricultural products.

TABLE 9.8

Marine Polysaccharide Edible Coatings for Agricultural Products

Product	Film Material	Benefit	Refs.
Apple	Alginate/CaCl$_2$	Inhibits native microflora during storage, enhances firmness	Moldão-Martins et al.[164]
Apple	Carrageenan	Maintains color, reduces surface dehydration during storage	Lee et al.[104]
Carrot	Alginate	Maintains initial quality	Vargas et al.[71]
Litchi fruit	Chitosan	Enhances shelf life, reduces browning	Jiang et al.[76]; Dong et al.[77]
Mushroom	Alginate	Reduces water evaporation, improves color	Nussinovitch and Kampf[95]
Melon	Alginate/CaCl$_2$	Enhances firmness	Rojas-Graü[92]; Rojas-Gratio et al.[161]
Clementine mandarin	Chitosan	Reduces water spot incidence and delays peel pigmentation	Fornes et al.[162]
Raspberry	Chitosan	Improves sensory quality	Han et al.[163]
Strawberry	Chitosan	Enhances texture and nutritional quality	Han et al.[163]
Peach, pear, kiwifruit	—	Reduces respiration rate and ethylene release, maintains quality	Du et al.[66]
Butternut squash	Chitosan	Maintains quality	Ponce et al.[60]
Papaya	Chitosan	Prevents deterioration and preserving	González-Aguilar et al.[67]
Mango	Chitosan	Delays ripening and decay during storage at 15°C and 85–90% RH	Zhu et al.[68]
Sweet potato	Chitosan	Maintains color	Waimaleongora-Ek et al.[73]
Banana	Chitosan	Extends shelf life	Kittur[144]

9.6.5.2 Seafood

Seafood products are highly perishable due to bacterial growth, as they contain relatively large quantities of free amino acids and volatile nitrogenous bases, which support bacterial growth. Increasing demand for high-quality fresh seafood has intensified the search for new methods and technologies for better fish preservation. The antimicrobial properties of chitosan coating can help enhance the shelf life of fresh fishery products.[78] Salmon, for example, is a high-quality-fat fish with considerable nutritional and economic importance. Much of the fresh salmon is sold to the consumer as whole or gutted salmon, but significant amounts are also sold as fillets. Dipping fresh salmon in chitosan solution has been found to reduce aerobic mesophilic and psychrophilic cell counts, extending the quality of the fish up to 6 days under refrigerated conditions. Because of its antioxidant activity, chitosan coatings can also reduce lipid oxidation in fresh fish. Glazing skinless pink salmon fillets with 1% (w/w) chitosan delayed lipid oxidation during 8 months of frozen storage. Fillets glazed with chitosan solution exhibited significantly higher yield and thaw yield than fillets glazed with lactic acid or distilled water.[38]

Chitosan coatings inhibited the growth of spoilage microorganisms on raw refrigerated shrimp, thus extending its shelf life. Gelatin-based edible films enriched with chitosan and an extract of either oregano (*Origanum vulgare*) or rosemary (*Rosmarinus officinalis*) in combination with high-pressure treatment at 300 MPa for 15 minutes at 20°C enhanced the shelf life of cold-smoked sardines (*Sardina pilchardus*). The chitosan and plant extract provided both antimicrobial as well as antioxidant activities, again enhancing the shelf life of the product. The combination of high pressure with edible films yielded the best results in terms of controlling lipid oxidation and microbial growth.[79] Chitosans from snow crab (*Chinoecetes opilio*) having various molecular weights and hence apparent viscosities were used to coat Atlantic cod (*Gadua morhua*) and herring (*Clupea harengus*). Coatings with a viscosity in the range of 50 to 360 cPs resulted in a 29 to 40% reduction in moisture loss in the fish fillets after storage for up to 12 days at 4°C. The coating also significantly reduced lipid oxidation in the products.[80] A chitosan fish oil coating has been shown to enhance the shelf life of lingcod (*Ophiodon elongates*) fillets. The fillets were treated with a solution containing 10% (w/w) chitosan fish oil (91% eicosapentaenoic and decosahexaenoic omega-3 fatty acids). The fresh fish fillets were vacuum-impregnated in the coating solution at 100 mmHg for 10 minutes followed by atmospheric restoration for 15 minutes and storage at 2°C for 3 weeks and –20°C for 3 months. The chitosan fish oil coating increased the total lipid and omega-3 fatty acid contents of the fish about threefold, reduced lipid oxidation in both fresh and frozen samples, and decreased the drip loss of frozen samples by about 28%. The coating resulted in a reduction in total plate and psychrotrophic counts in cold stored and frozen stored samples. The color of the fillets was not affected.[81]

9.6.5.3 Meat Products

Applying antimicrobial chitosan films inhibits surface spoilage bacteria in processed meats and enhances the shelf life of meat products.[82–84] Chitosan films reduced lipid oxidation in precooked beef patties during storage at 4°C; the coating was as effective as polyvinylchloride film in reducing moisture loss.[47,85] Coating with chitosan film in combination with gamma irradiation can enhance the refrigerated shelf life of meat products, as the coating controls microbial contamination and the development of rancidity, and the gamma irradiation reduces the initial microbial load in the sample. No viable bacteria or fungi were detected in chitosan-coated, irradiated, intermediate-moisture meat products; in contrast, similar products that were not subjected to gamma radiation showed visible fungal growth within 2 weeks. The chitosan-coated products also showed lower rancidity during storage for up to 4 weeks, which could be attributed to the low barrier properties of the film and suggests a potential role for chitosan coatings in the preparation of safe and stable irradiated intermediate-moisture meats.[86]

9.6.5.4 Poultry

Coating eggs with chitosan can provide a protective barrier against moisture and CO_2 transfer from the albumen through the egg shell, thus extending the shelf life of eggs. Coating with low-molecular-weight (470 kDa) chitosan was found to be effective in preventing weight loss. The Haugh unit and yolk index values indicated that the albumen and yolk quality of coated eggs can be preserved for up to 5 weeks at 25°C, which is at least 3 weeks longer than observed for the control, uncoated eggs. Overall consumer acceptability of coated eggs did not differ from that for control and commercial eggs.[87] Chitosan coatings containing sorbitol as a plasticizer may offer enhanced protection of eggs.[88]

Chitosan–lysozyme coating can control the growth of *Salmonella enteritidis* and reduce undesirable changes in the quality of on hard-boiled eggs. The eggs were coated with the film and inoculated with *Listeria monocytogenes* or *S. enteritidis* at 10^4 CFU/g and stored for 4 weeks at 10°C. At the end of the 4-week storage, *S. enteritidis* on chitosan–lysozyme-coated eggs had about 4-log fewer colony-forming units than the uncoated control eggs. Coatings reduced the populations of coliform and total plate counts and completely inhibited mold growth during 6 weeks of storage, and coatings reduced the weight loss of eggs at the end of 10 weeks of storage. The pH of coated eggs remained stable throughout the storage period, while that of the control eggs increased from 7.6 to 8.6. Color change in the chitosan–lysozyme-coated eggshells was less than for the control eggs.[105]

9.6.5.5 Dairy Products

Chitosan–lysozyme (CL) film-forming solutions at a pH of 4.4 to 4.5 were prepared by incorporating 0% or 60% lysozyme (per dry weight of chitosan) into solution. The prepared film was used to package sliced Mozzarella cheese preinoculated with *Listeria monocytogenes, Escherichia coli,* or *Pseudomonas fluorescens* at 10^4 CFU per g or with mold and yeast at 10^2 CFU per g. The presence of lysozyme in the film produced a greater antimicrobial effect on *P. fluorescens* and *L. monocytogenes* in the cheese compared to the sample packaged in chitosan film in the absence of lysozyme. Mold growth was completely inhibited in cheese packaged in chitosan–lysozyme films. The antimicrobial packaging also resulted in some reduction in yeast populations.[81]

9.6.5.6 Miscellaneous

Chitosan can also have food biotechnology-related applications. Chitosan purified from crab shell was used as the matrix for the immobilization of α-galactosidase, which was cross-linked with glutaraldehyde, giving a maximum 72% yield of the enzyme. The immobilized enzyme showed better thermal and storage stability than the free counterpart. The immobilized enzyme was used to hydrolyze raffinose oligosaccharides in soy milk in a continuous stirred batch reactor, which resulted in a reduction of the oligosaccharide content of the soy milk by 77%. The hydrolyzed soy milk was free from flatus-inducing factors such as raffinose and stachyose and therefore could serve as an alternative to cow's milk for those who are lactose intolerant.[89]

9.7 Alginate

Alginate is a versatile polysaccharide used in films and coatings for food products, for the microencapsulation of nutraceuticals, and as a carrier of various nutraceuticals and drugs. The high affinity of alginate for water makes it useful for controlling moisture in food products. Film-forming solutions of alginates are transparent, colorless, and noncoagulable on heating and have a wide range of viscosity. The ability of alginate to gel in the presence of divalent and trivalent cations is utilized to make alginate films. Among the cations, Ca^{2+} is more effective as a gelling agent than other polyvalent cations (see Chapter 4). A two-step procedure is generally used to apply alginate coatings. In the first step, the food products are dipped in or sprayed with an aqueous sodium alginate solution. In the second step, the treated food is dipped in a solution of $CaCl_2$, which induces gelation of the alginate. The resulting calcium alginate coating helps reduce moisture loss from the coated food and improves the texture. A number of studies have reported the barrier properties of alginate films. Alginate films at 55% RH exhibited tensile strength and elongation at break values of 49 MPa and 5.2%, respectively, with a WVP of 9.7×10^{-7} g·m/Pa·hr·m^2. Waxes or saturated lipids are capable of reducing WVP.[90]

The barrier properties of films, in general, were influenced by the content of guluronic and mannuronic acids in the alginate, presence of plasticizer, relative humidity (RH), presence of cations, etc. Alginate films containing dissimilar amounts of guluronate (G) and mannuronate (M)—M/G ≈ 0.45 and M/G ≈ 1.5—were soaked in a CaCl$_2$ solution for up to 20 minutes and were then evaluated for WVP. The M/G ≈ 0.45 films proved to be better moisture barriers compared to the M/G ≈ 1.5 films. The WVP of the M/G ≈ 0.45 and M/G ≈ 1.5 films decreased as the time of immersion in the calcium increased. As the relative humidity increased, tensile strength decreased and elongation at break increased for all films. This effect was more pronounced on films that contained plasticizers and had lower tensile strength values. Plasticizers did not increase elongation at break at 58% RH, but at 78% RH and above glycerol and sorbitol showed a significant increase in elongation at break values compared to films containing polyethylene glycol (PEG) as the plasticizer. The presence of PEG provided lower tensile strength and elongation at break values, and glycerol showed the highest tensile strength and elongation at break values among all plasticizers. The WVP in the presence of sorbitol was the lowest, in contrast to that in the presence of PEG.[91] Sodium-alginate-based films containing the antimicrobial agents lysozyme, nisin, grapefruit seed extract (GFSE), or ethylenediaminetetraacetic acid (EDTA) exhibited antimicrobial activities against a number of microorganisms. Film containing GFSE–EDTA showed the maximum inhibitory effect.[51] Incorporating garlic oil up to 0.4% resulted in a decrease in tensile strength and elongation at break and a slight increase in WVP.[55] The addition of sunflower oil with essential fatty acids (omega-3 and omega-6) at 0.025 to 0.125% (w/v) significantly reduced the WVP of alginate film.[92]

Alginate (2% w/v)-based edible coating containing 0.63% glycerol and the antibrowning agent *N*-acetylcysteine (1%) can control browning of fresh-cut apples during storage. The combination of glycerol and ascorbic acid at 1 to 2% and 1% (w/v) enhances the water barrier properties of the film. Furthermore, the addition of 1% (w/v) ascorbic acid enhanced the antioxidant activity of papaya throughout storage; the ascorbic acid in the alginate edible coatings helped to preserve the natural ascorbic acid content in fresh-cut papaya, thus helping to maintain its nutritional quality throughout storage. No substantial changes in the respiratory rate and ethylene production of the coated papayas were observed.[93] It has been suggested that alginate-based edible coatings could be used as carriers of probiotic bacteria in papaya. When 2% (w/v) alginate containing glycerol (0.6 to 2.0%), *N*-acetylcysteine (1%), ascorbic acid (1%), and/or citric acid (1%) were used to coat fresh-cut apple and papaya cylinders, the WVP of the film was 0.31×10^{-9} g·m/m^2·s·Pa. At 25°C, the water solubility of the film was 0.74 to 0.79 and the swelling ratio of the film was 1.6 to 2.0.

When fresh-cut papaya and apple were coated with 2% (w/v) alginate containing viable bifidobacteria, the WVP values of the alginate were 6.31 and 5.52×10^{-9} g·m/m²·s·Pa for the probiotic coatings of papaya and apple, respectively. Values > 10^6 CFU/g for *Bifidobacterium lactis* Bb-12 were maintained for 10 days during refrigerated storage of fresh-cut fruits, demonstrating the feasibility of alginate coatings carrying and supporting viable probiotics on fresh-cut fruit. The combination of coating followed by osmotic dehydration (OD) has been suggested as a method for the preservation of pineapple with better quality characteristics. The fruit was coated with 0.5 to 5.0% (w/v) sodium alginate solution by dipping for 60 s and 120 s and then drying at 50°C for 10 and 40 min. OD in the sucrose solution was carried out for coated and uncoated samples Studies on water loss, solid gain, performance ratio, and weight reduction suggest that the coating may control the problem of solid gain without affecting water removal during OD. The application of coating before OD can also help retain nutrients and flavor. The coated osmotically dehydrated fruits after rehydration could be added to ice cream, yogurt, and confectionary products.[94] Alginate (1%) coatings extend shelf life and preserve texture of coated mushrooms by reducing the rate of water evaporation. Alginate-coated mushrooms, stored at 20°C or at 4°C, had higher *L* values, indicating a lighter color and better appearance compared to the uncoated controls. At room temperature, the rate of water evaporation from the coated vegetables decreased. These positive effects were more pronounced in the cold-stored mushrooms.[95]

Gelation of alginate in the presence of calcium can be used to improve the texture of fruits. Processing operations may result in a dramatic loss of firmness in fruit tissues due to the action of pectic enzymes. Subcellular compartmentalization is disrupted at the cut surfaces, and the mixing of substrates and enzymes, which are normally separated, can initiate reactions that normally do not occur.[165] The most common way to control the softening phenomenon in fresh-cut fruits is to treat them with calcium salts. Calcium ions interact with pectic polymers to form a cross-linked network that increases mechanical strength, thus delaying senescence and controlling physiological disorders in fruits and vegetables. The combination of calcium with alginate can favor better texture of some fruits. Dipping in alginate solution followed by spraying or dipping in a calcium chloride solution could enhance the firmness of apple and melon.[92]

Alginate-based coatings prevented moisture loss and lipid oxidation in fishery products to some extent. A calcium alginate coating incorporating nisin could be used as antimicrobial packaging to maintain the quality of fresh fish. Northern snakehead (*Channa argus*) was coated with a calcium alginate coating incorporating nisin (1000 IU/mL) and ethylenediaminetetraacetic acid (150 μg/mL) prolonged the quality of fresh fish at 4 ± 1°C. The coating predominantly reduced chemical spoilage, reflected in total volatile base nitrogen, trimethylamine, pH, lipid oxidation, and water loss, and improved the overall sensory scores of fish fillets.[96]

Calcium alginate coatings containing lysozyme and nisin could be used to control the growth of *Listeria monocytogenes* and *Salmonella anatum* on the surface of ready-to-eat smoked salmon at refrigerated temperatures. The smoked fish samples were dipped into calcium alginate containing lysozyme, isolated from either oysters or hen eggs, and nisin; they were then stored at 4°C for 35 days. The growth of organisms at the end of storage were in the range of 2.2 to 2.8 log CFU/g, much lower compared to the untreated samples.[97] Coating beef cuts with the commercial alginate coating known as Flavor-Tex® reduced moisture loss during storage at 5°C.[98,99] A bitter taste is a concern with alginate coating that could possibly be addressed by incorporating catechin.[75]

9.8 Carrageenan

Carrageenan is a versatile material for biodegradable films. The κ- and ι-carrageenans form gels in the presence of monovalent and divalent cations (see Chapter 4) that can be used for films. Incorporating glycerine as a plasticizer favorably modified the tensile strength, flexibility, and elongation properties of carrageenan films.[100] Edible films made of ι-carrageenan were found to have good mechanical properties; they stabilized the emulsions and reduced oxygen transfer. When film-forming κ-carrageenan (or agar) dispersions were subjected to microwave irradiation in the presence of methyl methacrylate (MMA) and potassium persulfate (KPS), the prepared films were significantly hydrophobic compared to those of the parent polysaccharides. Morphological studies showed that the carrageenan film had sets of pores distributed randomly.[31]

The synergistic effects on various bacteria and pathogens of lysozyme and nisin in sodium-alginate- and κ-carrageenan-based biopolymer films were evaluated. Films were prepared using 2% (w/v) alginate and 1% (w/v) carrageenan with 0.75% plasticizer containing equal proportions of polyethylene glycol and glycerol together with 100 μg nisin per mL and 0.1% lysozyme solutions. Sodium-alginate-based films exhibited a greater extent of inhibition than κ-carrageenan-based films with the same antimicrobial compound additions. Tensile strength and elongation at break values were significantly less in films with added antimicrobials.[51] Environmentally friendly films exhibiting both antibacterial and antioxidative properties have been produced from chitosan and tetrahydrocurcuminoids (THCs). The curcuminoids were prepared from natural curcumin extracted from turmeric roots. The resulting films exhibited high antioxidant activity.[59] Table 9.9 lists some applications of marine macromolecules for antimicrobial packaging.

TABLE 9.9

Marine Macromolecules as Antimicrobial Packaging

Biopolymer	Antimicrobial Agents	Food Preserved	Refs.
Agar	Nisin	Poultry	Gennadios et al.[99]; Fan et al.[108]
Alginate	Lysozyme and nisin	Control of spoilage bacteria in foods	Park et al.[57]; Duan et al.[81]; Dutta et al.[97]
	Glucose oxidase and nisin	Fish, skim milk, beef, poultry	Lu et al.[96]; Gennadios et al.[99]
Chitosan	—	Fresh fish	Jeon et al.[80]; Gennadios et al.[99]
	Sorbate, bacteriocin	Various pathogens	Hwang et al.[101]
	Sodium benzoate and potassium sorbate; acetic/propionic acid	Butter	Lu et al.[96]
	Plant extracts such as rosemary and oleoresins	Butter squash	Ponce et al.[60]
	—	Dairy products	Coma et al.[50]
	Film containing oregano	Bologna	Coma et al.[50]; Zivanovic et al.[56]
Carrageenan	Lysozyme and nisin	Control of spoilage bacteria in foods	Lu et al.[96]

Carrageenan-based coatings have been applied for a long time to a variety of foods, including fresh and frozen meat, poultry, and fish, to prevent spoilage and surface dehydration.[99] Films prepared with 2% (w/v) κ-carrageenan and containing 0.1% KCl and 0.75% PEG or glycerol as a plasticizer controlled lipid oxidation in vacuum-packed mackerel mince stored for 25 days at temperatures ranging from +20° to –15°C. The film did not prevent weight loss of the mince during storage.[101] Locust bean gum and other mannans are often added to carrageenan gels to prevent syneresis and increase elasticity. The carrageenan coating is capable of preventing moisture loss and freezer burn in intermediate moisture foods. Carrageenan film also reduced lipid oxidation in beef patties during storage at 4°C and reduced moisture loss in the product. In this respect, the film was as effective as polyvinyl chloride (PVC) film.[102] A patent application described dipping precooked meat into an aqueous calcium carrageenan dispersion prior to freezing.[103]

Carrageenan has been used as a carrier of various additives for active packaging. Combined with antibrowning agents, 0.5% (w/v) carrageenan extended the shelf life of minimally processed apple slices by 2 weeks when stored at 3°C by maintaining color and reducing microbial counts. The antibrowning treatment involved dipping the fruit in one or more solutions of citric, ascorbic, or oxalic acids as well as organic acid plus calcium chloride mixtures. Mixtures of acid and calcium chloride had a synergistic effect on color. The treated apple slices also maintained acceptable sensory scores for firmness, flavor, and overall preference when stored up to 14 days.[104]

A combination of 0.5% carrageenan, 3% glycerol, and 5% sorbitol increased the foldability of potato starch edible paper up to 60 times. The film was reasonably transparent, with a transparency value to 76%.[107] Carrageenan could replace polyethylene as a coating of paper used to package oily or greasy foods. Both carrageenan-coated papers and films were highly impermeable to lipids, with κ-carrageenan-coated paper showing maximum impermeability followed by ι- and λ-carrageenan films. Lipid impermeability increased as the thickness of the κ-carrageenan layer increased. Carrageenan-coated papers, 4 and 5 kg per ream (278 m²), showed lipid resistance comparable to that of polyethylene-coated papers. Carrageenan-coated papers having more than 4 kg per ream were shown to have adequate lipid barrier properties for use in the packaging of greasy foods.[106] Carrageenan is also used in the manufacture of soft nongelatin capsules to carry antimicrobials and antioxidants and to reduce moisture loss, oxidation, or disintegration.[106]

9.9 Agar

Agar films are biodegradable, transparent, clear, homogeneous, and flexible and can be easily handled. The water vapor permeabilities of agar film are comparable to those of most of the polysaccharide-based films and with some protein-based films. Depending on the environmental moisture pressure, the WVP of the films varies but remains constant when the relative humidity is above 84%.[108] The biodegradability of the film was demonstrated by the fact that, when subjected to natural weathering exposure in a humid tropical climate for 90 days, the film underwent changes in mechanical, thermal, structural, and morphological properties as a result of photodegradation process. These changes altered the crystallinity of the polysaccharide, causing contraction that led to formation of microfractures, which promoted microbial attack. Accelerated weathering exposure of agar films showed that outdoor climate parameters played an important role in their degradation.[109]

9.10 Microbial Polysaccharides

Microbial polysaccharides have been shown to be amenable to film development. Pullulan, a water-soluble glucan produced by the yeast-like fungus *Aureobasidium pullulans*, has been well studied for film development and its applications. The films are clear, nontoxic, impermeable to oxygen, biodegradable, and edible. They readily dissolve in water and melt in the mouth. As edible food coatings, pullulan films as thin as 5 to 60 μm can be formed from a solution of 5 to 10 g of the glucan in 150 mL distilled water, which is cast onto a Teflon®-coated glass plate followed by drying at room temperature for a day. Recent studies have reported a tensile strength value of 67 MPa, elongation at break of 11%, and WVP of 4.4×10^{-7} g·m/m²·hr·Pa at 55% RH and 20°C.[90] The physical properties of the films can be modified by adding gelling agents, such as agar and gellan. Blending pullulan with alginate or carboxymethylcellulose (CMC) up to a total polymer concentration of 17 to 33% (w/w) reduced the film solubilization time in water. The addition of glycerol reduced tensile strength, increased elongation at break, and weakened water barrier properties but enhanced the solubilization in water. The oxygen resistance of pullulan film is well suited for control of the oxidation of fats and vitamins in foods. The pullulan film can be used for coating or packaging dried foods, including nuts, noodles, confectioneries, vegetables, fish, and meat. The films are resistant to oil and grease and can be printed on with edible inks. They also can be heat sealed to form single-serving packages that dissolve in water.

Application of a pullulan, sorbitol, and stearic ester coating on strawberries resulted in significant changes in the internal fruit atmosphere composition that were beneficial for extending the shelf life of this fruit. The coated fruit showed much higher levels of CO_2, a large reduction in internal O_2, better firmness and color retention, and a reduced rate of weight loss. In contrast, similar studies on whole kiwifruits showed increased levels of internal ethylene, which caused acceleration of fruit ripening during storage. Pullulan has been shown to form a stable emulsion with turmeric oleoresin, which could be microencapsulated to protect the resin from loss of quality due to exposure to light, heat, oxygen, and alkaline conditions. Gum arabic and maltodextrin provided the wall material for encapsulation of turmeric oleoresin prepared by spray drying an emulsion containing an optimum concentration of 10% of the oleoresin.[110] The exopolysaccharide of a *Pseudomonas* sp. was recently used to produce a film. The film properties could be improved by incorporating plasticizers and blending with other biopolymers.[111] Information on the film-making properties of polysaccharides from marine microorganisms is sparse; however, it is fair to assume that these compounds are capable of forming films that are both edible and biodegradable.

9.11 Marine Polysaccharides as Encapsulation Matrices

Marine polysaccharides, particularly those of seaweed origin, have major potential for the delivery of nutraceuticals, proteins, enzymes, vitamins, antioxidants, and probiotics because of their favorable gel-forming properties.[28]

Alginate and chitosan have been developed for encapsulation of probiotics, particularly lactobacilli, which have played a crucial role in the production of fermented products. Dairy products, such as yogurt, fermented milks, and cheeses, are conventional carriers of probiotics. Encapsulation techniques have been used to preserve their activity in several food matrices. Microencapsulation with alginate improved the viability of probiotic organisms in freeze-dried yogurt stored for 6 months at 4° and 21°C. Sodium alginate (1%) along with 3% fructooligosaccharides and a peptide favored survival of the probiotics *Lactobacillus acidophilus, L. casei, Bacillus bifidum*, and *Bifidobacterium longum*, which remained viable in the microcapsules.[112,113]

Chitosan in the form of hydrogel and composite gel matrices can be a good carrier of nutraceuticals (see Chapter 3). Astaxanthin was microencapsulated in a chitosan matrix cross-linked with glutaraldehyde by multiple emulsion/solvent evaporation. A powdered product was obtained containing microcapsules with a diameter of 5 to 50 µm and which improved the stability of the pigment in the microcapsules during storage for 8 weeks. The microencapsulated pigment did not suffer any chemical changes under the storage conditions examined.[114]

Polyelectrolyte hydrogels of xanthan gum and chitosan microscapsules can be used for the encapsulation and controlled release of food ingredients, cells, enzymes, and therapeutic agents. Capsules were completely cross-linked under all of the conditions studied when the initial xanthan solution concentration was 1.5% (w/v). Changes in the conformation of the chitosan chains as chitosan pH approached 6.2 were found to be important in achieving capsule network structures with different cross-linking densities.[115]

A novel interfacial engineering technology based on production of multilayer membranes around oil droplets was found to be effective for producing spray-dried encapsulated tuna oil. Chitosan in combination with maltodextrin (MD) or whey protein isolate (WPI) was used recently to encapsulate tuna oil. The process involved a combination of emulsification, ultrasonic atomization, and freeze-drying at the optimum ratios of 1:10 chitosan to MD or 1:1 chitosan to WPI. The combination of chitosan and MD had the highest emulsion stability. The eicosapentaenoic acid (EPA) and docosahexaenoic acid (DHA) contents of the encapsulated powder were 240 mg/g. The capsules had low moisture content and water activity and an acceptable appearance.[116] In an earlier study, tuna oil-in-water emulsion was stabilized in lecithin–chitosan membranes. The emulsions were prepared using a layer-by-layer electrostatic deposition method that involved adsorbing cationic chitosan onto the surface of anionic lecithin-stabilized droplets. The addition of corn syrup solids increased the stability of secondary emulsions. The powders had high oil retention levels (>85%). The structure of the microcapsules was unaffected by drying temperatures ranging from 165 to 195°C. The powdered tuna oil produced by this method has good physicochemical properties and dispensability, suggesting potential for its widespread utilization as a food additive.[117] Microcapsules containing neem (*Azadirachta indica* A. Juss.) seed oil were prepared by encapsulation of the natural liquid pesticide in a polyelectrolyte complex of κ-carrageenan and chitosan in an optimum carrageenan-to-chitosan ratio of 1:36.[118]

Entrapment within spheres of calcium alginate gels and, to a lesser extent, potassium carrageenan and agarose stands out as the most promising technique for immobilizing living organisms such as bacteria, cyanobacteria, fungi, and plant and animal cells. The food matrix has been found relevant to the absorption of vitamins. The absorption efficiency of highly lipophilic food microconstituents, including fat-soluble vitamins (A, E, D, and K), carotenoids, and phytosterols, depends on such factors as the presence of fat and the type of food matrix. The possible uses of such systems in industry, medicine, and agriculture are numerous.[14] Flavors can also be encapsulated in alginate matrices. The viscosity and uronic acid composition influence the properties of the film and microcapsules prepared from alginates. Further, the Ca^{2+} content in the mesosphere also influences the rate of release of encapsulated material. In microspheres having a defined size range, encapsulation efficiency and release rates are highly dependent on viscosity and the extent of Ca^{2+}–alginate interaction. Small microsphere fractions had excellent encapsulation efficiencies but showed faster release of the material. Viscosity appeared to exert a predominant influence on the microsphere properties.[119]

The lipid globules incorporated as emulsion-based alginate films can serve as the carriers of active volatile molecules. Favorable interactions among flavor and alginate molecules affect barrier properties, oxygen permeability, aroma compound permeability, liquid aroma permeability, and surface properties of emulsified alginate films, suggesting the involvement of emulsion-based edible films as a matrix and the ability to protect encapsulated aroma compounds.[120] The efficiency of alginate encapsulation can be modified by the addition of pectin. A composite matrix consisting of alginate and pectin increased the efficiency of folic acid encapsulation. The loading efficiency of the vitamin varied from 55 to 89%, depending on the composition of the polymers in the capsules. Folic acid was fully retained in the freeze-dried capsules after 11 weeks of storage at 4°C.[121] Calcium alginate capsules containing shark liver oil were prepared by ionotropic gelation of alginate solutions. Because the capsules were permeable to the oil, they were coated with a membrane of chitosan–alginate polyelectrolyte complex to reduce permeability. Encapsulation efficiencies as high as 87% (w/w) were obtained by using 6% (w/v) alginate solution. The oil content in the capsules was higher than 65% (w/w) at an optimal concentration of 10% (w/v) of alginate; however, above this level the oil was exuded after 48 hours. The stability of the capsule was adversely affected at pH 7.4 or higher.[122] A proteinase preparation (Flavourzyme®) was encapsulated using κ-carrageenan for cheese making. Cheese treated with the encapsulated enzyme showed higher rates of proteolysis than the control cheese throughout ripening. Differences in textural and sensory quality between treated and control cheeses were consistent with release of proteinases from the capsules.[112]

Carrageenan provides a very powerful matrix for the encapsulation of flavor compounds. Lipid globules incorporated in emulsion-based films containing up to 90% fat on a dry weight basis can be the carriers of flavor compounds. Neither the fat globule size nor the film water permeability was significantly modified by heat treatment, and no phase separation occurred due to heat.[20] Use of ι-carrageenan film for flavor encapsulation leads to interactions between aroma compounds and the ι-carrageenan, resulting in favorable modifications of the surface structure of the film and hence its permeability, making the film an ideal matrix to protect encapsulated flavors. These organic compounds are often classified based on their polarity in the following descending order: acids, alcohols, esters, ethers, aldehydes, ketones, aromatic hydrocarbons, unsaturated, and saturated aliphatic hydrocarbons.[120–124]

To date, the principal marine polysaccharides of interest for material production have been chitosan and alginate. A technology has already been developed to adhere chitosan to flexible packaging films to avoid migration and reduce bacterial activity.[36] The commercial alginate coating marketed under the trade name Flavor-Tex®, referred to earlier, was developed in the 1970s. The formulation includes sodium alginate with maltodextrin in the first solution and $CaCl_2$ with carboxymethylcellulose in the second solution.

9.12 Multicomponent Edible Films

Multicomponent or composite films and coatings consist of blends of various polymers, polysaccharides, proteins, and/or lipids. The purpose of developing multicomponent films is to improve the barrier and mechanical properties of films to broaden their potential applications. Improved properties of these films are due to the combination of specific characteristics of each component of the film; for example, polysaccharide coatings are generally less permeable to gases and more permeable to water vapor, as mentioned earlier in this chapter. Incorporating a protein or lipid in the film gives it a tight structure by inter- or intramolecular folding, resulting in lower water vapor permeability. Current research on multicomponent systems is focused on optimizing film composition, methodologies for their preparations, evaluating barrier properties, identifying the mechanism of mass transfer, determining the interactions of various films with food components, and evaluating their applications in the food industry. Many multicomponent films studied to date involve a lipid as a moisture barrier and a high polar polymer such as polysaccharide or protein as a structural matrix.[123,124]

TABLE 9.10

Water Vapor Permeability and Solubility of Chitosan-Starch-Based
Edible Films

Film Formulation	Water Vapor Permeability ($\times 10^{10}$ g/m·s·Pa)	Solubility (%)
Blend of 1% chitosan + 2% tapioca starch (film thickness, 0.197 mm)	28 ± 0.3	25 ± 6
Blend of 1% chitosan + tapioca starch + 0.05% potassium sorbate (film thickness, 0.197 mm)	6.7 ± 1.0	24 ± 1
Tapioca starch alone (film thickness, 0.163 mm)	12.1 ± 1.0	35 ± 2
Corn starch + chitosan	$3.76–4.54 \times 10^{-1}$	—

Source: Adapted from Vásconez, M.B. et al., *Food Res. Int.*, 42, 762, 2009; Garcia, M.A. et al., *Starch/Stärke*, 58, 453, 2006.

A composite film of agar and starch has enhanced barrier properties. Incorporation of agar at the 5% level increased the tensile strength of the starch film from 5.33 to 11.76 MPa and the elongation at break from 32.5 to 58.33%.[125] The high WVP of agar-based film can be significantly reduced when arabinoxylan is incorporated, and blending with cassava starch increases the surface wettability of agar films. Adding agar also enhances the elongation and tensile strength of cassava starch films; however, the mechanical properties of agar-based films degrade when the starch or arabinoxylan is added.[126]

Table 9.10 shows the water vapor permeability and solubility of chitosan-starch-based edible films. The problem with the high WVP of chitosan films could be addressed by incorporating lipid, protein, starch, or other polysaccharides into the film. Mixed films consisting of chitosan and lauric acid, poly-3-hydroxybutyric acid (PHB), or oleic acid have lower WVP values: the higher the oleic acid content, the lower the WVP. The addition of oleic acid also contributed to a significant increase in gloss and translucency and a decrease in the tensile strength, elongation at break, and elastic modulus of the composite films.[127] Composite films of chitosan and proteins such as sodium caseinate gave superior barrier properties, displaying improved tensile strength (19.6 MPa) and increased impact strength (35.6 GPa) in comparison with films of chitosan alone, which had corresponding values of 17 MPa and 26.6 GPa, respectively.[40]

Films of chitosan and gelatin containing polyol as a plasticizer were prepared by casting their aqueous solutions at 60°C and evaporating at 22° or 60°C. An increase in the plasticizer up to 50% (v/v) resulted in a considerable decrease of elasticity and tensile strength (up to 50%) of the original values, whereas the elongation at break increased up to 150%. Evaporation at 22°C led to a higher crystallinity of gelatin, which resulted in significant decrease of CO_2 and O_2 permeability in the chitosan–gelatin blends. An increase in the total plasticizer (water or polyol) content of these blends resulted in an increase in their gas permeabilities.[128]

The biodegradable composite films of chitosan and starch have a homogeneous matrix, stable structure, and interesting water barrier and mechanical properties. Smooth surfaces and homogeneous and compact film structures were observed from microstructure studies using scanning electron microscopy (SEM). The addition of glycerol reduced film opacity and increased the solubility of composite films. The WVP of films plasticized with glycerol ranged between 3.76 and 4.54 × 10^{-11} g/m·s·Pa, lower than those of the single-component films. The films were resistant, and their flexibility increased with glycerol addition. Tensile strength values of composite films were comparable to those of low-density and high-density polyethylene but lower than that obtained for cellophane; however, composite biodegradable films showed lower elongation at break values than the synthetic commercial ones.[129]

A composite film of rice starch and chitosan had lower WVP values than chitosan film. The introduction of chitosan increased the crystalline peak structure of the starch film; however, a very high chitosan concentration resulted in phase separation between the polymers.[130] Films of chitosan and hydroxypropyl guar gum (HGG) were prepared using a conventional solvent-casting technique. With a 60% content of HGG in the film, the maximum tensile strength and breaking elongation values of the film were 58.9 MPa and 17.25%, respectively. The blended film had greater thermal stability and optical transparency in comparison with pure chitosan films.[131]

Composite films of chitosan and hydroxypropylmethylcellulose (HPMC) that also contain lipids have been prepared. Incorporating stearic acid into the composite film reduced the water sensitivity of the film. Cross-linking of composite chitosan–HPMC with citric acid led to a 40% reduction in solubility in water. Whereas chitosan improved the WVP of the HPMC film, the addition of stearic acid reduced it.[132] Xanthan–chitosan microcapsules have been developed by complex coacervation for effective delivery of bioactive components.[115]

Table 9.11 shows the water vapor permeabilities of some composite films of alginate and chitosan. Fibers of blends of alginate and N-succinyl–chitosan (SCS) were prepared by spinning their mixture solution through a viscose-type spinneret into a coagulating bath containing aqueous $CaCl_2$. The fibers demonstrated good miscibility between the alginate and SCS. The fibers containing 30% (w/w) SCS had maximum tensile strength, breaking elongation, and water-retention properties.[133] Sodium alginate–chitosan composite membranes were prepared using a casting and solvent evaporation technique. $NaHCO_3$ was used as an additive to form pores in the interior of the composite membranes and glycerol was introduced as a plasticizer. The average pore size, water uptake capacity, and oxygen permeation property of the

TABLE 9.11

Water Vapor Permeabilities of Alginate and Chitosan Composite Films at 25°C

Basic Film	Composite Film Component	Water Vapor Permeability (g/mm²/day)	Thickness (mm)	Relative Humidity (%)
Starch, alginate, and glycerine in the ratio of 5.2:2.1:0.63	None	5.1	0.11	50–100
	Glycerine + lecithin + lauric acid (2.1)	2.2	0.13	50–100
	Glycerine + lecithin + palmitic acid (2.1)	2.9	0.10	50–100
	Glycerine + lecithin + stearic acid	1.3	0.12	50–100
Chitosan	None	314.6	0.03	—
	Chitosan + lauric acid	154.3	0.03	—
	Chitosan + palmitic acid	233.3	0.03	—
	Chitosan + butyric acid	526.7	0.03	—
	Chitosan AM (medium molecular weight)	477.3	0.03	—

Source: Adapted from Wu, Y. et al., *Adv. Food Nutr. Res.*, 44, 347, 2002; Wong, D.W.S., *J. Agric. Food Chem.*, 40, 540, 1992.

composite membranes could be adjusted by altering the ratio of NaHCO₃ in the alginate solution. The composite membranes showed high water uptake capacity, suitable mechanical strength, excellent oxygen permeability, and good biocompatibility.[134]

Transparent alginate and pectin composite films containing 10% glycerol had acceptable mechanical properties, low solubility, and limited swelling. Whereas increasing the glycerol concentration increased film solubility in water, moisture content, volumetric swelling, and flexibility, it decreased resistance to tensile stress.[135] Composite films of polyvinyl alcohol (PVA) and sodium alginate were prepared by casting aqueous solutions. The blended films exhibited improved thermal stability and mechanical properties.[136] Blending pullulan with alginate up to a total polymer concentration of about 17 to 33% reduced the film solubilization time in water. The addition of glycerol reduced tensile strength, increased elongation at break, and weakened water barrier properties but enhanced solubilization in water.[90]

Extrusion technology can be employed to develop composite films, as shown in the case of sodium alginate and gelatin with glycerol as the plasticizer, prepared using extrusion temperatures ranging from 120° to 135°C. The film had optimal tensile strength, Young's modulus, puncture resistance, color, turbidity, and vapor transfer rate.[21] An enzymatic technique utilizing transglutaminase of either mammalian or microbial origin has been used to enhance cross-linking of protein ingredients such as whey protein, ovalbumin, or gelatin with chitosan to improve barrier properties. A mushroom tyrosinase can catalyze gel formation of gelatin–chitosan blends, and

TABLE 9.12

Gelatin and Sodium Alginate Composite Film Properties

Parameter	Value
Thickness	0.364 mm
Tensile strength	3.595 MPa
Young's modulus	0.069 MPa
Puncture resistance	5.367 kg
Water vapor transmission rate	42.897 g H_2O/day/100 m^2

Source: Adapted from Liu, L. et al., *J. Food Agric. Environ.*, 3, 51, 2005. With permission from WFL Publisher.

the strength of the composite gels can be adjusted by altering the gelatin and chitosan compositions.[137] Table 9.12 summarizes the properties of composite films of gelatin and alginate.

Carrageenans have also been used for composite films. The addition of κ-carrageenan and gellan increased the melting point of fish gelatin gels. Also, the polysaccharides at a level of 2% enhanced the tensile strength and WVP of the fish gelatin films.[138] Edible films were prepared from powders of whey protein concentrate (WPC) and alginate and carrageenan. Films formed from co-dried powders had lower WVP and higher tensile strength, elastic modulus, and elongation than equivalent films formed from the dry blended powders. Films containing alginate had lower WVP and higher tensile strength, elastic modulus, and elongation values than films containing pectin, carrageenan, or konjac flour.[139]

Composite films based on combinations of κ-carrageenan (molecular weight, 5.1 × 10^5 Da) and chitosan (molecular weight, 1.71 × 10^5 Da), as well as ascorbic acid, were compared for their physical properties with films of pure κ-carrageenan or chitosan. Tensile strength was highest in pure κ-carrageenan films combined with 2% ascorbic acid. Composite films showed intermediate levels of tensile strength but lower values of elongation and permeability than pure films.[140] Preparation of composite film of carrageenan and corn zein consists of immersing preformed κ-carrageenan films into 10 to 95% corn zein in ethyl alcohol with polyethylene glycol and glycerol at 20% and 24% of zein (w/w), respectively. Increases in corn zein concentration produce proportionate decreases in WVP, water solubility, tensile strength, and swelling. Carrageenan films coated with corn zein showed heat sealing properties, even though their sealing strength was less than half that of corn zein film.[106]

9.12.1 Applications of Multicomponent Films

The enhanced barrier and functional properties of multilayer films can have many applications in the packaging of a variety of products and as carriers of such compounds as flavor and nutraceuticals, among others.[47] Chitosan–tapioca starch composite films exhibited antimicrobial properties against

Zygosaccharomyces bailii which were enhanced by the incorporation of sorbate in the film.[141] Chitosan–HPMC-based films inhibited the growth of *Listeria monocytogenes*.[132] Composite films of chitosan and polylactic acid (PLA) demonstrated antifungal activity against three mycotoxin-producing fungal strains: *Fusarium proliferatum*, *F. moniliforme*, and *Aspergillus ochraceus*.[142]

A composite edible film of chitosan and yam starch was found to have antimicrobial properties and hence could be a viable alternative for controlling microbiological growth in certain agricultural products. Samples of minimally processed carrot slices were immersed into coatings based on 4% yam starch (w/w) containing 2% glycerol (w/w) and an optimal concentration of 1.5% chitosan. After the treatment, the samples were placed in expanded polystyrene trays, wrapped in polyvinylchloride film and stored at 10°C for 15 days. During storage, all the samples had fewer than 100 CFU/g for *Staphylococcus aureus* and a most probable number of less than 3 CFU/g for *Escherichia coli*. Because of the antimicrobial activity of chitosan, chitosan–starch-coated samples showed reductions in mesophilic aerobes, mold and yeast, and psychrotrophic counts compared to the control. Chitosan in the coating inhibited lactic acid bacteria throughout the storage period.[143]

Chitosan-based composite film has been shown to prolong the shelf life of bananas.[144] A biodegradable laminate of chitosan–cellulose and polycaprolactone, developed in Japan, is a suitable film for the modified atmosphere packaging of head lettuce, cut broccoli, whole broccoli, tomatoes, and sweet corn at 10 to 25°C.[145]

As discussed earlier in this chapter, polyelectrolyte hydrogels of xanthan gum and chitosan microcapsules were formed by complex coacervation. The cross-linking density was found to be less dependent on the chitosan concentration than xanthan and the pH of chitosan. The capsules were completely cross-linked under all of the conditions studied when the initial xanthan solution concentration was 1.5% (w/v). The microcapsules could be used for encapsulation and controlled release of food ingredients, cells, enzymes, and therapeutic agents.[114]

A robust hydrogel system with superior absorbency and pH resistance has been developed incorporating agar and sodium alginate. The grafted polymer had a swelling capacity of 24 g/g in acidic medium when compared with a value of 14 g/g of the nongrafted film. The copolymer hydrogel may be useful in health, personal care, and agricultural applications.[146]

Marine polysaccharides including chitosan and κ-carrageenan could also be used to enhance the properties of polypropylene (PP) film using a simple casting method in the presence of plasticizers such as glycerol. When high glossy surfaces were observed on the coated films with chitosan and κ-carrageenan, the sucrose-plasticized chitosan coating gave the highest gloss of 142.7 gauge units. The type of biopolymers had a noticeable effect on the color of the coated films. Chitosan- and κ-carrageenan-coated PP films showed greater transparency, tensile strength, and elongation than the other coated films. Incorporation of nisin resulted in antimicrobial activity.[58] Food packaging films comprised of varying proportions of chitosan and polyethylene resin could be commercially prepared by extruding a mixture of the components. The chitosan particles of such films are evenly dispersed and distributed in the matrix, offering excellent elongation and tensile strength.[147]

9.13 Nanotechnology

Nanoparticle technology takes advantage of the unique properties of submicron-sized particles. The technique allows the creation and use of structures, devices, and systems that have novel properties and functions because of their small or intermediate size (1 to 100 nm in diameter). Nanotechnology in the field of food science can be applied to the development of new materials with novel functional properties, devising better delivery systems for drugs and nutraceuticals, and detecting pathogens, among others.[123] In the area of packaging, the technology can be used to reinforce films through the dispersion of nanometer-sized particles in the polymer matrix, thereby improving mechanical, thermal, optical, and physiochemical properties, in addition to improving mechanical and barrier properties. Consequently, natural biopolymer-based nanocomposite packaging materials with biofunctional properties have a huge potential for application in the active food packaging industry.[106,148,149]

Polysaccharides have important applications as nanomaterials for biopackaging systems. Traditionally, mineral fillers such as clay and silica are incorporated in films of these compounds to improve performance. Recently, the preparation of a nanoclay and carbohydrate film was reported. In this process, carbohydrates are pumped together with clay layers through a high shear cell to produce a film that has substantially reduced water vapor permeability. Introduction of the dispersed clay layers into the biopolymer matrix structure greatly improves the mechanical strength of the film. There is potential to produce such nanoforms of alginate, chitosan, and other polysaccharides.[150]

A chitosan-based nanocomposite films using nanoclay particles such as Cloisite® 30B, Nano-Silver, and silver ions was prepared by the solvent casting method. For this, a chitosan film solution was prepared by dissolving 2% chitosan (w/v) in a 1% acetic acid solution with 2% glycerol. The clay solution was prepared by dispersing layered nanoclay particles in the same solvent as that used for the film-forming solution. The clay solution was added to the polymer solution dropwise, and the resulting mixture was subjected to high shear mixing and ultrasonic treatment. The resulting solution was allowed to dry in ambient or elevated temperature conditions to make a free-standing film.[106] The ability of chitosan to interact strongly with milk proteins opens up the possibility to make novel hybrid nanoparticles.[151]

Unique nanoparticles of size 50 to 100 nm were obtained from chitin after consecutive acid hydrolysis and mechanical ultrasonication treatments. The chitin nanoparticles exhibited lower crystallinity when compared to conventional chitin. Chitin nanoparticles obtained by physical cross-linking between tripolyphosphate and protonized chitosan were used as filler in a glycerol plasticized-starch matrix for use in food packaging. The preparation showed improved tensile strength, storage modulus, glass transition temperature, water vapor barrier, and thermal stability due to the filler and matrix interactions; however, higher chitin nanoparticle loads (8%) resulted in the aggregation of chitin nanoparticles in the composites.[152]

Electrospinning allows the fabrication of nanofibers with diameters below 100 nm. Their unique properties are derived from a high orientation of polymers within the fibers that leads to mechanically superior properties. Further, the small dimensions of the fibers result in very high specific surface areas and associated properties. These fibers can serve as carriers of nutraceuticals, antioxidants, antimicrobials, and flavors.[153] Chitosan nanoparticles could be efficient carriers of potent antioxidants. Quercetin, for example, is an abundant flavonoid in plants that has numerous biological activities and is widely used as a potent antioxidant. Being sparingly soluble in water and subject to degradation in aqueous intestinal fluids, the absorption of quercetin is limited upon oral administration. Quercetin-loaded nanoparticles were prepared based on the ionic gelation of chitosan with tripolyphosphate anions; the technique was found to be useful in improving the bioavailabilty of quercetin.[154] A detailed description of food nano delivery systems based on polysaccharides and other polymers has been provided recently, with particular references to analytical techniques that can be used for the identification and characterization of these delivery systems in food products.[155]

9.14 Conclusion

The apparent lack of interest on the part of the food industry in using bio-based materials for packaging is perhaps due to a lack of knowledge about the materials themselves and their compatibility with existing packaging technology, an inability to absorb the additional costs of using bio-based materials in large-scale product packaging, and a reluctance to face legislative hurdles. Growing problems with synthetic packaging, increased consumer interest in biopackaging, and the availability of these bio-based materials at more favorable prices may tilt the balance in favor of bio-based materials for the packaging of food products.[156] The potential exists to use bio-based materials, including marine polysaccharides, as food packaging for diverse food product categories (e.g., agricultural products, seafood, meat and dairy products, ready-to-eat meals, beverages, snacks, dried foods, frozen products). It is likely that these polysaccharides will most likely find immediate application in foods requiring short-term chilled storage, such as fruits and vegetables. Globally, research and development activities in the area of food biopackaging have intensified over the last decade to find solutions to existing problems with food packaging.[157] The use of active edible coatings on an industrial scale is expected to develop in Europe due to implementation of the European Union Framework Regulation 2004/1935/EC, which authorizes the concept of active packaging with intentional active agent release.[158] Nanotechnology can deliver a wide spectrum of opportunities for the food and packaging industries; however, at present, many of these may be difficult to adopt commercially because of techno-economic problems. Nevertheless, the food industry is beginning to realize the full potential of nanotechnology in such areas as food processing, packaging, nutraceutical delivery, and safety and sensing.[159,160]

The development of new technologies to improve the delivery properties of edible films and coatings is a major issue for future research. Further research should be focused on providing more realistic information that can be applied to commercial applications of edible and biodegradable films, including those from marine sources, as the food industry is looking for edible films and coatings that could be used on a broad spectrum of foods and add value to their products while increasing their shelf life.[161] Marine biopolymers such as chitosan and alginate have increasing roles to play in development of edible and biodegradable films on a commercial scale.

References

1. Reineccius, G. A., Flavor encapsulation, in *Edible Films to Improve Food Quality*, Krochter, J. M. et al., Eds., CRC Press, Boca Raton, FL, 1994, pp. 105–120.
2. Marsh, K. and Bugusu, B., Food packaging: roles, materials and environmental issues, *J. Food Sci.*, 72, R39, 2007.
3. Dutta, A., Raychaudhuri, U., and Chakraborty, R., Biopolymers for food packaging, *Indian Food Ind.*, 25, 32, 2006.
4. Leistner, L. and Goms, L. M., Food preservation by hurdle technology, *Trends Food Sci. Technol.*, 6, 41, 1995.
5. Brody, A. L., Edible packaging, *Food Technol.*, 56, 65, 2005.
6. López-Rubio, A. et al., Overview of active polymer-based packaging technologies for food applications, *Food Rev. Int.*, 20, 357, 2004.
7. Ozdemir, M. and Floros, J. D., Active food packaging technologies, *Crit. Rev. Food Sci. Nutr.*, 44(3), 185, 2004.

8. Cutter, C. N., Microbial control by packaging: a review, *Crit. Rev. Food Sci. Nutr.*, 42, 151, 2002.

9. Raybaudi-Massilia, R. M. et al., Control of pathogenic and spoilage microorganisms in fresh-cut fruits and fruit juices by traditional and alternative natural antimicrobials, *Comp. Rev. Food Sci. Food Safety*, 8, 157, 2009.

10. Suppakul, P. et al., Active packaging technologies with an emphasis on antimicrobial packaging and its applications, *J. Food Sci.*, 68, 408, 2003.

11. Cha, D. S. and Chinnan, M. S., Biopolymer-based antimicrobial packaging: a review, *Crit. Rev. Food Sci. Nutr.*, 44, 223, 2004.

12. USFDA, *Listing of Food Additive Status*, U.S. Food and Drug Administration, Washington, D.C., 2010 (www.fda.gov/Food/FoodIngredientsPackaging/FoodAdditives/FoodAdditiveListings/ucm091048.htm).

13. Yoga, U. et al., Applications of microencapsulation in the food industry, *Bev. Food World*, 36(7), 28, 2009.

14. McClements, D. J. et al., Structural design principles for delivery of bioactive components in nutraceuticals and functional foods, *Crit. Rev. Food Sci. Nutr.*, 49, 577, 2009.

15. Garg, M. L. et al., Means of delivering recommended levels of long chain n-3 polyunsaturated fatty acids inhuman diets, *J. Food Sci.*, 71, R66, 2006.

16. Krochta, M. and DeMulder-Johnston, C., Edible and biodegradable polymer films: challenges and opportunities, *Food Technol.*, 51(2), 61, 1997.

17. Siracusa, V. et al., Biodegradable polymers for food packaging: a review, *Trends Food Sci. Technol.*, 19, 634, 2008.

18. Tharanathan, R. N., Biodegradable films and composite coatings: past, present and future, *Trends Food Sci. Technol.*, 14, 71, 2003.

19. Lopez-Rubio, A., Gavara, R., and Lagaron, J. M., Bioactive packaging: turning foods into healthier foods through biomaterials, *Trends Food Sci. Technol.*, 17, 567, 2006.

20. Karbowiak, H. et al., Influence of thermal process on structure and functional properties of emulsion-based edible films, *Food Hydrocoll.*, 21, 879, 2007.

21. Liu, L., Kerry, J. F., and Kerry, J. P., Selection of optimum extrusion technology parameters in the manufacture of edible biodegradable packaging films derived from food-based polymers, *J. Food Agric. Environ.*, 3, 51, 2005.

22. Park, H. J., Development of advanced edible coatings for fruits, *Trends Food Sci. Technol.*, 10, 254, 1999.

23. Rojas-Gratio, M. A., Soliva-Fortuny, R., and Martin-Belloso, O., Edible coatings to incorporate active ingredients for fresh fruits: a review, *Trends Food Sci. Technol.*, 20, 438, 2009.

24. Bourlieu, C. et al., Edible moisture barriers: how to assess their potential and limits in food products shelf-life extension?, *Crit. Rev. Food Sci. Nutr.*, 49, 474, 2009.

25. Lin, D. and Zhao, Y., Innovations in the development and applications of edible coatings for fresh and minimally processed fruits and vegetables, *Comp. Rev. Food Sci. Food Safety*, 6, 60, 2007.

26. Ayranci, E. and Tunc, S., A method for the measurement of the oxygen permeability and the development of edible films to reduce the rate of oxidative reactions in fresh foods, *Food Chem.*, 80(3), 423, 2003.

27. Guilbert, S. et al., Prolongation of the shelf-life of perishable food products using biodegradable films and coatings, *LWT Food Sci. Technol.*, 29, 10, 1996.

28. Karbowiak, T. et al., From macroscopic to molecular scale investigations of mass transfer of small molecules through edible packaging applied at interfaces of multiphase food products, *Innov. Food Sci. Emerg. Technol.*, 10, 116, 2009.

29. Cutter, C. N., Opportunities for bio-based packaging technologies to improve the quality and safety of fresh and further processed muscle foods, *Meat Sci.*, 74, 131, 2006.

30. Souza, B. W. S. et al., Influence of electric fields on the structure of chitosan edible coatings, *Food Hydrocoll.*, 24, 330, 2010.

31. Prasad, K., Meena, R., and Siddhantha, A. K., A new approach to functionalize agar and κ-carrageenan based thin films with improved barrier properties, *J. Polym. Mater.*, 25, 373, 2008.

32. Butler, B. L. et al., Mechanical and barrier properties of edible chitosan films as affected by composition and storage, *J. Food Sci.*, 61, 953, 1996.

33. Krajewska, B., Membrane based processes performed with use of chitin/chitosan materials, *Sep. Purif. Technol.*, 41, 305, 2005.

34. Nadarajah, K. et al., Sorption behaviour of crawfish chitosan films as affected by chitosan extraction process and solvent types, *J. Food Sci.*, 71, E33, 2006.

35. Wiles, J. L. et al., Water vapor transmission rates and sorption behavior of chitosan films, *J. Food Sci.*, 65, 1175 2000.

36. Kim, K. M. et al., Properties of chitosan films as a function of pH and solvent type, *J. Food Sci.*, 71, E119, 2006.

37. Baskar, D. and Sampath Kumar, T. S., Effect of deacetylation time on the preparation, properties and swelling behavior of chitosan films, *Carbohydr. Polym.*, 78, 767, 2009.

38. Sathivel, S. et al., The influence of chitosan glazing on the quality of skinless pink salmon (*Oncorhynchus gorbuscha*) fillets during frozen storage, *J. Food Eng.*, 83, 366, 2007.

39. Jia, D., Fang, Y., and Yao, K., Water vapor barrier and mechanical properties of konjac glucomannan–chitosan–soy protein isolate edible films, *Food Bioproducts Proc.*, 87(1), 7, 2009.

40. Pereda, M. et al., Water vapor absorption and permeability of films based on chitosan and sodium caseinate, *J. Appl. Polym. Sci.*, 111, 2777, 2009.

41. Srinivasa, P. C. et al., Effect of plasticizers and fatty acids on mechanical and permeability characteristics of chitosan films, *Food Hydrocoll.*, 21, 7, 2007.

42. Srinivasa, P. C. et al., Drying kinetics and properties of chitosan films prepared under different drying conditions, *J. Food Eng.*, 63, 79, 2003.

43. Castro, S., A Chitosan and/or Porous Solids Film for Biomedical and/or Food Uses, a Process for the Obtention Thereof and Use of the Same, European Patent No. MXPA04001347 (A), 2004 (http://v3.espacenet.com/publicationDetails/biblio?DB=EPODOC&adjacent=true&locale=en_EP&FT=D&date=20040915&CC=MX&NR=PA04001347A&KC=A).

44. de Britto, D. and de Assis, O. B. G., Synthesis and mechanical properties of quaternary salts of chitosan-based films for food application, *Int. J. Biol. Macromol.*, 41, 198, 2007.

45. Pereda, M. et al., Water vapor absorption and permeability of films based on chitosan and sodium caseinate, *J. Appl. Polym. Sci.*, 111, 2777, 2009.

46. Hellander, I. M. and Murmiaho-Lassila, E. I., Chitosan disrupts the barrier properties of the outer membrane of Gram-negative bacteria, *Int. J. Food Microbiol.*, 71, 235, 2001.

47. Natarajah, N. et al., Sorption behavior of crawfish chitosan films as affected by chitosan extraction process and solvent types, *J. Food Sci.*, 71, E33, 2006.
48. Sai, G. J. and Su, W. H., Antibacterial activity of shrimp chitosan against *Escherichia coli*, *J. Food Prot.*, 62, 239, 1999.
49. Takeda, H., Antibacterial activity of chitosan sheet, *Chitin Chitosan Res.*, 12, 192, 2006.
50. Coma, V., Deschamps, A., and Martial-Gros, A., Chitosan polymer: antimicrobial activity assessment on dairy-related contaminants, *J. Food Sci.*, 68, 2788, 2003.
51. Dutta, P. K. et al., Perspectives for chitosan based antimicrobial films in food applications, *Food Chem.*, 114, 1173, 2009.
52. Cha, D. S. et al., Release of nisin from various heat-pressed and cast films, *LWT Food Sci. Technol.*, 36, 209, 2003.
53. Fernandez-Saiz, P., Lagar, J. M., and Ocio, M. J., Optimization of the film-forming and storage conditions of chitosan as an antimicrobial agent, *J. Agric. Food Chem.*, 57, 3298, 2009.
54. Ouattara, B. et al., Diffusion of acetic and propionic acids from chitosan-based antimicrobial packaging films, *J. Food Sci.*, 65, 768, 2000.
55. Pranoto, Y., Rakshit, K., and Salokhe, V. M., Mechanical, physical and antimicrobial characterization of edible films based on alginate and chitosan containing garlic oil, *Dev. Chem. Eng. Mineral Proc.*, 13, 617, 2005.
56. Zivanovic, S., Chi, S., and Draughon, F. A., Antimicrobial activity of chitosan films enriched with essential oils, *J. Food Sci.*, 70, M45, 2005.
57. Park, S. I., Daeschel, M. A., and Zhao, Y., Functional properties of antimicrobial lysozyme-chitosan composite films, *J. Food Sci.*, 69, M215, 2004.
58. Hong, S. I. et al., Properties of polysaccharide-coated polypropylene films as affected by biopolymer and plasticizer types, *Packaging Technol. Sci.*, 18, 1, 2005.
59. Portes, E. et al., Environmentally friendly films based on chitosan and tetrahydrocurcuminoid derivatives exhibiting antibacterial and antioxidative properties, *Carbohydr. Polym.*, 76, 578, 2009.
60. Ponce, A. G. et al., Antimicrobial and antioxidant activities of edible coatings enriched with natural plant extracts: *in vitro* and *in vivo* studies, *Postharvest Biol. Technol.*, 49, 294, 2008.
61. Twu, Y.-K. et al., Preparation and sorption activity of chitosan/cellulose blend beads, *Carbohydr. Polym.*, 54, 425, 2003.
62. Pichavant, F. et al., Fat resistance properties of chitosan-based paper packaging for food applications, *Carbohydr. Polym.*, 61, 259, 2005.
63. Aider, M., Chitosan application for active bio-based films production and potential in the food industry [review], *LWT Food Sci. Technol.*, 43, 837, 2010.
64. Campeniello, D. et al., Chitosan: antimicrobial activity and potential applications for preserving minimally processed strawberries, *Food Microbiol.*, 25, 992, 2000.
65. Ghaouth, A. et al., Use of chitosan coatings to reduce water loss and maintain quality of cucumber and bell pepper, *J. Food Proc. Pres.*, 15, 339, 1992.
66. Du, J. et al., Effects of chitosan coating on the storage of peach, Japanese pear, and kiwifruit, *J. Jap. Soc. Hort. Sci.*, 66, 15, 1997.
67. González-Aguilar et al., Effect of chitosan coating in preventing deterioration and preserving the quality of fresh-cut papaya "Maradol," *J. Food Sci. Agric.*, 89, 15, 2009.

68. Zhu, X. et al., Effects of chitosan coating on postharvest quality of mango fruits, *J. Food Proc. Pres.*, 32, 770, 2008.
69. Srivastava, P. C. et al., Storage studies of mango packed using biodegradable chitosan films, *Storage Technol.*, 215, 504, 2002.
70. Olivas, G. I. and Barbosa-Canovas, G. V., Edible coatings for fresh-cut fruits, *Crit. Rev. Food Sci. Nutr.*, 45, 657, 2005.
71. Vargas, M. et al., Effect of chitosan-based edible coatings applied by vacuum impregnation on quality preservation of fresh-cut carrot, *Post-Harvest Biol. Technol.*, 51, 263, 2009.
72. Casareigo, A. et al., Chitosan coating surface properties as affected by plasticizer, surfactant and polymer concentrations in relation to the surface properties of tomato and carrot, *Food Hydrocoll.*, 22, 1452, 2008.
73. Waimaleongora-Ek, P. et al., Selected quality characteristics of fresh-cut sweet potatoes coated with chitosan during 17-day refrigerated storage, *J. Food Sci.*, 73, S418, 2008.
74. Fuchs, S. J. et al., Effect of edible coatings on postharvest quality of fresh green asparagus, *J. Food Proc. Pres.*, 32, 951, 2008.
75. Yasuyo, K. and Yoshiharu, M., Chitosan-Containing Food and Drink, Japanese Patent No. JP2007110982 (A), 2007.
76. Jiang, Y., Li, J., and Jiang, W., Effects of chitosan coating on shelf life of cold-stored litchi fruit at ambient temperature, *LWT Food Sci. Technol.*, 38, 757, 2005.
77. Dong, H. et al., Effects of chitosan coating on quality and shelf life of peeled litchi fruit, *J. Food Sci.*, 64, 355, 2004.
78. Cakii, S. et al., Application of Chitosan and Its Phosphatic Derivatives for Quality Preservation of Fresh Fish Stored at Refrigerated Temperature, paper presented at the 2nd Joint Trans-Atlantic Fisheries Technology Conference, Quebec City, October 29–November 1, 2006.
79. Gómez-Estaca, J. et al., Effect of functional edible films and high pressure processing on microbial and oxidative spoilage in cold-smoked sardine (*Sardina pilchardus*), *Food Chem.*, 105, 511, 2007.
80. Jeon, Y. J., Kamil, J. Y. V. A., and Shahidi, F., Chitosan as an edible invisible film for quality preservation of herring and Atlantic cod, *J. Agric. Food Chem.*, 50, 5167, 2002.
81. Duan, J. et al., Antimicrobial chitosan-lysozyme (CL) films and coatings for enhancing microbial safety of mozzarella cheese, *J. Food Sci.*, 72, M355, 2007.
82. Cuttler, C. N., Opportunities for bio-based packaging technologies to improve the quality and safety of fresh and further processed muscle foods, *Meat Sci.*, 74, 131, 2006.
83. Quintavalla, S. and Vicini, L., Antimicrobial food packaging in meat industry, *Meat Sci.*, 62, 373, 2002.
84. Ouattara, B. et al., Inhibition of surface spoilage bacteria in processed meats by application of antimicrobial films prepared with chitosan, *Int. J. Food Microbiol.*, 62, 139, 2000.
85. Wu, Y. et al., Moisture loss and lipid oxidation for precooked beef patties stored in edible coatings and films, *J. Food Sci.*, 65, 300, 2000.
86. Rao, M. S., Chander, R., and Sharma, A., Development of shelf-stable intermediate moisture meat products using active edible chitosan coating and irradiation, *J. Food Sci.*, 70, M325, 2005.

87. Bhale, S. et al., Chitosan coating improves shelf life of eggs, *J. Food Sci.*, 68, 2378, 2003.
88. Kim, S. H. et al., Plasticizer types and coating methods affect quality and shelf life of eggs coated with chitosan, *J. Food Sci.*, 71, S111, 2008.
89. Dhananjay, S. K. and Mulinani, V. H., Optimization of immobilization process on crab shell chitosan and its application in food processing, *J. Food Biochem.*, 32, 521, 2008.
90. Tong, Q., Xiao, Q., and Lim, L.-T., Preparation and properties of pullulan–alginate–carboxymethylcellulose blend films, *Food Res. Int.*, 41, 1007, 2008.
91. Olivas, G. I. and Barbosa-Cánova, G. V., Alginate–calcium films: water vapor permeability and mechanical properties as affected by plasticizer and relative humidity, *LWT Food Sci. Technol.*, 41, 359, 2008.
92. Rojas-Graü, M. A., Alginate and gellan-based edible coatings as carriers of antibrowning agents applied on fresh-cut Fuji apples, *Food Hydrocoll.*, 21, 118, 2007.
93. Tapia, M. S. et al., Use of alginate- and gellan-based coatings for improving barrier, texture and nutritional properties of fresh-cut papaya, *Food Hydrocoll.*, 22, 1493, 2008.
94. Singh, C., Sharma, H. K., and Sarkar, B. C., Influence of process conditions on the mass transfer during osmotic dehydration of coated pineapple samples, *J. Food Proc. Pres.*, 34(4), 700, 2010.
95. Nussinovitch, A. and Kampf, N., Shelf-life extension and conserved texture of alginate-coated mushrooms (*Agaricus bisporus*), *LWT Food Sci. Technol.*, 26, 469, 1993.
96. Lu, F. et al., Alginate–calcium coating incorporating nisin and EDTA maintains the quality of fresh northern snakehead (*Channa argus*) fillets stored at 4°C, *J. Sci. Food Agric.*, 89, 848, 2009.
97. Dutta, S. et al., Control of *Listeria monocytogenes* and *Salmonella anatum* on the surface of smoked salmon coated with calcium alginate coating containing oyster lysozyme and nisin, *J. Food Sci.*, 73, M67, 2008.
98. Williams, S. K. et al., Evaluation of a calcium alginate film for use on beef cuts, *J. Food Sci.*, 43, 292, 1978.
99. Gennadios, A. et al., Application of edible coatings on meats, poultry and seafoods: a review, *LWT Food Sci. Technol.*, 30, 337, 1997.
100. Briones, A. V. et al., Tensile and tear strength of carrageenan film from Philippine *Eucheuma* species, *Mar. Biotechnol.*, 6, 148, 2004.
101. Hwang, K. T., Rhim, J. W., and Park, H. J., Effects of κ-carrageenan-based film packaging on moisture loss and lipid oxidation of mackerel mince, *Korean J. Food Sci. Technol.*, 29, 390, 1997.
102. Wu, Y. et al., Moisture loss and lipid oxidation for precooked beef patties stored in edible coatings and films, *J. Food Sci.*, 65, 300, 2000.
103. Shaw, C. P., Secrist, J. L., and Tuomy, J. M., Method of Extending the Storage Life in the Frozen State of Precooked Foods and Products Thereof, U.S. Patent No. 4, 196, 219, 1980.
104. Lee, J. Y. et al., Extending shelf-life of minimally processed apples with edible coatings and antibrowning agents, *LWT Food Sci. Technol.*, 36, 323, 2003.
105. Kim, K. W. et al., Edible coatings for enhancing microbial safety and extending shelf life of hard-boiled eggs, *J. Food Sci.*, 73, M227, 2008.
106. Rhim, J. W. and Ng, P. K. G., Natural biopolymer-based nanocomposite films for packaging applications, *Crit. Rev. Food Sci. Nutr.*, 47, 411, 2007.

107. Conggui, C. et al., The application of konjac glucomannan and carrageenan in developing edible wrapping paper made from potato starch, *Food Sci. China*, 25, 98, 2004.
108. Fan, T. D. et al., Functional properties of edible agar-based and starch-based films for food quality preservation, *J. Agric. Food Chem.*, 52, 973, 2005.
109. Palegrin, Y.-F. et al., Degradation of agar films in a humid tropical climate: thermal, mechanical, morphological, and structural changes, *Polym. Degrad. Stabil.*, 92, 244, 2007.
110. Kshirsagar, K. C. et al., Efficacy of pullulan in emulsification of turmeric oleoresin and its subsequent microencapsulation, *Food Chem.*, 113, 1139, 2009.
111. Freitas, F. et al., Characterization of an extracellular polysaccharide produced by a *Pseudomonas* strain grown on glycerol, *Bioresource Technol.*, 100, 859, 2009.
112. Kailasapathy, K. and Lam, S. H., Application of encapsulated enzymes to accelerate cheese ripening, *Int. Dairy J.*, 15, 6, 2005.
113. Chen, K. N. et al., Optimization of incorporated prebiotics as coating materials for probiotic microencapsulation, *J. Food Sci.*, 70, M260, 2005.
114. Ciapara, H. I. et al., Microencapsulation of astaxanthin in a chitosan matrix, *Carbohydr. Polym.*, 56, 41, 2004.
115. Argin-Soysal, S., Kofinas, P., and Martin, Y., Effect of complexation conditions on xanthan–chitosan polyelectrolyte complex gels, *Food Hydrocoll.*, 23, 202, 2009.
116. Klaypradit, W. and Huang, Y.-W., Fish oil encapsulation with chitosan using ultrasonic atomizer, *LWT Food Sci. Technol.*, 41, 1133, 2008.
117. Klinkesorn, U. et al., Characterization of spray-dried tuna oil emulsified in two-layered interfacial membranes prepared using electrostatic layer-by-layer deposition, *Food Res. Int.*, 39, 447, 2006.
118. Devi, N. and Maji, T. K., A novel microencapsulation of neem (*Azadirachta indica* A. Juss.) seed oil (NSO) in polyelectrolyte complex of κ-carrageenan and chitosan, *J. Appl. Polym. Sci.*, 113, 1576, 2009.
119. Lee, H.-Y. et al., Influence of viscosity and uronic acid composition of alginates on the properties of alginate films and microspheres produced by emulsification, *J. Microencaps.*, 23, 912, 2006.
120. Hambleton, A. et al., Protection of active aroma compound against moisture and oxygen by encapsulation in biopolymeric emulsion-based edible films, *Biomacromolecules*, 9, 1058, 2008.
121. Madziva, H., Kailasapathy, K., and Phillips, M., Alginate–pectin microcapsules as a potential for folic acid delivery in foods, *J. Microencaps.*, 22, 343, 2005.
122. Peniche, C. et al., Formation and stability of shark liver oil loaded chitosan/calcium alginate capsules, *Food Hydrocoll.*, 18, 865, 2004.
123. Shefer, A. and Shefer, S., Novel encapsulation system provides controlled release of food ingredients, *Food Technol.*, 57(11), 40, 2003.
124. Fabra, M. J. et al., Influence of interactions on water and aroma permeabilities of ι-carrageenan–oleic acid–beeswax films used for flavour encapsulation, *Carbohydr. Polym.*, 76, 325, 2009.
125. Wu, Y. et al., Development and application of multicomponent edible coatings and films: a review, *Adv. Food Nutr. Res.*, 44, 347, 2002.
126. The, D. P., Debeaufort, F., Voilley, A., and Luu, D., Biopolymer interactions affect the functional properties of edible films based on agar, cassava starch and arabinoxylan blends, *J. Food Eng.*, 90, 548, 2009.

127. Vargas, M. et al., Characterization of chitosan–oleic acid composite films, *Food Hydrocoll.*, 23, 536, 2009.
128. Arvanitoyannis, I. S., Nakayama, A., and Aiba, S., Chitosan and gelatin based edible films: state diagrams, mechanical and permeation properties, *Carbohydr. Polym.*, 37, 371, 1998.
129. Garcia, M. A., Pinotti, A., and Zaritzky, N. E., Physicochemical, water vapor barrier and mechanical properties of corn starch and chitosan composite films, *Starch/Stärke*, 58, 453, 2006.
130. Bourtoom, T. and Chinnan, M. S., Preparation and properties of rice starch–chitosan blend biodegradable film, *LWT Food Sci. Technol.*, 41, 1633, 2008.
131. Zhio, C. et al., Study of blend films from chitosan and hydroxypropyl guar gum, *J. Appl. Polym. Sci.*, 90, 1991, 2003.
132. Moiler, H. et al., Antimicrobial and physicochemical properties of chitosan–HPMC-based films, *J. Agric. Food Chem.*, 52, 6585, 2004.
133. Fan, L. et al., The novel alginate/N-succinyl-chitosan antibacterial blend fibers, *J. Appl. Polym. Sci.*, 116, 2151, 2010.
134. Ma, L., Yu, W., and Ma, X., Preparation and characterization of novel sodium alginate/chitosan two-ply composite membranes, *J. Appl. Polym. Sci.*, 106, 397, 2007.
135. da Silva, M. A. et al., Alginate and pectin composite films crosslinked with Ca^{2+} ions: effect of the plasticizer concentration, *Carbohydr. Polym.*, 77, 736, 2009.
136. Gaykara, T. and Serkan, D., Preparation and characterization of blend films of poly (vinyl alcohol) and sodium alginate, *J. Macromol. Sci.*, 43, 1113, 2006.
137. Di Pierro, P., Transglutaminase-catalyzed preparation of chitosan–ovalbumin films, *Enz. Microbiol. Technol.*, 40, 437, 2007.
138. Yudi Pranoto, Y., Lee, C. M., and Park, H. J., Characterizations of fish gelatin films added with gellan and κ-carrageenan, *LWT Food Sci. Technol.*, 40, 766, 2007.
139. Coughlan, N. B. et al., Combined effects of proteins and polysaccharides on physical properties of whey protein concentrate-based edible films, *J. Food Sci.*, 69, E271, 2004.
140. Park, H. J. et al., Mechanical and barrier properties of chitosan-based biopolymer films, *Chitin Chitosan Res.*, 5, 19, 1999.
141. Vásconez, M. B. et al., Antimicrobial activity and physical properties of chitosan–tapioca starch based edible films and coatings, *Food Res. Int.*, 42(7), 762, 2009.
142. Sebastien, F., Novel biodegradable films made from chitosan and poly(lactic acid) with antifungal properties against mycotoxinogen strains, *Carbohydr. Polym.*, 65, 185, 2006.
143. Durango, A. M., Soares, N. F. F., and Andrade, N. J., Microbiological evaluation of an edible antimicrobial coating on minimally processed carrots, *Food Control.*, 17, 336, 2006.
144. Kittur, F. S. et al., Polysaccharide based composite coating formulations for shelf-life extension of fresh banana and mango, *Eur. Food Res. Technol.*, 213, 306, 2001.
145. Makino, Y. and Hirata, T., Modified atmosphere packaging of fresh produce with a biodegradable laminate of chitosan–cellulose and polycaprolactone, *Postharvest Biol. Technol.*, 10, 247, 1997.
146. Meena, R. et al., Development of a robust hydrogel system based on agar and sodium alginate blend, *Polym. Int.*, 57, 329, 2008.

147. Ho, H. J., Process for Preparation of Food Packaging Film Containing Chitosan, European Patent No. US2008097003 (A1), 2008.
148. Sorrentino, A. S., Gorrasi, G., and Vittoria, V., Potential perspectives of bio-nanocomposites for food packaging applications, *Trends Food Sci. Technol.*, 18, 84, 2007.
149. Sherman, L. M., Chasing nanocomposites, *Plastics Technology Online*, 2005 (www.ptonline.com/articles/200411fa2.html).
150. Andersson, C., New ways to enhance the functionality of paperboard by surface treatment: a review, *Packaging Technol. Sci.*, 21, 339, 2008.
151. Singh, H., Ye, A., and Thompson, A., Nanoencapsulation systems based on milk proteins and phospholipids, in *Micro/Nanoencapsulation of Active Food Ingredients*, Huang, Q. et al., Eds., American Chemical Society, Washington, D.C., pp. 131–142.
152. Chang, P. R. et al., Fabrication and characterisation of chitosan nanoparticles/plasticised-starch composites, *Food Chem.*, 120, 736, 2010.
153. Min, B. M. et al., Chitin and chitosan nanofibers: electrospinning of chitin and deacetylation of chitin nanofibers, *Polymer*, 45, 7137, 2004.
154. Zhang, Y. et al., Physicochemical characterization and antioxidant activity of quercetin-loaded chitosan nanoparticles, *J. Appl. Polym. Sci.*, 107, 891, 2008.
155. Luykx, D. M. A. M. et al., A review of the analytical methods for the identification and characterization of nano delivery systems in foods, *J. Agric. Food Chem.*, 56, 8231, 2008.
156. Weber, C. J., Ed., Production and application of biobased packaging materials for the food industry, *Food Add. Contam.*, 19, 172, 2002.
157. Daraba., A., Future trends in packaging: edible, biodegradable coats and films, *J. Environ. Prot. Ecol.*, 9, 652, 2008.
158. Guillard, V. et al., Food preservative content reduction by controlling sorbic acid release from a superficial coating, *Innov. Food Sci. Emerg. Technol.*, 10, 108, 2009.
159. Weiss, J., Takhistov, P., and McClements, J., Functional materials in food nanotechnology: IFT Scientific Status Summary, *J. Food Sci.*, 71, R107, 2006.
160. Chau, C.-F. and Yen, G.-C., A general introduction to food nanotechnology, *FoodInfo Online Features* October 31, 2008 (http://www.foodsciencecentral.com/fsc/ixid15445).
161. Rojas-Gratio, M. A., Soliva-Fortuny, R., and Martin-Belloso, O., Edible coatings to incorporate active ingredients for fresh fruits: a review, *Trends Food Sci. Technol.*, 20, 438, 2009.
162. Fornes, F. et al., Low concentration of chitosan coating reduce water spot incidence and delay peel pigmentation of Clementine mandarin fruit, *J. Sci. Food Agric.*, 85, 1105, 2005.
163. Han, C. et al., Edible coatings to improve storability and enhance nutritional value of fresh and frozen strawberries (*Fragaria × ananassa*) and raspberries (*Rubus ideaus*), *Postharvest Biol. Technol.*, 33, 67, 2004.
164. Moldão-Martins, M. et al., The effects of edible coatings on postharvest quality of the "Bravo de Esmolfe" apple, *Eur. Food Res. Technol.*, 217, 325, 2003.
165. Toivonen, P. M. A. and Brummell, D. A., Biochemical bases of appearance and texture changes in fresh-cut fruit and vegetables, *Postharvest Biol. Technol.*, 48(1), 1–14, 2008.

10

Safety and Regulatory Aspects

10.1 Introduction

Consumers around the world are becoming more aware of how certain foods can adversely affect their health, and concern about the nutritive value of processed foods is increasing, resulting in a rising demand for food products that can supply adequate quantities of essential nutrients. Also, frequent reports of foodborne health hazards and food recalls have raised questions about the safety of processed foods. Regulatory agencies closely monitor food quality and draft needs-based legislations and specifications with a view to protecting consumers. Global food supply chains have reacted to these concerns by implementing systems designed to improve product quality and safety. Accordingly, Good Manufacturing Practice (GMP) protocols must be observed in the food production and supply industries, and Good Nutritional Practice (GNP) protocols have been suggested to protect the nutritional quality of food and integrate it with food safety regulations; GNP is based on a model that covers as many as nine good practices along the food supply chain.[1,2] This chapter briefly discusses the regulatory and safety aspects of marine polysaccharides as food additives.

10.2 Safety of Food Additives

A *food additive* has been defined by the Codex Alimentarius Commission (CAC) of the Food and Drug Administration (FAO) as "any substance not normally consumed as a food by itself and not normally used as a typical ingredient of the food, the intentional addition of which to food for a technological (including sensory) purpose in the manufacture, processing, preparation, treatment, packing, packaging, transport or holding of such food may reasonably be expected to result directly or indirectly, in it or its byproducts becoming a component of or otherwise affecting the characteristics of such foods."[3] The *acceptable daily intake* (ADI) is an estimate by the Joint Expert

Committee on Food Additives of the FAO of the amount of a food additive, expressed on a body weight basis, that can be ingested daily over a lifetime without appreciable health risk. The primary objective of establishing maximum use levels for food additives in various food groups is to ensure that the intake of an additive from all of its uses does not exceed its ADI. The *maximum use level* of an additive is the highest concentration of the additive determined to be functionally effective in a food or food category and agreed to be safe by the CAC. It is generally expressed as milligrams of additive per kilogram of food. Food additives used in accordance with this standard should be of appropriate food-grade quality and should at all times conform with the applicable regulations. The CAC regularly updates its lists of various additives and their maximum use levels.[3] The safety of food additives is evaluated by toxicological testing. The U.S. Food and Drug Administration (FDA) bases the number and types of toxicological tests required for new food additives on several critical types of information, including the chemical structure and level of use of the product. Typically, a minimum battery of tests is necessary depending on the *level of concern* assigned to the product.[5] The protocol for toxicological testing is given in Table 10.1.

10.3 Regulation of Food Additives

In September 1955, the Joint FAO/WHO Conference on Food Additives initiated the first international system to regulate the safety of food additives. Since then, more than 600 substances have been evaluated and provided with specifications for purity by the Joint FAO/WHO Expert Committee on Food Additives (JECFA).[5] The International Numbering System for Food Additives (INS), prepared by the Codex Committee on Food Additives, was designed to identify food additives intended for use in one or more member countries. The criteria for INS inclusion include: (1) the compound must be approved by a member country as a food additive, (2) the compound must be toxicologically cleared for use by a member country, and (3) the compound must be required to be identified on the final product label by a member country. The INS numbers for some polysaccharide food additives including those of marine origin are provided in Table 10.2.

In the United States, the basic food law is the Federal Food, Drug, and Cosmetic Act (FD&C Act) of 1938. This Act gives the FDA primary responsibility for the safety and wholesomeness of the food supply in the country. Three important amendments strengthened this Act: (1) Miller Pesticide Amendment of 1954, which provided for the establishment of safe tolerances (permissible amounts) for pesticide residues on raw agricultural commodities; (2) Food Additives Amendment of 1958, which required premarketing clearances for substances intended to be added and for substances occurring

TABLE 10.1

Toxicological Evaluations of Food Additives

Concern Level	Type of Evaluations
1	Acute toxicity
	Short-term feeding study (at least 28 days' duration)
	Short-term tests for carcinogenic potential (Ames mutagenicity assay, cell transformation test, *in vitro* chromosome aberration test)
2	Subchronic feeding study (at least 90 days' duration) in a rodent species
	Subchronic feeding study (at least 90 days' duration) in a non-rodent species
	Multigeneration reproduction study with a teratology phase in a rodent species
	Short-term tests for carcinogenic potential (*in vivo* acytogenetics, mammalian cell gene mutation assay)
3	Carcinogenicity studies in two rodent species
	Chronic feeding study in a rodent species (at least one year's duration)
	Chronic feeding study in a non-rodent species (at least one year's duration)
	Multigeneration reproduction study with at least two generations with a teratology phase in a rodent species
	Short-term tests of carcinogenic potential

TABLE 10.2

International Numbering System/European Council Numbers for Polysaccharide Food Additives and Their Functions

Compound	INS/E No.	Functions
Marine polysaccharides		
Alginic acid	400	Thickening, gelling agent, stabilizer, and emulsifier
Sodium alginate (1999)	401	Thickening, gelling agent, stabilizer, and emulsifier
Potassium alginate	402	Thickening and stabilizing agent
Agar	406	Thickening agent and stabilizer
Carrageenan including furcellaran	407	Thickening, gelling agent, stabilizer, and emulsifier
Polysaccharides of non-marine origin		
Xanthan gum	415	Thickener, foaming agent, stabilizer, and emulsifier
Gellan (1999)	418	Thickening and stabilizing agent
Curdlan (2001)	424	Thickening and stabilizing agent
Pullulan (2009)	1204	Thickening and stabilizing agent

Note: Additives permitted for use in foods by Codex General Standards for Food Additives (CODEX STAN 192-1995). Year listed is in parentheses.

in foods during processing, storage, or packaging; and (3) Color Additive Amendment of 1960, which regulates the listing and certification of color additives. The Food Additives Amendment to the FD&C Act classifies substances that are added to food into four regulatory categories:

1. *Food additives*—Substances that have no proven track record of safety and must be approved by the FDA before they can be used

2. *Generally Recognized as Safe (GRAS)*—Substances for which use in food has a proven track record of safety based either on a history of use before 1958 or on published scientific evidence, and that need not be approved by the FDA prior to being used

3. *Prior sanctioned*—Substances that were assumed to be safe by either the FDA or the U.S. Department of Agriculture before 1958, to be used in a specific food (e.g., the preservative nitrate can be used in meat because it was sanctioned before 1958 but it cannot be used on vegetables because they were not covered by the prior sanction)

4. *Color additives*—Dyes that are used in foods, drugs, cosmetics, and medical devices and must be approved by the FDA before they can be used

Since 1958, when the Food Additives Amendment was adopted, scientific techniques have been developed to evaluate the safety and carcinogenicity of substances in the food supply. About 200 substances were exempted from testing requirement because were judged by experts to be GRAS under the conditions of their use in foods at the time. A GRAS substance is one that has a long, safe history of common use in foods or that is determined to be safe based on proven science. Some substances may be GRAS for one use but not for others. GRAS or prior-sanctioned status does not guarantee the safety of a substance. If new data suggest that a substance under either of these categories may be unsafe, the FDA may take action to remove the substance from food products or require the manufacturer to conduct studies to evaluate the newly raised concern. The Office of Regulatory Affairs and its components are responsible for inspecting the full range of FDA-regulated products, both before they are marketed and afterward.[6,7] Most GRAS substances have no quantitative restrictions as to use, although their use must conform to Good Manufacturing Practices.

Food hydrocolloids, integral processing aids, are regulated either as a food additive or as a food ingredient (with the exception of gelatin, a proteinous hydrocolloid).[7] The Food Additives Status List is maintained by the FDA Center for Food Safety and Applied Nutrition (CFSAN) under an ongoing program known as the Priority-Based Assessment of Food Additives (PAFA). The PAFA database contains administrative, chemical, and toxicological information on over 2000 substances directly added to food, including substances regulated by the FDA as direct, secondary direct, and color additives, as well as GRAS and prior-sanctioned substances.

An inventory of more than 3000 substances often referred to as the *Everything Added to Food in the United States* (EAFUS) database includes ingredients added directly to food that the FDA has either approved as food additives or listed or affirmed as GRAS. Information on GRAS ingredients can be found in the GRAS Notice Inventory (http://www.fda. gov/Food/FoodIngredientsPackaging/GenerallyRecognizedasSafeGRAS/ GRASListings/default.htm), Listing of Food Additive Status (http:// www.fda.gov/Food/FoodIngredientsPackaging/FoodAdditives/Food AdditiveListings/ucm091048.htm), or Color Additive Status List (http:// www.fda.gov/ForIndustry/ColorAdditives/ColorAdditiveInventories/ ucm106626.htm). Discoveries and developments in chemistry such as those in nanotechnology, will continue to present challenges to food regulators.[8]

The European Union harmonizes food additive regulations within its member countries through its Council Directive 78/663/EEC of July 25, 1978, laying down specific criteria on purity for emulsifiers, stabilizers, thickeners, and gelling agents for use in foodstuffs. Clearance of food hydrocolloids by the European Council was given in 1995 under Directive 95/2/EC for food additives other than colors and sweeteners under the Miscellaneous Additives Directive. The Institute for Health and Consumer Protection of the Joint Research Centre (JRC), a directorate of the European Commission, gives guidelines on the use of materials and articles intended to come into contact with foodstuffs. The Community Reference Laboratory for Food Contact Materials of the JRC addresses the needs of Member State laboratories and provides the means for enforcing compliance measures, methods, and reference substances.

The regulation of preservatives is based on Council Directive 64/54/EEC of November 5, 1963, on the approximation of the laws of the Member States concerning the preservatives authorized for use in foodstuffs intended for human consumption. This directive established a list of agents that are fully accepted for use in the Member States. These compounds are designated in Annex I of EU Directive 80/597/EEC with an appropriate serial number (e.g., E400 = alginic acid). Standardization of labeling procedures includes providing details of trade, name, manufacture, designated number, etc. The approved list has two parts: substances whose primary function is preservative and those whose primary functions are other than preservative but which have a secondary preservative effect. Separate labeling laws require that the presence of additives, including preservatives, be declared. Permitted preservatives must be safe and technologically effective.[8]

In the European Union, safety evaluations are carried out by the Scientific Committee for Food. Technological effectiveness is assessed by experts from the governments of Member States, the food and chemical industries, and the European Commission. Specific purity criteria were established in EU Directive 65/66/EEC of January 26, 1965, laying down specific criteria of purity for preservatives authorized for use in foodstuffs intended for human consumption. This directive specifies the criteria that be satisfied before an additive may be approved, including guidance on what constitutes "need," appropriate safety evaluations (and reevaluations, as necessary). Agreed conditions of use ensure that acceptable daily intakes are not compromised; where these are sufficiently high, the concept of *quantum satis* might be applied.[9]

The Scandinavian counties and Belgium divide foods and additives into classifications that are permitted, whereas the United Kingdom and the Netherlands use separate regulations for individual additives. In the United Kingdom, the schedule of Emulsifiers and Stabilizers in Food Regulations, promulgated by the Ministry of Agriculture, Fisheries, and Food, includes alginic acid, plant gums such as guar, agar, and other emulsifiers. Japan has their own specifications, which include many of the food additives particular to Japan. Manufacturers of hydrocolloids are required to file petitions with agencies such as the Food Chemicals Codex, U.S. Pharmacopoeia, European Pharmacopoeia, and Japan Pharmacopoeia. These organizations establish methods to identify specific products and standards of purity for pharmaceutical and drug use. Such standards are necessary for future expansion of the use of polysaccharides, including those of marine origin.[10,11]

Regulatory authorities require the labeling of *all* ingredients in food products, including seafood items, along with their particular technological functions and EU or FDA identification codes. Following the general rules on food additives, additive producers are obliged to present methods to analyze the additive in the final food product.

10.4 Polysaccharides

Polysaccharides, similar to other commercial food additives, must satisfy certain safety criteria in order to protect the health of consumers. As pointed out in Chapter 8, a number of polysaccharides from terrestrial microbes have received regulatory approval and are being used commercially for food product development. Dextran, from *Leuconostoc mesenteroides*, was the first microbial extracellular polysaccharide (EPS) to receive regulatory approval, in 1947. Later, three EPSs from other non-marine microorganisms were also approved as food additives by the FDA: xanthan (isolated from *Xanthomonas campestris*), gellan (*Sphingomonas paucimobilis*), and curdlan (*Agrobacterium* spp. Biovar. 1 or *A. radiobacter*).[12] In the United States, xanthan is permitted as a food ingredient under food additive regulations controlled by the FDA. The polysaccharide is on the list of GRAS compounds. The use of xanthan is permitted for maintenance of viscosity; suspension of particulate matter; emulsion and freeze–thaw stabilities of meat sauces, gravies, and meats; canned, frozen, or refrigerated meat salads; canned or frozen meat stews; canned chili or chili with beans; pizza topping mixes and batter; and breading at concentrations sufficient for purposes in accordance with 21 CFR 172.5.[6,9] An acceptable daily intake for xanthan has been recommended by the Joint FAO/WHO Expert Committee on Food Additives. In addition to animal feeding studies, evaluations on the dietary effects of xanthan in human volunteers have demonstrated an absence of adverse effects and changes in enzymatic and toxicological indicators.[12] The compound was accepted as a food additive (E415) by the European Union in 1974. It is estimated that about 60% of the xanthan currently produced is food grade.[13]

Gellan gum was approved for use in food in Japan in 1988, later in the United States, and more recently in Europe.[14] The approved uses of gellan are as a gelling, stabilizing, and suspending agent.[13,14] In the United States, curdlan was approved in 1996 as Pureglucan™ and the product was launched as a formulation and processing aid, stabilizer, and thickener or texture modifier for foods.

No toxicity or carcinogenicity of curdlan has been observed in animal studies and *in vitro* tests, including acute, subchronic, and chronic toxicity studies and reproduction and carcinogenicity studies. No evidence of any toxicity or carcinogenicity or of any effects on reproduction has been observed. The only effects seen in these studies were reductions in weight gain at higher dietary concentrations due to replacement of part of the diet by curdlan; there was an effect on the body weights of the pups with the 15% diet, which was shown in additional studies to be due to the reduced food availability in the animals at this dose level. There was no evidence of adverse effects on the nutritional status of the animals nor on the absorption of minerals.[15] Curdlan was approved as early as 1989 and commercialized for food uses in Korea, Taiwan, and Japan.[15] In Japan, a number of polysaccharides from microorganisms, not necessarily of marine origin, have found commercial applications; these compounds are regarded as natural products, although EPSs from marine microorganisms have yet to receive approvals.

Suppliers are responsible for ensuring that these microbial polysaccharides are safe, and adequate testing is required to determine that the polysaccharide is not an irritant and lacks sensitizing activity under standard test conditions. The user must also ensure that the polymer is safe for the proposed application. In the case of potential food additives, the product must also be shown to be free of adverse effects when fed to animals over several generations.[15] Table 10.3 lists some of the permitted additives according to European Parliament and Council Directive No 95/2/EC of 20 February 20, 1995, on food additives other than colors and sweeteners.

10.5 Marine Polysaccharides

The Food and Agriculture Organization (FAO) of the United Nations, among its other responsibilities, is concerned with the proper management of living marine resources, including fisheries and seaweeds. With regard to marine algae, these interests include developing better catalogs of worldwide commercial seaweed resources according to species, improved methods of assessing and managing these wild resources, understanding the impact of

TABLE10.3

Selected Polysaccharide Additives Permitted by the European Council

Polysaccharides	Source	Functions, Food Products, and Permitted Levels
Alginic acid (E400); sodium alginate (E401); potassium alginate (E402); ammonium alginate (E403); calcium alginate (E404); propylene glycol alginate (E405)	Large brown seaweeds, such as *Laminaria hyperborea, Ascophyllum nodosum,* and *Macrocystis* spp.	*Functions*—Emulsifier, suspending, stabilizer, gelling agent, thickener. *Products*—Jam, jellies, marmalades; sterilized, pasteurized, and ultra-heat treatment (UHT) cream; low-calorie cream, pasteurized low-fat cream; weaning foods for infants and young children in good health. *Permitted levels*—10 g/kg (individual or in combination); 0.5 g/kg in weaning foods (individual or in combination)
Agar (E406)	Mainly *Gelidium, Pterocladia,* and *Gracilaria*	*Functions*—Emulsifier, stabilizer, gelling agent, thickener. *Products*—Ice creams, milk shakes, instant desserts, custard tarts, suspending agent in soft drinks, spreads, partially dehydrated and dehydrated milk, tinned goods, glazes for meats. *Permitted level*—*quantum satis*
Carrageenan (E407)	Mainly *Eucheuma, Betaphycus, Kappaphycus,* and *Chondrus crispus*	*Functions*—Emulsifier, stabilizer, gelling agent, thickener. *Products*—Ice creams, milk shakes, instant desserts, custard tarts, suspending agent in soft drinks, spreads. *Permitted level*—0.3 g/L in infant formulae

Source: European Parliament and Council Directive 95/2/EC of February, 20, 1995, on food additives other than colors and sweeteners.

seaweed harvesting on other commercial resources for which seaweeds form a habitat, and determining the contribution of macroalgae to those marine food chains leading to commercial fish populations. The FAO is interested in seaweed cultivation methods (mariculture), wild stock harvesting, and their processing, marketing, and trade and relies on established experts such as the International Seaweed Association.[11]

10.5.1 Chitin and Chitosan

Chitin and chitosan products fall within the lowest level of concern for toxicological testing (see Table 10.1); nevertheless, some tests in categories 2 or 3 might also be necessary based on anticipated use levels and lack of available toxicological data. Being naturally present in living organisms, chitin and its deacetylated derivative chitosan are considered safe. The available literature on chitin and chitosan suggests a low order of toxicity, based on chemical structure and animal studies. Like several high-molecular-weight food polymers of natural origin such as cellulose and carrageenan, chitin and chitosan are not expected to be digested or absorbed from the human gastrointestinal tract. The human gastrointestinal tract does not have the ability to degrade the β-(1,4)-glycosidic linkage. The apparent lack of enzymes to degrade the β-(1,4)-glycosidic linkage in the human gastrointestinal tract and the high molecular weights of chitin and chitosan suggest that they would be excreted unchanged in the feces without significant absorption. This expected lack of absorption would preclude significant systemic toxicity. To date, chitosan would appear to be well tolerated clinically; however, its prolonged use in diets may have to be monitored to ensure that it does not disturb the intestinal flora or interfere with the absorption of micronutrients, particularly lipid-soluble vitamins and minerals, or have any other negative effect.[16,17]

The safety of chitooligomers prepared by the enzymatic depolymerization of chitosan has been reported in a short-term mice feeding study; the oral maximum tolerated dose of the chitooligomers was more than 10 g/kg body weight. No mutagenicity was observed, as judged by the Ames test, mouse bone marrow cell micronucleus test, and mouse sperm abnormality test. A 30-day feeding study did not show any abnormal symptoms and clinical signs or deaths in rats. No significant differences were found in body weight, food consumption, food availability, hematology values, clinical chemistry values, or organ/body weight ratios. No abnormality of any organ was found during histopathological examination.[18]

Some chitosan derivatives have also been tested for safety. Carboxymethyl derivatives of chitosan used to enhance the postharvest shelf life of coated fruits and vegetables have been evaluated for safety. Aqueous preparation of the compounds at 1% (w/w) were used to coat pellets of feed administered to albino rats. The preparations were also given orally (1 mL, 2% aqueous solution) to the animals. After 4 weeks of feeding, no significant changes were observed in body weight gain, weight of vital organs, or the hematology and histopathology of the animals, thus indicating that the coating formulations are safe.[17]

In the United States, no petition for the use of chitin or chitosan in food has been submitted; however, the 1994 Dietary Supplement Health and Education Act permitted their use as food supplements without premarket approval as long as no health claims are made.[19] The use of chitin and chitosan as ingredients in foods or pharmaceutical products, however, will require standardization of identity, purity, and stability. Manufacturers should consider filing petitions with agencies such as Food Chemical Codex, U.S. Pharmacopoeia, European Pharmacopoeia, and Japan Pharmacopoeia. These organizations establish methods to identify specific products and standards of purity for pharmaceutical and drug use. Such standards will be necessary for future expansion of the use of chitin and chitosan. Chitin and chitosan have been approved for pesticide and seed treatments, as fertilizer, and as animal feed additives. The U.S. Environmental Protection Agency has approved the use of commercially available chitosan for wastewater treatment up to a maximum level of 10 mg/L.[18,19] Chitosan produced by Primex® of Norway has received GRAS status from the FDA and is recognized as a functional food.[20]

In Japan, chitosan was approved as a food additive in 1983 and placed on the List of Food Additives Other Than Chemical Synthetics by the Japanese Ministry of Health, Labour, and Welfare Public Health Bureau, Food Chemistry Division. In the list of approved thickeners and stabilizers, the source of chitin is indicated as material obtained from the acid treatment of shells of crustaceans. In Japan, it is added to foods such as noodles, potato crisps, and biscuits. Based on its definition of functional foods, chitin and chitosan possess most of the required attributes related to enhancement of immunity, prevention of illness, delaying of aging, and recovery from illness. Chitin and chitosan are also approved in Canada for various food applications. In the European market, chitosan is sold in the form of dietary capsules to assist weight loss. In Norway, chitin is permitted as a food additive. In view of the various derivatives of chitin and chitosan that have found innumerable applications in the food and pharmaceutical industries, there is a need for standardization and specifications for all materials derived from chitin.[16,19,21]

10.5.2 Glucosamine

Glucosamine, a natural amino sugar, is the end product of the hydrolysis of chitosan (see Chapter 3). It is also found in large concentrations in certain foods such as milk, eggs, liver, yeast, and molasses. In the body it exists as components of mucopolysaccharides, mucoproteins, and mucolipids and is synthesized from L-glutamine and glucose. Glucosamine can be absorbed easily into the human intestine and has low toxicity.[22] Heating aqueous solutions of glucosamine to 150°C at a pH of 4 to 7.5 results in the formation of furfurals. Heating at pH 8.5 causes the generation of flavor components such as pyrazines, 3-hydroxypyridines, pyrrole-2-carboxaldehyde, furans, and acetol, among others. As ammonia is liberated from glucosamine, it initiates the ring opening of furfurals to form 5-amino-2-keto-3-pentenals. Intramolecular condensation of these intermediates between the amino group and the carbonyl groups leads to the formation of 3-hydroxypyridines and pyrrole-2-carboxaldehyde. These results are important with respect to heat processing foods containing. Glucosamine is an over-the-counter medicine in Japan that comes in combination with chondroitin sulfate.[22,23]

10.5.3 Seaweed and Seaweed Polysaccharides

Seaweeds have been used safely in the Far East for several centuries; however, for food product development, it is important that seaweed products meet industrial and technical specifications and consumer safety regulations.[24] Food regulatory authorities have specifically recommended seaweed species for use as raw materials for the extraction of the commercial gums. A number of seaweeds are approved by FAO/WHO Codex Food Standards as sources of commercial gums. The FDA regulates specific seaweeds used for extraction of the hydrocolloids. Kelp is classified by the FDA as a natural substance and extractive and GRAS when prepared under Good Manufacturing Practices. In Europe, seaweeds have received approval as food ingredients. France has extended approval for seaweeds as vegetables and condiments, thus opening new opportunities for the food industry. Processed eucheuma seaweed (PES) is now permitted as a food additive in all countries under INS 407 (E407a in the EU). The Joint FAO/WHO Expert Committee on Food Additives (JEFCA) allocated to PES a nonspecified acceptable daily intake (ADI). In North America, since 1990, PES has been approved and labeled as carrageenan. As additives, seaweeds are allowed in products such as vinegar and boiled vegetables, including mushrooms, at levels ranging from 200 to 350 mg per kg, whereas in canned or fermented vegetables up to 1000 mg/g is approved.[3] An ADI of 70 mg/kg body weight for prolylene glycol alginate was set by JECFA in 1993.

10.5.3.1 Alginate

The nontoxic nature of algin has been established. Food-grade-quality alginates comply with relevant international and national purity specifications. Alginic acid; its sodium, potassium, ammonium, and calcium salts; and propylene glycol alginates have been given INS numbers 400 to 405 (E400 to E405). To meet the varied needs of customers a range of product grades are available with differing viscosities, gelling properties, and particle sizes.[25]

10.5.3.2 Agar

The first phycocolloid used by humans, agar was one of the first food ingredients approved as GRAS by the FDA. It has passed all the toxicological, teratological, and mutagenic testing required by the FDA. The consumption of agar for several centuries and the many toxicological studies performed have confirmed the safety of the product. Under the code INS 406 (E406), agar is a permitted thickening agent/stabilizer for food that is authorized in all countries without limitation of daily intake.[25]

10.5.3.3 Carrageenan

Carrageenans have received particular attention in view of certain early adverse reports of gastrointestinal and immunological consequences as a result of consumption of food products containing the polysaccharides. Carrageenans are not degraded to any extent in the gastrointestinal tract and are not absorbed by species such as rodents, dogs, and non-human primates. Available data on long-term bioassays do not provide evidence of carcinogenic, genotoxic, or tumor-promoting activity by carrageenans. Like many dietary fibers, however, there is significant cecal enlargement in rodents when carrageenan is administered at high doses, but this does not appear to be associated with any toxicological consequences to the rodent.

Feeding studies on carrageenans from *Eucheuma spinosum* fed to guinea pigs, monkeys, and rats through the diet showed that there was little or no absorption of high-molecular-weight carrageenans by guinea pigs or rats. The various toxicological studies related to orally administered food-grade carrageenan have been summarized;[26,27] however, substantial amounts of low-molecular-weight (40,000 Da) and intermediate-molecular-weight (150,000 Da) carrageenans were found in the livers of these animals. Urinary excretion of carrageenan was limited to low-molecular-weight materials of 20,000 Da or lower. Qualitative and quantitative evidence indicated that there was an upper limit to the size of carrageenan molecules absorbed.[28]

Whereas carrageenans themselves are safe, the degradation products of carrageenans (poligeenans) having molecular weights of 20 to 30 kDa have been shown to exhibit toxicological properties. Foods containing high-molecular-weight carrageenans generally do not contain poligeenans (formerly referred to as degraded carrageenan). Poligeenans are not considered food additives.[29] Carrageenans are susceptible to degradation by bacterial carrageenases, including those of marine habitats; for example, *Cytophaga* bacteria decompose carrageenans. These bacteria are Gram-negative, facultative, anaerobic, nonflagellate organisms that form spreading colonies; they contain carotenoid pigments and have phosphatase activities. The bacteria also decompose agar, casein, gelatin, and starch.[30]

A κ-carrageenase (EC 3.2.1.83)-producing *Vibrio* sp. was isolated from the surface of a marine alga. The enzyme, having a molecular weight of 35 kDa, specifically hydrolyzed κ-carrageenan to neocarrabiose and neocarratetrose sulfates. The extracellular enzyme had a maximum activity at pH 8.0 and 40°C, with P_i and K_m values of 9.2 and 3.3 mg/mL, respectively.[31] Another bacterial strain that degrades various sulfated galactans (carrageenans and agar) was isolated from the marine red alga *Delesseria sanguinea*. This extracellular enzyme has a molecular weight of 40 kDa and optimal activity at pH 7.2. Seven marine bacteria that degraded carrageenans from *Eucheuma spinosum* were isolated, and the degradation products of the carrageenans could be separated and characterized by rapid size-exclusion chromatography.[32]

Various nonenzymatic processes can also degrade carrageenans. These processes include autoclaving, microwave treatment, and ultrasonication in the presence of acetate, citrate, lactate, malate, and succinate. Autoclaving in the presence of citrate or malate at 110 to 120°C for 2 hours generates five to seven types of oligosaccharides with a depolymerization rate of approximately 23.0%.[33] These findings must be considered when developing heat-processed foods containing carrageenans. The prolonged cooking of natural carrageenan with other ingredients that may be acidic should always be avoided, as there may be some degradation of carrageenan under these conditions.

When evaluating the safety of carrageenans, the following facts should be considered. Carrageenan at doses normally used in foods is not harmful, as shown by animal feeding studies. The form most susceptible to decomposition is κ-carrageenan; ι-carrageenan is the least. These decomposition products (molecular weights of 10,000 Da or less) may be harmful, as they can cause irritation and ulceration of the digestive tract of nonruminants such as rhesus monkeys, and they are absorbed and retained in the liver of these animals. Because the breakdown of carrageenan results in a loss of some of the desirable functional properties of the carrageenan, it is in the best interests of food processors to prevent such decomposition.

Carrageenan (INS 407, E407) is a permitted food additive in all countries. In the United States, carrageenan is generally recognized as safe for use in food when used in accordance with Good Manufacturing Practices. The FDA lists carrageenans on the GRAS list for food additives and considers natural carrageenan to be safe as a food additive, but the agency suggests measuring the molecular weights of samples of gums prior to their use in foods to be sure that degraded products are not used.[34] The Joint FAO/WHO Expert Committee on Food Additives has recommended an acceptable daily intake of 0 to 75 mg carrageenan per kg body weight. JECFA is of the view that, based on the information available, it is inadvisable to use carrageenan or processed euchema seaweed in infant formulae.[36] The Scientific Committee on Food (SCF) of the European Union, which recently evaluated research data on the biological effects of carrageenan, observed that the ADI level recommended by JECFA could be maintained.[36,37] The SCF endorsed a molecular weight distribution limit on carrageenan that is more restrictive than in the United States. The SCF acknowledged that "there is no evidence that exposure to low molecular weight carrageenan from the use of food-grade carrageenan is occurring"; however, it advised against the use of carrageenan in formula for infants that are fed from birth. The SCF had no objection to its use for older infants as follow-on milk or in weaning foods, for which carrageenan may be added up to 0.3 g/L milk.[37] The regulation of carrageenan is not uniform internationally, and controversy over the use of carrageenan has not been completely resolved.[36–38] EU regulations recognize the structural and functional similarities of carrageenan and furcellaran and classify them as E407. Table 10.4 summarizes FDA-approved food uses of marine polysaccharides, and Table 10.5 shows marine polysaccharides on the Food Additives Status List of the FDA.

10.6 Regulatory Aspects of Polysaccharide-Based Edible Films

Regulatory authorities classify edible films and coatings as food products, food ingredients, food additives, food contact substances, or food packaging materials. Because these films and coatings are an integral part of the edible

TABLE 10.4

USFDA-Approved Uses of Marine Polysaccharides in Food Products

Polysaccharide	Function	Product	Concentration
Agar–agar	Stabilize and thicken	Thermally processed canned jelly food products	0.25% of finished product
Algin	Stabilize and thicken	Breading mix, sauces	Sufficient for purpose in accordance with 21 CFR 172.5
Mixture of sodium alginate, CaCO₃, calcium lactate (lactic acid), or glucono-δ-lactone (GDL)	Bind meat pieces	Restructured meat food products	Maximum limit: Sodium alginate, 1%; $CaCO_3$, 0.2%; lactic acid/calcium lactate or GDL, 0.3% of product formulation; ingredients/mix to be added dry
Carrageenan	Extend and stabilize product	Breading mix, sauces	Sufficient for purpose in accordance with 21 CFR 172.5
Carrageenan	Prevent purging of brine solution	Cured pork products	Not to exceed 1.5% of formulations; not permitted in combination with other binders approved for use in cured pork products
Carrageenan (locust bean gum, xanthan gum blend)	Prevent purging of brine solution	Cured pork products	In combination not to exceed 0.5% of product formulation; not permitted in combination with other binders approved for use in cured pork products
Mixture consisting of water, sodium alginate, CaCl₂, sodium carboxymethyl-cellulose, and corn syrup solids	Reduce cooler shrinkage and help protect surface	Freshly dressed meat carcasses	Not to exceed 1.5% of hot carcass weight; chilled weight may not exceed hot weight

Note: Approval of substances for use in the preparation of products per Title 9 (Animals and Animal Products) CFR §318.7; updated January 1 each year.

TABLE 10.5

Marine Polysaccharides on the USFDA Food Additives Status List

Polysaccharide	Description	Applications
Agar–agar	MISC, GRAS/FS, GRAS	In baked goods and baking mixes, 2%; in confectionery/frosting, 1 to 2%; in soft candies, 0.25%
Ammonium alginate	MISC, REG	Boiler water additive
Potassium alginate	GRAS	Stabilizer and thickener: <0.1% in confections and frostings, <0.7% in gelatins and puddings, <0.25% in processed fruits and fruit juices, <0.01% in all other food categories
Alginic acid/algin	GRAS/FS	Cheeses, frozen desserts, jellies, preserves
Sodium alginate	STAB, GRAS/FS	Cheeses, frozen desserts: <0.5% finished product
Calcium or potassium alginate	GRAS	Confectioneries, gelatin, pudding, processed foods
Carrageenan and its NH_4/K/Na/Ca salts, *Gigartina* extracts	STAB, REG, GMP, REG/FS	<0.8% in finished cheese; also added with Polysorbate 80 at a maximum level of 500 ppm
Furcellaran and its K/Na/Ca salts	MISC, EMUL, STAB, REG/FS, GMP	Ice cream

Note: EMUL, emulsifier; FS, substances permitted as optional ingredient in a standardized food; GMP, in accordance with good manufacturing practices; GRAS, generally recognized as safe; GRAS/FS, substances generally recognized as safe in foods but limited in standardized foods where the standard provides for its use; MISC, miscellaneous; REG, food additives for which a petition has been filed and a regulation issued; REG/FS, food additives regulated and included in a specific food standard; STAB, stabilizer.

Source: USFDA, *Listing of Food Additive Status*, U.S. Food and Drug Administration, Washington, D.C. (http://www.fda.gov/Food/FoodIngredientsPackaging/Food Additives/FoodAdditiveListings/ucm091048.htm).

portion of food products, they should observe all regulations required for food ingredients.[39] In Europe, the ingredients that can be incorporated into edible coating formulations are regarded as food additives and are listed within the list of additives for general purposes. To maintain product safety and edibility, all film-forming components, as well as any functional additives in the film-forming materials, should be food-grade nontoxic materials, and all process facilities should meet the high standards of hygiene as required by the European Union.[40] Specific purity criteria for food additives are addressed by Commission Directive 2008/84/EC of August 27, 2008, laying down specific purity criteria on food additives other than colours and sweeteners.[42,43]

Because edible coatings could have ingredients with a functional effect, inclusion of such compounds should be noted on the product label. In Europe, the use of food additives must always be indicated on the packaging

TABLE 10.6

Commercial Status of Important Marine Polysaccharides

Polysaccharide	Commercial Status
Agar	Limited, capital-intensive market; raw materials limited; increasing competition from other gelling agents either used singly or in combinations such as carrageenan/locust bean gum (LBG) or xanthan/LBG or curdlan; use in bacteriological medium is established
Alginate	Mature product, with no new major applications; major commercial product, propylene glycol alginate; supply more fragile; many benefits from positive image as a marine product
Carrageenan	Multifunctional uses greatly expanded with acceptance in meat and poultry products; accepted stabilizer in the growing dairy industry; labeled as natural food stabilizer by some food companies; semi-refined carrageenan approved in the United States and Western Europe; market more competitive and less able to bear added value; niche markets for producers of expensive alcohol-precipitated refined carrageenan; some consumer concerns; cost and availability of raw materials are major problems
Chitosan	Water treatment, cosmetics, food and beverages, healthcare, agrochemicals, biotechnology, etc.; more than 50 companies involved in the business

Source: Seisun, D., in *Gums and Stabilizers for the Food Industry 11*, Williams, P.A. and Phillips, G.O., Eds., Royal Society of Chemistry, Cambridge, U.K., 2002, pp. 3–9; GIA, *Chitin & Chitosan*, Global Industry Analysts, Inc., February 1, 2007, 215 pp. (http://www.market research.com/map/prod/1473672.html); Bixler, H.J. and Porse, H., *J. Appl. Phycol.*, 2010 (DOI 10.1007/s10811-010-9529-3).

label according to their category (e.g., antioxidant, preservative, colorant) with either their name or E number.[43] In the United States, the FDA requires that any compound included in a formulation should be generally recognized as safe or regulated as a food additive and used within specified limitations.[43,44]

10.7 Commercial Status

Table 10.6 summarizes the commercial utilization of some important marine polysaccharides. The U.S. marine biotechnology market has surpassed $1 billion. The non-U.S. market is over $2.2 billion and is growing faster than the U.S. market, with an annual average growth rate of 4.7%. The non-U.S. market is projected to experience an average annual growth rate of 6.4%.[45]

References

1. Raspor, P. and Jevsnik, M., Good nutritional practice from producer to consumer, *Crit. Rev. Food Sci. Nutr.*, 48, 276, 2008.
2. Beulens, A. J. M., Food safety and transparency in food chains and networks: relationships and challenges, *Food Control*, 16, 481, 2005.
3. *Codex General Standard for Food Additives*, Codex STAN 192-1995, Codex Alimentarius Commission, Rome, 1995.
4. Bend, J. et al., *Evaluation of Certain Food Additives and Contaminants*, Technical Report Series No. 947, World Health Organization, Geneva, 2007.
5. *Safety Evaluation of Certain Food Additives and Contaminants*, prepared by the 68th Meeting of the Joint FAO/WHO Expert Committee on Food Additives (JECFA), WHO Food Additives Series 59, World Health Organization, Geneva, 2008.
6. Rados, C., FDA law enforcement: critical to product safety, *FDA Consumer*, January/February, 2006.
7. Armstrong, D. A., Food chemistry and U.S. food regulations, *J. Agric. Food Chem.*, 57, 8180, 2009.
8. Howlett, J. F., The regulation of preservatives in the European Community, *Food Add. Cont.*, 9, 607, 1992.
9. Gamvros, R. J. and Blekas, G. A., Legal aspects and specifications of biopolymers used in foods, *Develop. Food Sci.*, 41, 419, 2003.
10. Caddy, J. A. and Fisher, W. A., FAO interests in promoting understanding of world seaweed resources, their optimal harvesting, and fishery and ecological interactions, *Hydrobiologia*, 124, 111, 1985.
11. Heinze, T., Barsett, H., and Ebringerova, A., *Polysaccharides: Structure, Characterisation and Use*, Springer-Verlag, Heidelbrg, Germany, 2005.
12. Jarvis, I., Healthy xanthan gum market attracts new competitors, *Chem. Market Report.*, 265, 14, 2004.
13. Sworn, G., Gellan gum, in *Handbook of Hydrocolloids*, Phillips, C. O. and Williams, P. A., Eds., CRC Press, Boca Raton, FL, 2000, pp. 117–134.
14. Giavasis, I., Harvey, L. M., and McNeil, B., Gellan gum, *Crit. Rev. Biotechnol.*, 20, 177, 2000.
15. Spicer, F. J. F. et al., A toxicological assessment of curdlan, *Food Chem. Toxicol.*, 37, 455, 1999.
16. Weiner, M. L., An overview of the regulatory status of the safety of chitin and chitosan as food and pharmaceutical ingredients, in *Advances in Chitin and Chitosan*, Elsevier, London, 1992, pp. 663–672.
17. Ramesh, H. P., Viswanatha, S., and Tharanathan, R. N., Safety evaluation of formulations containing carboxymethyl derivatives of starch and chitosan in albino rats, *Carb. Polym.*, 58, 435, 2004.
18. Qin, C. et al., Safety and evaluation of short-term exposure to chitoligomers from enzymatic preparation, *Food Chem. Toxicol.*, 44, 855, 2006.
19. Subasinghe, S., Chitin from shellfish waste: health benefits overshadowing industrial uses, *Infofish Int.*, 3, 58, 1999.
20. Koide, S. S., Chitin–chitosan: properties, benefits, and risks, *Nutr. Res.*, 18, 1091, 1998.

21. Preuss, H. G. and Kaats, G. R., Chitosan as a dietary supplement for weight loss: a review, *Curr. Nutr. Food Sci.*, 2, 297, 2006.
22. Shu, C.-K., Degradation products formed from glucosamine in water, *J. Agric. Food Chem.*, 46, 1129, 1998.
23. Bhaskar, N., Hokkaido University, personal communication, 2009.
24. Mabeau, S. and Fleurence, J., Seaweed in food products: biochemical and nutritional aspects, *Trends Food Sci. Technol.*, 4, 103, 1993.
25. Marinalg International, www.marinalg.org/.
26. Cohen, S. M. and Ito, N., A critical review of the toxicological effects of carrageenan and processed *Eucheuma* seaweed on the gastrointestinal tract, *Crit. Rev. Toxicol.*, 32, 413, 2002.
27. Tobacman, J. K., Review of harmful gastrointestinal effects of carrageenan in animal experiments, *Environ. Health Perspect.*, 109, 983, 2001.
28. Michel, C. and Farlane, G. T., Digestive fates of soluble polysaccharides from marine macroalgae: involvement of the colonic microflora and physiological consequences for the host, *J. Appl. Bacteriol.*, 80, 349, 1996.
29. Pittman, K. A., Golberg, L., and Coulston, F., Carrageenan: the effect of molecular weight and polymer type on its uptake, excretion and degradation in animals, *Food Cosmet. Toxicol.*, 14, 85, 1976.
30. Sarwar, G., Sakata, T., and Kakimoto, D., Isolation and characterization of carrageenan-decomposing bacteria, *J. Gen. Appl. Microbiol.*, 29, 145, 1983.
31. Araki, T., Higashimoto, Y., and Morishita, T., Purification and characterization of κ-carrageenase from a marine bacterium, *Vibrio* sp. CA-1004, *Fish. Sci.*, 65, 937, 1999.
32. Joo, D. S. and Cho, Y. S., Preparation of carrageenan hydrolysates from carrageenan with organic acid, *J. Korean Soc. Food Sci. Nutr.*, 32, 42, 2003.
33. Kuntsen, S. H. et al., A rapid method for the separation and analysis of carrageenan oligosaccharides released by ι- and κ-carrageenase, *Carbohydr. Res.*, 33, 101, 2001.
34. FDA, Food Additives Permitted for Direct Addition to Food for Human Consumption, 21 CFR 172, Subpart C, Product Quality Control, U.S. Food and Drug Administration, Washington, D.C., 2006.
35. Watson, D. B., Public health and carrageenan regulation: a review and analysis, *J. Appl. Phycol.*, 20, 505, 2008.
36. Food and Agriculture Organization/World Health Organization, 68th Meeting of the Joint FAO/WHO Expert Committee on Food Additives (JECFA), Geneva, June 19–28, 2007.
37. Scientific Committee on Food, European Commission, Health & Consumer Protection Directorate-General, *Opinion of the Scientific Committee on Food on Carrageenan*, SCF/CS/ADD/EMU/199 Final, February 21, 2003.
38. Carthew, P., Safety of carrageenan in foods, *Environ. Health Perspect.*, 110, A176, 2002.
39. Brody, A. L., Strupinsky, E. R., and Kline, L. R., *Active Packaging and Food Applications*, Technomic, Lancaster, PA, 2001.
40. Chen, M. C., Yeh, G. H. C., and Chiang, B. H., Antimicrobial and physicochemical properties of methylcellulose and chitosan, *Trends Food Sci. Technol.*, 12, 1, 2005.
41. Guilbert, S., Cug, B., and Gontard, N., Innovations in edible and/or biodegradable packaging, *Food Addit. Contam.*, 14, 741, 1997.

42. European Parliament and Council Directive No. 95/2/EC of 20 February 1995 on food additives other than colours and sweeteners (http://ec.europa.eu/food/fs/sfp/addit_flavor/flav11_en.pdf).
43. Rojas-Graü, M. A., Soliva-Fortuny, R., and Martín-Belloso, O., Edible coatings to incorporate active ingredients to fresh-cut fruits: a review, *Trends Food Sci. Technol.*, 20, 438, 2009.
44. de Kruijf, N. et al., Active and intelligent packaging: applications and regulatory aspects, *Food Add. Cont.*, 19(Suppl.), 141, 2002.
45. BCC Research, *Biomaterials from Marine Sources*, Report No. BIO046B, February 2003 (http://www.bccresearch.com/report/BIO046B.html).
46. Seisun, D., Overview of the hydrocolloid market, in *Gums and Stabilizers for the Food Industry 11*, Williams, P. A. and Phillips, G. O., Eds., Royal Society of Chemistry, Cambridge, U.K., 2002, pp. 3–9.
47. GIA, *Chitin & Chitosan*, Global Industry Analysts, Inc., February 1, 2007, 215 pp. (http://www.marketresearch.com/map/prod/1473672.html).

Section III

Biomedical Applications

11

Biomedical Applications of Marine Polysaccharides: An Overview

11.1 Introduction

Polysaccharides are emerging ingredients in biomedical applications because they are biodegradable, water soluble, and functionally active. Polysaccharides for such applications are available from diverse sources, including agricultural and marine resources. Recent interest in polysaccharide-based materials for biomedical use can also be attributed to new possibilities for chemical modifications that enhance their functional activities for specific purposes. These strategies involve combinations of polysaccharides with other polymers and applications of nanotechnology. Some of the major applications of polysaccharides in the biomedical field include controlled drug delivery, tissue regeneration, wound dressing, dental implants, blood plasma expanders, vaccines, and nonviral gene delivery, among others. Polysaccharide-based delivery systems that carry molecules of interest within their networks have been developed for the biomedical and pharmaceutical sectors to transport drugs and other bioactive compounds to targeted sites. Tissue engineering has the goal of regenerating tissue using novel compatible biomaterials. Interest in these fields is indicated by several recent articles.[1-4]

The remarkable ability of polysaccharides to form hydrogels means they are capable of absorbing a great amount of water once immersed in biological fluids and assuming a structure similar to extracellular matrix or biological tissue. These hydrogels, which are highly hydrated polymer networks, allow cells to adhere, proliferate, and differentiate, essential in the treatment of diseased or injured tissues and organs. These gels can protect drugs from hostile environments and release them in response to particular environmental stimuli such as pH and temperature. The hydrogel-forming property also makes them ideal materials for cellular scaffolds, coatings, and devices for the treatment of various diseases. Hydrogel scaffolds of natural polysaccharides are useful in tissue engineering, and being able to inject them via a needle without a loss of rheological properties further adds to their applicability. Chemical modifications of these hydrogels, such as the insertion of sulfate groups, improve their biocompatibility.[5,6] Innovative manufacturing technologies have resulted in new strategies for the stabilization of sensitive drugs and the development of novel approaches to site-specific carrier targeting.[7] In addition, nanotechnology enhances the potential of marine polysaccharides to be used in the areas of food and pharmaceutical sciences (see Chapter 9).

11.2 Marine Polysaccharides for Biomedical Applications

Polysaccharides from marine sources, similar to their terrestrial counterparts, offer diverse therapeutic functions derived from the fact that most of them are biocompatible, biodegradable to harmless products, nontoxic, physiologically inert, and capable of forming hydrogels due to their remarkable hydrophilicity, which helps them to bind proteins and other compounds. Of the various marine polysaccharides, alginate, fucoidan, and chitosan have established themselves as promising materials for a variety of uses in medicine. Chemical modifications involving the combination of polysaccharides with other polymers and compounds allow the development of novel functionalities for these macromolecules.[8]

11.2.1 Crustacean Polysaccharides: Chitin and Chitosan

The applicability of chitin is restricted due to its poor solubility. On the other hand, chitosan is more versatile in its applications in medicine and pharmacology, apart from agriculture, biotechnology, and food (see Chapter 6). These applications are made possible because of the characteristic structural features of chitosan, particularly its deacetylated nature, net cationic charge, and the presence of multiple reactive amino groups in the molecule which help the polymer to interact with water and other compounds (see Chapter 3).[9,10] Chitosan has diverse medical applications, including hemodialysis membranes, artificial skin, hemostatic agents, hemoperfusion columns, and drug delivery systems. The property of chitosan to form gels at a slightly acid pH gives chitosan its antacid and antiulcer activities. Chitosan exhibits anticholesterolemic and antiuricemic properties when administered orally. Oral administration of chitosan also suppresses serum cholesterol levels and hypertension. The hypocholesterolemic mechanism of chitosan is due to its ability to bind fatty acids, bile acids, phospholipids, uric acid, and the toxic gliadin fraction. Chitosan does not depress serum iron and hemoglobin. It has no influence on the human intestinal microorganisms but lowers the putrefaction metabolites. Chitin and chitosan oligosaccharides, when intravenously injected, enhance antitumor activity by activating macrophages.[11–15] Interesting applications of chitosan in health care are discussed below.

11.2.1.1 Chitosan as Drug Delivery Matrix

Chitosan is considered to be the drug carrier for the 21st century.[16] The polysaccharide enhances the dissolution properties of poorly soluble drugs and aids in the transdermal delivery of drugs. It also prevents drug irritation in the stomach. For effective drug delivery, chitosan can be used in the form of microspheres, microparticles, nanoparticles, granules, gels, or films. Chitosan microspheres are useful for the controlled release of antibodies, antihypertensive agents, anticancer agents, protein and peptide drugs, vaccines, and nutraceutical compounds. Chitosan and its derivatives also are promising nonviral vectors for gene delivery.[17–20] The drug-carrying ability of chitosan can be enhanced through derivatization and complex formation with other polymers; for example, a novel dual cross-linked complex gel bead for oral delivery of protein drugs has been reported recently. The composite capsules are composed of carboxymethyl chitosan and alginate; the beads are capable of withstanding the acidity of gastric fluids without liberating substantial amounts of loaded protein, and they retard protein release in the intestine, suggesting their efficacy as carriers for oral protein drug delivery.[21]

A biodegradable, glucose-sensitive *in situ* gelling system that utilizes chitosan for the delivery of insulin has been developed. The glucose-sensitive gel responded well to varied glucose concentrations *in vitro*. The gel released the entrapped insulin in a pulsatile manner in response to the glucose concentration *in vitro*.[22] In another study, calcium alginate beads were coated with chitosan using ionotropic gelation. The bead particles, which ranged in size from 200 to 400 μm, exhibited excellent muco-adhesive properties. The release of drugs from the beads was dependent on the composition of the beads, the component polymer, and its possible interactions.[23] Drugs encapsulated in chitosan have immense potential to control colon-based diseases. Large amounts of enzymes are present in the human colon; they are secreted by colon bacteria that ferment the polysaccharides.[24,25] A composite collagen–chitosan membrane has potential for the treatment of periodontal defects in dentistry.[21]

Chitosan–carrageenan composite nanoparticles show promise as carriers of therapeutic macromolecules, with potential applications in drug delivery, tissue engineering, and regenerative medicine. The nanoparticles can be obtained in a hydrophilic environment, without the use of organic solvents or other aggressive technologies for their preparation. These nanocarriers vary in size from 350 to 650 nm and positive zeta potentials of 50 to 60. Using ovalbumin as the model protein, nanoparticles demonstrated excellent capacity for the controlled release of the protein for 3 weeks.[26] A biocompatible gel of chitosan and β-glycerol phosphate (GP) material is a promising vehicle for a variety of cell encapsulation and injectable tissue-engineering applications.[27]

11.2.1.2 Wound Healing

The bacteriostatic and fungistatic properties of chitosan are particularly useful for wound treatment. Because of this, chitosan has found use as a wound healing agent in skin ointments. Chitosan implanted in animal tissues encourages wound healing and hemostatic activities. Transdermal films containing chitosan can slowly release drugs into the blood. The sulfated derivatives of chitin and chitosan have anticoagulant and lipolytic activities in animal blood. It is possible to modulate the wound healing process using N-carboxybutyl chitosan, which favors an ordered reconstruction of dermal architecture, while collagen provides a valid scaffold for organizing cell and stromal matrices. Hemostasis is immediately obtained after the application of most of the commercial chitin-based dressings to traumatic and surgical wounds. Platelets are activated by chitin with redundant effects and superior performances compared with known hemostatic materials. Chitin and chitosan activate the angiogenesis necessary to support physiologically ordered tissue formation, and chitosan possesses bioadhesive properties that make it useful in sustained-release formulations. Chitosan has been found to encourage nerve growth, and chitosan-coated hydroxyapatite microspheres and granules reduce bleeding and hasten healing with hard tissue growth in dental and orthopedic applications. The antibacterial and antifungal activities of chitosan prevent bacterial and fungal infection.[28] Biocompatible wound dressings derived from chitin are available in the form of hydrogels, xerogels, powders, composites, and films.[29,30] A novel biocompatible blended fiber was prepared by blending chitin with tropocollagen in an aqueous solution of acetic acid and methanol, which was spun through a viscose-type spinneret into a dilute aqueous ammonia solution containing ammonium sulfate at room temperature, which gave a white fiber of chitosan–tropocollagen.[31]

11.2.1.3 Tissue Engineering

Tissue engineering emerged in the late 1980s. Skin repair is an important aspect of tissue engineering, especially for extended third-degree burns, where current treatments are still insufficient in promoting satisfying skin regeneration. Biocompatible, biodegradable, and injectable compounds can serve as temporary skeletons to accommodate and stimulate new tissue growth. Hydrogels derived from natural proteins and polysaccharides are ideal scaffolds for tissue engineering because they resemble the extracellular matrices of the tissue comprised of various amino acids and sugar-based macromolecules. Chitosan-based scaffolds have potential use for promoting good tissue regeneration, as chitosan hydrogels maintain the correct morphology of chondrocytes and preserve their capacity to synthesize cell-specific extracellular matrices. Chitosan scaffolds incorporating growth factors and morphogenetic proteins have been developed for the purpose of rapid bone regeneration. Biomineralized alginate–chitosan microcapsules have been proposed as multifunctional scaffolds and delivery vehicles for the tissue regeneration of hard and soft tissues.[32] The potential of hydroxyapatite (HA)–chitin matrices to serve as tissue-engineered bone substitutes has been demonstrated. These nontoxic compounds favor bone regeneration along with biodegradation of the HA–chitin matrix. Although total joint replacement has become a common procedure in recent years, bacterial infection remains a significant postsurgery complication. One way to reduce the incidence of bacterial infection is to add antimicrobial agents to the bone cement used to fix the implant.[29,33–36]

The development of a biodegradable porous scaffold made from chitosan and alginate polymers with improved mechanical and biological properties has been reported. Bone-forming osteoblasts readily attached to the chitosan–alginate scaffold, proliferated well, and deposited calcified matrix. The hybrid scaffold had a high degree of tissue compatibility. Calcium deposition occurred as early as the fourth week after implantation. The scaffold can be prepared from solutions of physiological pH, which may provide a favorable environment for incorporating proteins with less risk of denaturation.[37]

Chitin fibers with improved strength can be obtained by making use of the nanostructures and mesophase properties of chitin. Hydrodynamic shaping and *in situ* cross-linking of hydrogel precursors are utilized in an efficient "hydrodynamic spinning" approach for synthesizing hydrogel fibers of different diameters. The material has significant potential for applications in tissue engineering.[38,39] Chitin and chitosan can be used as sutures because of their biocompatibility, biodegradability, and nontoxicity together with their antimicrobial activity and low immunogenicity. A process for the production of adsorbable surgical sutures has been developed by the Central Institute of Fisheries Technology in Cochin, India (K. Devasadan, pers. comm.). Table 11.1 summarizes general applications of chitin, chitosan, and some of their derivatives in health care, Table 11.2 shows salient properties of chitin and chitosan in biomedical applications, and Table 11.3 indicates the quality criteria required for chitosan in these applications.

11.2.1.4 Glucosamine

Glucosamine, the hydrolytic product of chitosan, is commonly consumed in combination with chondroitin sulfate from shark cartilage to treat arthritis and osteoporosis.[29]

11.3 Seaweed and Seaweed Polysaccharides

Seaweed has been reported to possess medicinal properties (see Chapter 7). Seaweed polysaccharides, in general, have interesting biological activities in that they are antibacterial, antiviral, antihyperlipidemic, anticoagulant, and

TABLE 11.1

— General Applications of Chitin, Chitosan, and Some Derivatives in Healthcare

Compound	Applications
Chitin	Wound dressing
	In vivo absorbable sutures
	Drug delivery
	Dialysis membrane
O-carboxymethyl chitin and O-hydroxypropyl chitin	Cosmetic ingredient
N-actylchitohexasacharide	Antitumor agent
Chitosan and its various derivatives	Artificial skin
	Blood anticoagulant and hemostatic materials
	Chitin and chitosan hydrogels for delivery of nutraceuticals
	Protein absorbents
	Drug delivery systems
	Hemodialysis membranes
	Immunostimulation, molecular recognition, and entrapment of growth factor
	Nanofiber scaffold for nerve tissue regeneration
	Hypocholesterolemic agents
	Wound-healing materials
	Nutrition (e.g., hypocholesterolemic agent, dietary fiber, weight reduction)
	Water treatment (remove metals, radioisotopes, pesticides)
	Cosmetics (shampoo, skin products)
N-hexanoylchitosan and N-octanoylchitosan	Antithrombogenic material for artificial blood vessels
	Contact lenses
	Blood dialysis membranes
	Artificial organs
N-carboxybutylchitosan	Wound dressing
5′-Methylpyrrolidinone chitosan	Dentistry
Sulfates of chitin and chitosan	Anticoagulant and lipolytic agents

Source: Adapted from Synowiecki, J. and Khatieb, N.A., *Crit. Rev. Food Sci. Nutr.*, 43, 145, 2003; Subasinghe, S., *Infofish Int.*, 3, 58, 1999.

antitumorigenic, and the potential for seaweed to ameliorate chronic renal failure in rats has been reported.[40] Sulfated polysaccharides from marine algae are particularly known for being antitumorigenic and for their anti-thrombin, cell recognition, cell adhesion, and receptor regulator functions.[41] Many seaweed species, including the Indian seaweed species *Eucheuma kappaphycus*, *Gracilaria edulis*, and *Acanthophora spicifera*, are effective antioxidants.[42] Table 11.4 lists biological activities and potential biomedical applications for polysaccharides derived from seaweeds, marine microalgae, and marine microorganisms.

TABLE 11.2

Salient Properties of Chitosan in Biomedical Applications

Applications	Salient Property
Wound healing, burn therapy	Chitosan forms tough, water-absorbent, biocompatible films that promote tissue growth.
Hemodialysis membranes	Chitosan–cellulose blended membranes using trifluroacetic acid as a cosolvent have improved dialysis properties in artificial kidney due to improved permeability.
Drug delivery matrix	Chitosan functions as an inexpensive carrier encapsulating nutraceuticals and drugs. Chitosan enhances dissolution properties of poorly soluble drugs, prevents drug irritation in the stomach, and aids transdermal delivery of drugs.
Removal of toxins	Chitosan-encapsulated activated charcoal has the potential to remove toxins and bilirubin.
Hemoperfusion	Chitosan and its oligomers can satisfy the requirements of specificity and blood compatibility; it can also be used as a selective adsorbent for antigen/antibodies.
Artificial cartilage scaffolds	Chitin and chitosan, when combined with chondroitin sulfate, support proteoglycan production to treat cartilage deficiency.
Anticholesterol drug	Chitosan reduces lipid absorption by trapping neutral lipids.
Dental bioadhesives, biodegradable sutures	A composite collagen–chitosan membrane has potential for the treatment of periodontal defects in dentistry.
Composite films and nanoparticles with alginate, fucoidan, β-lactoglobulin, etc.	Chitosan can be utilized in tissue engineering for drug delivery, skin recovery, scaffolds, etc.

Source: Adapted from Hirano, S. et al., *Biomaterials*, 21, 997, 2000; Muzarelli, R.A.A., *Carb. Polym.*, 76, 157, 2009.

11.3.1 Alginates

Alginates have been the focus of considerable research because of their versatile functional properties. The conventional biomedical uses of alginate primarily depend on its thickening, gel-forming, and stabilizing properties. In the medical field, alginates serve as vectors for drug delivery, dental impressions, absorbent in dressings, antireflex therapies, and more. Novel composites based on calcium sulfate blended with alginate and materials based on alginate and acrylic polymers, such as alginate–polyethylene glycol copolymers, can be used for a variety of biomedical applications (e.g., reconstructive scaffolds).[29] Typical examples of biomedical applications of alginate are discussed below.

TABLE 11.3

Requirements for Pharmaceutical-Grade Chitosan Products

Parameter	Requirement
Viscosity	Specify
Acid content	Specify
Moisture	Specify
Solubility, turbidity	>99.9%
Heavy metals	<27 ppm
Microorganisms	<1 CFU/g
Color	Colorless, grayish, cream
Ash	2–3%
Appearance	Powder, flakes
Grain size	Less than 25% should be >2 mm in size

Source: Adapted from Steinbüchel, A. and Rhee, S.K., Eds., *Polysaccharides and Polyamides in the Food Industry: Properties, Production, and Patents*, Vol. 1, p. 163, 2005. Copyright Wiley-VCH Verlag GmbH & Co. KGaA. Reproduced with permission.

TABLE 11.4

Biological Activities of Polysaccharides from Seaweeds, Marine Microalgae, and Marine Microorganisms and Their Applications

Polysaccharide	Function
Alginate	Delivery of drugs, including proteins
	Dental impressions
	Absorbent in dressings
	Antireflex therapies
	Scaffold in reconstructive processes
	Wound dressings
Carrageenans	Antiinflammatory and immune responses
	Antiviral activities
Fucoidans	Anticoagulant
	Antitumor activity
	Immunomodulating activity
	Hypoglycemic activity
	Hypolipidemic activity
	Antiinflammatory activity
	Antiviral activity
	Ameliorates chronic renal failure
Ulvan	Anticoagulant activity
	Prevention of ischemic cardiovascular diseases
Polysaccharides microalgae and corals	Anticoagulant activity
	Antithrombotic activity
	Antiviral activity
Polysaccharides from marine microorganisms	Antitumor activity
	Immunostimulatory activity

10.3.1.1 Wound Dressing

Wound dressings that can be formed *in situ* offer several advantages over the use of preformed dressings such as conformability with the wound bed, ease of application, and improved patient compliance and comfort. Alginate gel dressings are particularly useful for bleeding wounds, because calcium alginate is a natural hemostat. The use of alginate as a wound dressing requires tight control of a number of material properties, including mechanical stiffness, swelling, degradation, cell attachment, and binding or release of bioactive molecules. Control over these properties can be achieved by chemical or physical modifications of the polysaccharide itself or the gels. During the sol–gel transition of alginate, channel-like pores are created, the dimensions of which can be influenced by the alginate gel concentration, its nature and conformation, the pH, or the temperature.[43] The use of an alginate–gelatin hydrogel containing borax for wound dressing has been reported; the composite matrix has the hemostatic effect of gelatin, the wound-healing-promoting feature of alginate, and the antiseptic property of borax. The hydrogel was found to have a fluid uptake of 90% of its weight, which would prevent the wound bed from accumulating exudates. In addition, the hydrogel can maintain a moist environment over the wound bed in a moderate to heavily exuding wound, which would enhance epithelial cell migration during the healing process. Using a rat model, it was demonstrated that within 2 weeks the wound covered with gel was completely filled with new epithelium without any significant adverse reactions.[44]

11.3.1.2 Drug Delivery

The role of alginate in drug delivery systems has been well documented. In drug formulations, the alginate gel can be prepared prior to use, or it can spontaneously form *in situ* in physiological fluids, under conditions of low pH, or in the natural presence of calcium ions in the site of administration. Alternatively, the gelling agent can be added as a part of the formulation or separately administered. Because of the relatively mild alginate gelation process, proteins can be loaded into and released from the polysaccharide gel matrices without loss of their biological activities. Alginates have also been investigated for use as insulin delivery systems for diabetic patients.[45]

Tablets are the most common dosage form, due to their convenience and ease of preparation. Alginate tablets can be prepared by direct compression, as well as by wet or dry granulation and coating with various techniques. Alginate tablets provide a targeted, slow release of drugs. The tablet properties are influenced by the chemical compaction of the alginate, its molecular weight, guluronic acid/mannuronic acid ratio, and the presence of salt. Alginate tablets have good elastic properties, although the compactibility of alginates has been reported to be lower than that of chitosan.[46] The compression properties indicate that sodium alginate tablets are more elastic than those made of potassium alginates; further, tablets containing alginates with low guluronic acid content were found to exhibit higher elasticity than tablets with alginates having a low mannuronic acid content. Hormone-producing cells can be encapsulated in alginate gels for delivery.[29,47,48]

Microcapsules of alginate (typically about 200 μm in size) are obtained by dropping an aqueous solution of the polysaccharide into a gelling solution that is either acidic (pH < 4) or, more usually, contains calcium chloride ($CaCl_2$) as the cross-linking agent (see Chapter 4). Drugs with unfavorable solid-state properties, such as low solubility, can be encapsulated in the gel matrix. Alternatively, smaller microspheres (<10 μm) can be produced by a water-in-oil emulsification process using an ultrasonicator. A surfactant agent is used to obtain a stable water-in-oil emulsion. An aqueous $CaCl_2$ solution is then added to the emulsion under stirring to allow ionotropic gelation of the particles.[49] Despite their tremendous potential, alginate devices are susceptible to rapid degradation at neutral pH and demonstrate low adhesion to mucosal tissues. The carrier properties of alginate could be improved through the incorporation of chitosan in the beads. These chitosan-treated alginate beads can be controlled release systems for small molecular drugs with high solubility. The drying technique used affects the bead properties.[50]

The oral administration of peptide or protein drugs requires protecting them from degradation in the gastric environment and improving their absorption in the intestinal tract. Chitosan–alginate beads loaded with a model protein, bovine serum albumin (BSA), were investigated to explore the temporary protection of protein against acidic and enzymatic degradation during gastric passage. The presence of chitosan in the coagulation bath during bead preparation resulted in increased entrapment of BSA. Release studies were done in simulated gastric fluid (SGF) at pH 1.2 and subsequently in simulated intestinal fluid (SIF) at pH 7.5 to mimic physiological gastrointestinal conditions. During incubation in simulated gastric fluid, the beads showed swelling and did not show any sign of erosion. After transfer to intestinal fluid, the beads eroded, burst, and released the protein. The presence of chitosan in the beads delayed the release of BSA, as the multilayer beads disintegrated very slowly. The enzymes pepsin and pancreatin did not change the characteristics of BSA-loaded chitosan–alginate beads. The results suggest that alginate beads reinforced with chitosan offer an excellent means for the controlled gastrointestinal passage of protein drugs.[51,52]

Alginate in combination with polyethylene glycol (PEG) is a well-studied delivery material for proteins. A chitosan–PEG–alginate microencapsulation process applied to proteins such as albumin was reported to be a good candidate for the oral delivery of bioactive peptides. While PEG exhibits useful properties such as low toxicity and immunogenicity together with the ability to preserve the biological properties of proteins, alginate acts as a coating membrane.[53] A spray-drying technique was applied to BSA–sodium alginate solutions to obtain spherical particles having a mean diameter less than

10 µm. The microparticles were hardened using first a solution of calcium chloride and then a solution of chitosan. The chitosan concentration and pH affected the BSA loading in the microparticles.[54]

Recently, alginate-bearing cyclodextrin molecules covalently linked on polymer chains for the sustained release of hydrophobic drugs have been developed. These cyclodextrin derivatives of alginate are promising as they exhibit the cumulative properties of size specificity of cyclodextrin and transport properties of the polymer matrix.[29]

11.3.1.3 Alginate Scaffolds for Tissue Engineering

Macroporous scaffolds are typically utilized in tissue engineering applications to allow the migration of cells throughout the scaffold and integration of the engineered tissue with the surrounding host tissue. A method to form macroporous beads from alginate incorporates gas pockets within alginate beads, and the gas bubbles are stabilized with various surfactants. Alginates in these scaffolds can have average molecular weights ranging from 5 to 200 kDa. These products support cell invasion *in vitro* and *in vivo*.[55] Similarly, novel composites based on calcium sulfate blended with alginate and materials based on acrylic polymers and alginate–polyethylene glycol copolymers can be used as scaffolds in reconstructive processes.[29] In a composite hydrogel derived from oxidized alginate and gelatin, the degree of cross-linking of the gel was found to increase with an increase in the degree of oxidation of alginate, whereas the swelling ratio and the degree of swelling decreased. The gel was found to be biocompatible and biodegradable.[56] The manufacture of highly stable and elastic alginate membranes with good cell adhesion and adjustable permeability has been reported. Clinical-grade, ultra-high-viscosity alginate was gelled by the diffusion of barium ions. The burst pressure of well-hydrated membranes depended on manufacture and storage conditions. NaCl-mediated membrane swelling can be prevented by partial replacement of salt with sorbitol. Properties of the film, such as hydraulic conductivity and mechanical stability, have been reported.[57]

11.3.2 Carrageenans

Carrageenans have been used by the pharmaceutical industry in the production of pills and tablets. In addition to their well-known biological activities related to inflammatory and immune responses, carrageenans are potent inhibitors of herpes and human papillomavirus (HPV) viruses. There are also indications that these polysaccharides may offer some protection against human immunodeficiency virus (HIV) infection. Moreover, chemical modifications of carrageenans can lead to derivatives with enhanced abilities to combat several diseases. Carrageenan and phospholipid fractions from the seaweed *Porphyra yezoensis* have been shown to exhibit antitumor activity.[58] In addition to alginate, carrageenans have been used for the production of tablets because of their elastic behavior.[59]

11.3.3 Fucoidans

As mentioned in Chapter 4, fucoidans are complex, heterogeneous polysaccharides containing sugars (particularly fucose) and high amounts of sulfate derived from marine brown algae. Fucoidans are present in the dietary brown seaweeds that are consumed frequently in Asian countries. It can also be extracted from marine invertebrates such as sea cucumber or sea urchin.[60] Fucoidans are reported to have antitumor, antimutagenic, immunomodulating, hypoglycemic, antiviral, hypolipidemic, and antiinflammatory properties. Being natural antioxidants, they have great potential for preventing free-radical-mediated diseases. Their anticoagulant activity is perhaps the most widely studied property of fucoidans. Fucoidans are potent thrombin and factor Xa inhibitors mediated by antithrombin or heparin cofactor II. They bind antithrombin, an inhibitor of blood coagulation.[61]

The marine green algae *Monostroma latissimum* was found to be high in rhamnose-containing sulfated polysaccharide, which displayed high anticoagulant activities.[62] The antithrombin activity of fucoidan from *Lessonia angustata* var. *longissima* is almost comparable to that of heparin. Similarly, native fucoidan with a molecular weight of 320,000 Da from *L. vadosa* showed good anticoagulant activity, whereas the depolymerized fraction, having a molecular weight of 32,000 Da, presented weak anticoagulant activity. Of the 16 species of Indian marine green algae screened for blood anticoagulant activity, *Caulerpa* species exhibited the highest activity, comparable to that of heparin.[63] Similarly, the red alga *Tichocarpus crinitus* cell wall contained sulfated galactans, which displayed high anticoagulant activities even at low concentrations.[64] These studies indicate that fucoidans can potentially replace the conventional anticoagulant heparin, which is prepared from mammalian mucosa.

Besides anticoagulant activity, fucoidans also exhibit antiviral activities both *in vivo* and *in vitro*. The polysaccharides exhibit low cytotoxicity compared with other antiviral drugs currently used in clinical medicine. Because they interfere with the molecular mechanisms of cell-to-cell recognition, fucoidans can be used to block cell invasion by various retroviruses such as HIV, herpes, cytomegalovirus, and African swine fever virus. A sulfated fucan having a molecular weight of 40 kDa was shown to be a selective inhibitor of herpes simplex virus type 1.[65] Fucoidan from brown algae showed antiprion activity and delayed disease onset when it was ingested after enteral prion infection. Daily uptake of fucoidans might be prophylactic against prion diseases caused by the ingestion of prion-contaminated materials. Fucoidan was found to inhibit proliferation and induce apoptosis in human lymphoma HS–Sultan cell lines. Fucoidans from *Laminaria saccharina, L. digitata, Fucus serratus, F. distichus,* and *F. vesiculosus* strongly blocked MDA-MB-231 breast carcinoma cell adhesion to platelets, an effect that could have critical implications in tumor metastasis.[58]

The formation of calcium oxalate (i.e., kidney stones) is a major disease that particularly affects the elderly. Abnormalities in oxalate metabolism have been suggested as a cause for the pathogenesis of stone disease, as an excessive excretion of oxalate leads to calcium oxalate crystal urea. Synthetic polysaccharides, such as low-molecular-weight heparin (LMWH), have been reported to have renoprotective effects. Fucoidans isolated from seaweed are similar to heparin; hence, the nephroprotective action of heparin derivatives and sodium pentosan polysulfate (SPP) could also be extended to fucoidans.[40] In view of these findings, significant potential is seen for the medical exploitation of fucoidans, as noted in a number of excellent recent reviews.[8,66–68]

11.3.4 Other Seaweed Polysaccharides

Ulvan isolated from *Ulva conglobata* demonstrates significant anticoagulant activity due to the direct inhibition of thrombin and the potentiation of heparin cofactor II.[69] Polysaccharides from *U. pertusa* have been reported to have potential for preventing ischemic cardiovascular and cerebrovascular diseases. The compounds studied contained 47% total carbohydrates (mainly composed of rhamnose, xylose, and glucose and smaller amounts of mannose, galactose, and arabinose), 23.2% uronic acids, 17.1% sulfate groups, 1.0% N, and 29.9% ash. The polysaccharides significantly lowered the levels of plasma total cholesterol, low-density lipoprotein cholesterol, and triglycerides and markedly increased serum high-density lipoprotein cholesterol compared with the hyperlipidemia control group.[70] Laminarin oligosaccharides and polysaccharides can be utilized to develop new immunopotentiating substances and functional alternative medicines.[40]

11.3.5 Microalgal Polysaccharides

Based on preliminary research, several potential therapeutic benefits have been identified for commercially produced microalgae. These benefits include the treatment or prevention of AIDS and cancer, cerebral, and vascular diseases. The microalga *Ostoc* has been reported to exhibit antitumor activity, and a variety of sulfated polysaccharides from the microalgae have anticoagulant and antithrombotic activities. They can also inhibit viral infections. A novel antiviral polysaccharide, nostoflan, isolated from *Nostoc flagelliforme*, exhibited potent activity against HSV-1, HSV-2, and human cytomegalovirus.[71–73] Significant biological activities have been attributed to the polysaccharide from *Gymnodinium* spp. An organism belonging to this genus produces a sulfated polysaccharide that shows antiviral activity against encephalomyocarditis virus. Microalgal polysaccharide have

TABLE 11.5

Therapeutic and Biotechnological Applications of Microbial Exopolysaccharides

Polysaccharide	Medical and Biotechnological Applications
Alginate	Hypoallergenic, wound-healing tissue encapsulation of nutrients
Emulsan	Hypoallergenic, wound-healing tissue encapsulation of nutrients
Gellan	Solidifying culture media, especially for studying marine microorganisms
Xanthan	In secondary and tertiary crude oil recovery, paints, pesticides, detergents, cosmetics, pharmaceuticals, printing inks (to control viscosity, settling, and gelation)
Cellulose	As artificial skin to heal burns or surgical wounds, as hollow fibers or membranes for specific separation technology, as acoustic membranes in audiovisual equipment
Curdlan	Immobilization matrix; curdlan with zidovudine (AZT) displays promising antiretroviral activity
Succinoglycan	Immobilization matrix
Dextran	As a blood plasma extender or blood flow improving agent and as cholesterol-lowering agent; in separation technology, as carrier in tissue and cell culture

Source: Adapted from Kumar, A.S. et al., *J. Basic Microbiol.*, 47, 103, 2008; Gibbs, P.A. and Seviour, R.J., in *Polysaccharides in Medicinal Applications*, Dumitriu, S., Ed., Marcel Dekker, New York, 1996, pp. 56–86.

been shown to be hypocholesterolemic agents in animals and to have antiviral activity against such animal viruses as HSV-1 and HSV-2, as well as retroviruses.[74]

11.3.6 Microbial Exopolysaccharides

Pharmaceutical applications of exopolysaccharides of non-marine origin such as bacterial alginate, emulsan, gellan, xanthan, curdlan, succinoglycan, and dextran are summarized in Table 11.5. Despite growing interest in identifying the biological activities (e.g., antitumor and immunostimulatory) of polysaccharides produced by marine bacteria, the biomedical and biotechnological potentials of these biopolymers from marine and deep-sea hydrothermal vent environments remain largely untapped. Bacterial alginates isolated from marine microorganisms have received some attention for diverse biotechnological applications.[75–77] Breakthroughs in the genetics of alginate-producing bacteria have opened up the prospect of polysaccharide engineering.[29]

10.3.7 Polysaccharides from Sponges

Diverse sulfated polysaccharides synthesized by species of marine sponges could possess interesting biological activities. Sulfated polysaccharides were extracted from four species of these organisms by papain digestion. Analysis of the purified polysaccharides revealed a species-specific variation in their chemical composition. The sulfated polysaccharides contained variable proportions of galactose, fucose, arabinose, and hexuronic acid with different degrees of sulfation. It has been suggested that these compounds play a role in the species-specific aggregation of sponge cells or in the structural integrity of sponges.[78] An exopolysaccharide extracted from the sponge *Celtodoryx girardae* had a unique molecular weight of about 800 kDa, contained significant amounts of sulfate, and displayed antiviral activity against HSV-1.[79]

11.4 Potentials of Nanotechnology

Nanotechnology offers the potential to enhance the therapeutic applications of marine polysaccharides. Hydrogel nanoparticles have attracted considerable attention in recent years as they offer significant promise as nanoparticulate drug delivery systems due to their unique properties obtained by combining the characteristics of a hydrogel system (hydrophilicity and extremely high water content) with a nanoparticle (very small size). Among the natural polymers, chitosan and alginate have been studied extensively for the preparation of hydrogel nanoparticles. Regardless of the type of polymer used, the release mechanism of the loaded agent from hydrogel nanoparticles is complex, resulting from three main vectors: drug diffusion, hydrogel matrix swelling, and chemical reactivity of the drug and matrix.[80]

Biodegradable nanotubes were fabricated through the layer-by-layer assembly technique of alternate adsorption of alginate and chitosan onto the inner pores of polycarbonate template with subsequent removal of the template. The assembled materials demonstrated good film-forming ability, low cytotoxicity, and good biodegradability and could readily be internalized into cancer cells.[81] Hydroxyl-containing antimony oxide bromide (AOB) nanorods were combined with biopolymer chitosan to form a hybrid biocompatible, crack-free, and porous chitosan–AOB composite film. The composite film could be used efficiently to entrap proteins for potential biomedical applications and for biosensors, biocatalysis, and bioelectronics.[82]

Another material with medical potential is nanofibers of chitosan membranes or mats developed through electron spinning of carboxymethyl chitosan.[83] A drug delivery system consisting of liposome–chitosan nanoparticle complexes has been developed.[84] The use of quaternary ammonium chitosan derivative nanoparticles (QCS NP) as bactericidal agents in polymethyl methacrylate (PMMA) bone cement has been found to be beneficial. A 10^3-fold reduction in the number of viable counts of *Staphylococcus aureus* and *S. epidermidis* upon contact with the surface of the CS NP- and QCS NP-loaded bone cements was achieved, and no cytotoxic effects were observed. Mechanical tests indicated that the addition of the CS and QCS in nanoparticle form retained the strength of the bone cement to a significant degree, suggesting a promising strategy for combating joint implant infection.[85] Multicomponent systems containing chitosan and β-lactoglobulin core shell nanoparticles were prepared as biocompatible carriers for the oral administration of nutraceuticals. Brilliant blue release experiments showed that the nanoparticles prepared with native β-lactoglobulin had properties favorable to resisting acid and pepsin degradation in simulated gastric conditions. When transferred to simulated intestinal conditions, the β-lactoglobulin shells of the nanoparticles were degraded by pancreatin.[86]

Uniform chitosan microspheres have been fabricated and weakly crosslinked for potential applications in colon-specific drug delivery. The effects of microsphere size, cross-linking density, and electrostatic interactions between the drug and chitosan on drug release were studied, employing model drugs of different acidities. When the drug was basic, all chitosan spheres exhibited 100% release within 30 minutes. As the acidity of the drug increased, the release slowed down and depended on the cross-linking density and microsphere size. The release of weakly acidic drug was most suppressed for large spheres (35 to 38 μm), while the small spheres (23 to 25 μm) with higher cross-linking exhibited the most retention of highly acidic drug, indicating that they are a promising candidate for colon-specific delivery.[87]

Microspheres referred to as fucospheres have been developed to serve as drug carriers for macromolecular drugs such as peptides and proteins; they are produced through the cross-linking of fucoidan with chitosan and range in size from 0.61 to 1.28 μm. Using 2.5% fucoidan, BSA was efficiently encapsulated at a maximum capacity of about 90%. The extent of drug release from the microsphere was dependent on the concentration of BSA, chitosan type, and the preparation method.[88]

11.5 Commercial Aspects

Alginate, fucoidan, and chitosan are the important marine polysaccharides that are the focus of biomedical research today.[8,90] The main commercial sources of seaweed polysaccharides for alginate are species of *Ascophyllum* (large brown alga in the family Fucaceae), *Laminaria* (a genus of 31 species of brown algae of the class Phaeophyceae, found mostly in Europe), *Lessonia* (seaweed largely obtained from France and South America), *Ecklonia* (a genus of kelp belonging to the family Lessoniaceae and found in South Africa), *Durvillaea* (a genus of brown algae of class Phaeophyceae, found in Australia, New Zealand, and South America), and *Macrocystis* (an immensely long blackish seaweed of the Pacific, found mostly in California).[89,90]

Recent years have witnessed marked growth in dietary chitosan commercial preparations offering both therapeutic and health benefits. Chitosan is sold in the form of dietary capsules to assist in weight loss in some European countries. Other dietary chitosan products include Fat Absorb™ and Fat Trapper. Chitosan combined with nutrients such as lecithin, vitamin C, vitamin E, garlic, and β-carotene are available in Japan. MinFAT™ is a fat trimmer marketed in Malaysia that claims to absorb 21 times its weight of fat.[13] Several brands of alginate-based wound dressings are available commercially.[1] Fucoidans are now being marketed as nutraceuticals and food supplements because of their significant biological functions.[91] Currently, microencapsulated marine omega-3 polyunsaturated fatty acid (PUFA) powder products are commercially available. Whereas progress has been made in the uses of marine macromolecules for the encapsulation and delivery of nutraceuticals, such as polyunsaturated fatty acids, glucosamine, and bioactive peptide, there is vast untapped potential for the use of these polysaccharides in practical clinical applications.

References

1. Paul, W. and Sharma, C. P., Polysaccharides: biomedical applications, in *Encyclopedia of Surface and Colloid Science*, Somasundaram, P., Ed., Taylor & Francis, London, 2006, pp. 5056–5067.
2. Rinaudo, M., Characterization and properties of some polysaccharides used as biomaterials, *Macromol. Symp.*, 245, 549, 2007.
3. Phillips, G. O. and Williams, P. A., Eds., *Handbook of Hydrocolloids*, 2nd ed., CRC Press, Boca Raton, FL, 2009.
4. Kamyshny, A. and Magdassi, S., Microencapsulation, in *Encyclopedia of Surface and Colloid Science*, Somasundaram, P., Ed., Taylor & Francis, London, 2006, pp. 3957–3969.
5. Peppas, N. A. et al., Hydrogels in pharmaceutical applications, *Eur. J. Pharm. Biopharm.*, 50, 27, 2000.
6. Barbucci, R. et al., Polysaccharides based hydrogels for biological applications, *Macromol. Symp.*, 204, 37, 2003.
7. Edwards, C. A. and Garcia, A. L., The health aspects of hydrocolloids, in *Handbook of Hydrocolloids*, 2nd ed., Phillips, G. O. and Williams, P. A., Eds., CRC Press, Boca Raton, FL, 2009, p. 50.
8. D'Ayata, G. G. et al., Marine derived polysaccharides for biomedical applications: chemical modification approaches, *Molecules*, 13, 2069, 2008.
9. Tolaimate, A. et al., On the influence of deacetylation process on the physicochemical characteristics of chitosan from squid chitin, *Polymer*, 41, 2463, 2000.
10. Prashanth, K. V. H. and Tharanathan, R. N., Chitin/chitosan: modifications and their unlimited application potential—an overview, *Trends Food Sci. Technol.*, 18, 117, 2007.
11. Muzzarelli, R. A. A. and Muzzarelli, C., Chitosan chemistry: relevance to the biomedical sciences, *Adv. Polym. Sci.*, 186, 151, 2005.

12. Shahidi, F. and Abuzaytoon, R., Chitin, chitosan and co-products: chemistry, production, applications and health effects, *Adv. Food Nutr. Res.*, 49, 94, 2005.
13. Subasinghe, S., Chitin from shellfish waste: health benefits overshadowing industrial uses, *Infofish Int.*, 3, 58, 1999.
14. Ravi Kumar, M. N. V. and Hudson, S. M., Chitosan, in *Encyclopedia of Biomaterials and Biomedical Engineering*, Wnek, G. E. and Bowlin, G. L., Eds., Informa Healthcare, New York, 2008, pp. 604–617.
15. Muzzarelli, R. A. A., Chitosan-based dietary foods, *Carbohydr. Polym.*, 29, 309, 1996.
16. Paul, W. and Sharma, C. P., Chitosan, a drug carrier for the 21st century: a review, *STP Pharma Sci.*, 10, 5, 2000.
17. Sato, T., Ishii, T., and Okahata, Y., *In vitro* gene delivery mediated by chitosan: effect of pH, serum, and molecular mass of chitosan on the transfection efficiency, *Biomaterials*, 22, 2075, 2001.
18. Molinaro, G. et al., Biocompatibility of thermosensitive chitosan-based hydrogels: an *in vivo* experimental approach to injectable biomaterials, *Biomaterials*, 23, 2717, 2002.
19. Knorr, E. and Lim, L. Y., Implantable applications of chitin and chitosan, *Biomaterials*, 24, 2339, 2003.
20. Shantha, K. L., Bala, U., and Rao, K. P., Tailor-made chitosans for drug delivery, *Eur. Polym. J.*, 31, 377, 1995.
21. Zheng, H. et al., Novel dual cross-linked complex gel bead based on carboxymethyl chitosan/alginate for oral delivery of protein drugs, PMSE 377, 233rd ACS National Meeting, March 25–29, 2007, Washington, D.C.
22. Kashyap, N. et al., Design and evaluation of biodegradable, biosensitive *in situ* gelling system for pulsatile delivery of insulin, *Biomaterials*, 28, 2051, 2007.
23. Elzathry, A. A. et al., Evaluation of alginate–chitosan bioadhesive beads as a drug delivery system for the controlled release of theophylline, *J. Appl. Polym. Sci.*, 111, 2452, 2009.
24. Sinha, V. R. and Kumria, R., Polysaccharides in colon-specific drug delivery, *Int. J. Pharmaceut.*, 224, 19, 2001.
25. Kosaraju, S. L., Colon targeted delivery systems: review of polysaccharides for encapsulation and delivery, *Crit. Rev. Food Sci. Nutr.*, 45, 251, 2005.
26. Grenha, A. et al., Development of new chitosan/carrageenan nanoparticles for drug delivery applications, *J. Biomed. Mat. Res.*, 92A, 1265, 2010.
27. Ahmadi, R., Joost, D., and de Bruijn, J. D., Biocompatibility and gelation of chitosan-glycerol phosphate hydrogels, *J. Biomed. Mat. Res.*, 86A, 824, 2008.
28. Wu, T. et al., Physicochemical properties and bioactivity of fungal chitin and chitosan, *J. Agric. Food Chem.*, 53, 3888, 2005.
29. D'Ayala, G. G. et al., Marine derived polysaccharides for biomedical applications: chemical modification approaches, *Molecules*, 13, 2069, 2008.
30. Yoshiharu, K. et al., New techniques and functions of chitosan products, *Chitin Chitosan Res.*, 12, 96, 2006.
31. Hirano, S. et al., Wet spun chitosan–collagen fibers, their chemical N-modifications, and blood compatibility, *Biomaterials*, 21, 997, 2000.
32. Muzarelli, R. A. A., Chitins and chitosans for the repair of wounded skin, nerve, cartilage and bone, *Carb. Polym.*, 76, 157, 2009.
33. Ge, Z. et al., Hydroxyapatite–chitin materials as potential tissue-engineered bone substitutes, *Biomaterials*, 25, 1049, 2004.

34. Francis, S. J. K. and Howard W. T. M., Application of chitosan-based polysaccharide biomaterials in cartilage tissue engineering: a review, *Biomaterials*, 21, 2589, 2000.
35. Boucard, N. et al., The use of physical hydrogels of chitosan for skin regeneration following third-degree burns, *Biomaterials*, 28, 3478, 2007.
36. Nettles, D. L., Elder, S. H., and Gilbert, J. A., Potential use of chitosan as a cell scaffold material for cartilage tissue engineering, *Tissue Eng.*, 8, 1009, 2002.
37. Li, Z. et al., Chitosan–alginate hybrid scaffolds for bone tissue engineering, *Biomaterials*, 26, 3941, 2005.
38. Hu, M. et al., Hydrodynamic spinning of hydrogel fibers, *Biomaterials*, 31(5), 863, 2010.
39. Restuccia, D. et al., New EU regulation aspects and global markets of active and intelligent packaging for food industry applications, *Food Control*, 21, 1425, 2010.
40. Veena, C. K. et al., Antioxidant activity of sulfated polysaccharide: beneficial role of sulfated polysaccharides from edible seaweed *Fucus vesiculosus* in experimental hyperoxaluria, *Food Chem.*, 101, 1552, 2007.
41. Ye, H. et al., Purification, anti-tumor and antioxidant activities *in vitro* of polysaccharide from the brown seaweed *Sargassum pallidum*, *Food Chem.*, 111, 428, 2008.
42. Ganesan, P., Kumar, C. S., and Bhaskar, N., Antioxidant properties of methanol extract and its solvent fractions obtained from selected Indian red seaweeds, *Biores. Technol.*, 99, 2717, 2008.
43. Augst, A. D., Kong, H. J., and Mooney, D. J., Alginate hydrogels as biomaterials, *Macromol. Biosci.*, 6, 623, 2006.
44. Balakrishnan, B. et al., Evaluation of an *in situ* forming hydrogel wound dressing based on oxidized alginate and gelatin, *Biomaterials*, 26, 6335, 2005.
45. Anitha, Y., Sharma, C. P., and Sony, P., *In vivo* absorption studies of insulin from an oral delivery system, *Drug Deliv.*, 8, 10, 2001.
46. Picker-Freyer, K. M. and Brink, K. D., Evaluation of powder and tableting properties of chitosan, *AAPS Pharm. Sci. Technol.*, 7, 75, 2006.
47. Jerry, N. et al., Evaluation of sodium alginate as drug release modifier in matrix tablets, *Int. J. Pharm.*, 309, 25, 2006.
48. Chan, L. W. et al., Mechanistic study on hydration and drug release behavior of sodium alginate compacts, *Drug Dev. Ind. Pharm.*, 33, 667, 2007.
49. Srivastavas, R. et al., Stabilization of glucose oxidase in alginate microspheres with photoreactive diazoresin nanofilm coatings, *Biotechnol. Bioeng.*, 91, 124, 2005.
50. Sezer, A. D. and Akbuga, J., Release characteristics of chitosan treated alginate beads. II. Sustained release of a low molecular drug from chitosan treated alginate beads, *J. Microencaps.*, 16, 687, 1999.
51. Anal, A. K. et al., Chitosan–alginate multilayer beads for gastric passage and controlled intestinal release of protein, *Drug Dev. Ind. Pharm.*, 29, 713, 2006.
52. Rayment, P. et al., Investigation of alginate beads for gastro-intestinal functionality. Part 1. *In vitro* characterization, *Food Hydrocoll.*, 23, 816, 2009.
53. T'nnesen, H. H. and Karlsen, J., Alginate in drug delivery systems, *Drug Dev. Ind. Pharm.*, 28, 219, 2002.
54. Coppi, G et al., Chitosan–alginate microparticles as a protein carrier, *Drug Dev. Ind. Pharm.*, 27, 393, 2001.

55. Eisell, P. et al., Porous carriers for biomedical applications based on alginate hydrogels, *Biomaterials*, 21, 1921, 2000.
56. Balakrishnan, B. and Jayakrishnan, A., Self-cross-linking biopolymers as injectable *in situ* forming biodegradable scaffolds, *Biomaterials*, 26, 3941, 2005.
57. Zimmermann, F. et al., Physical and biological properties of barium cross-linked alginate membranes, *Biomaterials*, 28, 1327, 2007.
58. Noda, H. et al., Antitumor activity of polysaccharides and lipids from marine algae, *Nippon Suisan Gakkaishi*, 55, 1265, 1989.
59. Campo, V. et al., Carrageenans: biological properties, chemical modifications and structural analysis—a review, *Carbohydr. Polym.*, 77, 167, 2009.
60. Ribeiro, A. C. et al., A sulfated alpha-L-fucan from sea cucumber, *Carbohydr Res.*, 255, 225, 1994.
61. Kuznetsova, T. A. et al., Anticoagulant activity of fucoidan from brown algae *Fucus evanescens* of the Okhotsk Sea, *Bull. Exp. Biol. Med.*, 136, 471, 2003.
62. Mao, W. et al., Chemical characteristic and anticoagulant activity of the sulfated polysaccharide isolated from *Monostroma latissimum* (Chlorophyta), *Int. J. Biol. Macromol.*, 44, 70, 1999.
63. Shanmugham, M. et al., Distribution of heparinoid-active sulfated polysaccharides in some Indian marine green algae, *Ind. J. Mar. Sci.*, 30, 222, 2001.
64. Babarnova, A. M. et al., Structure and properties of carrageenan-like polysaccharide from the red alga *Tichocarpus crinitus* (Gmel.) Rupr. (Rhodophyta, Tichocarpaceae), *J. Appl. Phycol.* 20, 1013, 2008.
65. Adhikari, U. et al., Structure and antiviral activity of sulphated fucans from *Stoechospermum marginatum*, *Phytochemistry*, 67, 2474, 2006.
66. Holtcamp, A. D. et al., Fucoidans and fucoidanases—focus on techniques for molecular structure elucidation and modification of marine polysaccharides, *Appl. Microbiol. Biotechnol.*, 82, 1, 2009.
67. Kusaykin, M. et al., Structure, biological activity and enzymatic transformation of fucoidans from brown seaweeds, *Biotechnol. J.*, 3, 904, 2008.
68. Li, H. et al., Fucoidan: structure and bioactivity, *Molecules*, 13, 1671, 2008.
69. Mao, W. et al., Sulfated polysaccharides from marine green algae *Ulva conglobata* and their anticoagulant activity, *J. Appl. Phycol.*, 18, 9, 2006.
70. Pengzhan, Y., Quanbin, Z. et al., Polysaccharides from *Ulva pertusa* (Chlorophyta) and preliminary studies on their antihyperlipidemia activity, *J. Appl. Phycol.*, 15, 21, 2003.
71. Plaza, M. et al., Innovative natural functional ingredients from microalgae, *J. Agric. Food Chem.*, 57, 7159, 2009.
72. De Phillippis, R. and Vincenzini, M., Exocellular polysaccharides from cyanobacteria and their possible applications, *FEMS Microbiol. Rev.*, 22, 151, 1998.
73. Raja, R. et al., A perspective on the biotechnological potential of microalgae, *Crit. Rev. Microbiol.*, 34, 71, 2008.
74. Matsunaga, T. et al., Marine microalgae, *Adv. Biochem. Eng./Biotechnol.*, 96, 965, 2005.
75. Rehm, B. H. A., Biosynthesis and applications of alginates, in *Encyclopedia of Biomaterials and Biomedical Engineering*, Bowlin, G. L. and Wnek, G., Eds., Informa Healthcare, New York, 2007, pp. 350–358.
76. Nichols, M. et al., Bacterial EPSs from extreme marine environments with special consideration of the Southern Ocean, sea ice, and deep-sea hydrothermal vents: a review, *Mar. Biotechnol.*, 7, 253, 2005.

77. Smelcerovic, A. et al., Microbial polysaccharides and their derivatives as current and prospective pharmaceuticals, *Curr. Pharm. Des.*, 14, 3168, 2008.
78. Zierer, M. S. and Mourão, P. A. S., A wide diversity of sulfated polysaccharides are synthesized by different species of marine sponges, *Carb. Res.*, 328, 209, 2000.
79. Rashid, A. M. et al., Isolation of a sulphated polysaccharide from a recently discovered sponge species (*Celtodoryx girardae*) and determination of its anti-herpetic activity, *Int. J. Biol. Macromol.*, 44, 286, 2009.
80. Hamidi, M. et al., Hydrogel nanoparticles in drug delivery, *Adv. Drug Deliv. Rev.*, 60, 1638, 2008.
81. Yang, Y. et al., Assembled alginate/chitosan nanotubes for biological application, *Biomaterials*, 28, 3083, 2007.
82. Lu, X., Wen, Z., and Li, J., Hydroxyl-containing antimony oxide bromide nanorods combined with chitosan for biosensors, *Biomaterials*, 33, 5740, 2006.
83. Akira, S., Electrospinning of chitosan and chitosan derivatives, *Chitin Chitosan Res.*, 12, 166, 2006.
84. Diebold, Y. et al., Ocular drug delivery by liposome–chitosan nanoparticle complexes (LCS-NP), *Biomaterials*, 28, 1553, 2007.
85. Shi, Z. et al., Antibacterial and mechanical properties of bone cement impregnated with chitosan nanoparticles, *Biomaterials*, 27, 2440, 2005.
86. Chen, L. and Subirade, M., Chitosan/β-lactoglobulin core shell nanoparticles as nutraceutical carriers. *Biomaterials*, 26, 6041, 2005.
87. Choy, Y. B. et al., Uniform chitosan microspheres for potential application to colon-specific drug delivery, *Macromol. Biosci.*, 8, 1173, 2008.
88. Sezer, A. D. and Akbua, J., Fucosphere—new microsphere carriers for peptide and protein delivery: preparation and *in vitro* characterization, *J. Microencaps.*, 23, 513, 2006.
89. Kardozo, K. H. M., Metabolites from algae with economical impact, *Comp. Biochem Physiol. C*, 146, 60, 2007.
90. Steinbüchel, A. and Marchessault, R. H., Eds., *Biopolymers for Medical and Pharmaceutical Applications: Humic Substances, Polyisoprenoids, Polyesters, and Polysaccharides*, Wiley–VCH, Hoboken, NJ, 2005.
91. Gouin, S., Microencapsulation: industrial appraisal and existing technologies and trends, *Trends Food Sci. Technol.*, 15, 330, 2004.
92. Synowiecki, J. and Khatieb, N. A., Production properties and some new applications of chitin and its derivatives, *Crit. Rev. Food Sci. Nutr.*, 43, 145, 2003.
93. Steinbüchel, A. and Rhee, S. K., Eds., *Polysaccharides and Polyamides in the Food Industry: Properties, Production, and Patents*, Vol. 1, Wiley-VCH, Hoboken, NJ, 2005, p. 163.
94. Kumar, A. S., Mody, K., and Jha, B., Bacterial exopolysaccharides: a perception, *J. Basic Microbiol.*, 47, 103, 2008.
95. Gibbs, P. A. and Seviour, R. J., Pullulan, in *Polysaccharides in Medicinal Applications*, Dumitriu, S., Ed., Marcel Dekker, New York, 1996, pp. 56–86.

Index

A

absorbents, 198
Acacia senegal, 31, 49
Acanthophora, 191
   A. spicifera, 336
acceptable daily intake (ADI), 309–310, 313
   carrageenans, 321
   processed eucheuma seaweed, 319
   xanthan, 314
acetals, 4
acetate, 6
acetic acid, 54, 69, 70, 71, 74, 75, 78, 81, 124,
   166, 271, 275, 277, 297, 334
Acetobacter pasteurianus, 148
Acetobacter xylinum, 146, 148
acetylcholinesterase (AChE), 193
Acinetobacter, 164
   A. calcoaceticus, 244
Acropora, 21
Actinomyces naeslundii, 146
active packaging, 262–264, 269, 275, 297, 299
acyl thiourea chitosan derivatives, 76
adenosine triphosphate (ATP), 12
adhesiveness, 34, 122, 219
adipates, 6
Aerobacter levanicum, 146
Aeromonas, 166
   A. hydrophila, 119, 163, 166
agar, 3, 16, 17, 55, 89, 90, 95–104, 191, 196, 197,
   198–202, 228, 246, 271, 288, 320
   alkali-treated, 96, 101
   –carrageenan blend, 201–202
   diet, 201
   extraction, 95–98
   films, 287, 292
   gelation, 99–101
   gel strength, 96–97, 98, 99, 102–103
   hydrocolloid, 16
   hydrogels, 296
   in baked products, 199
   interactions with food components,
      102–104, 198
   loss modulus, 41
   medical uses of, 202
   production of, 226
   quick-soluble, 201

   safety of, 319
   storage modulus, 41
   structure, 98
   sugar, and, 198
   sugar reactivity, 102–103
   swelling, 101
   vs. gelatin, 200
   yield, 96, 98
agaropectin, 98, 101
agarophytes, 99
agarose, 97, 98, 100–101, 290
   rheological properties of, 101
Agrobacterium, 139, 146, 314
agrowaste, 139, 147, 267
air–liquid interface, 33, 120
albumin, 340
Alcalase®, 66
Alcaligenes, 140, 146
alcohols, 4, 73, 76, 82, 141, 196, 210, 291
algae, 12, 153, 191; *see also* brown seaweed,
   green seaweed, macroalgae,
   microalgae, red seaweed, seaweed
   annual production, 225
   composition of, 90–92
algin, 104–111, 191, 202–206, 319
alginate, 3, 8, 17, 27, 28, 44, 50, 55, 73, 79, 89, 92,
   104–111, 135, 136, 139, 153, 191, 192, 196,
   197, 201, 202–206, 219, 228, 259, 265, 271,
   285, 288, 291, 299, 332, 346
   acid interactions, and, 54
   antioxidant activity, 192
   bacterial, 147, 239, 344
   beads, 109, 340
   biomedical applications of, 337–341
   calcium, and, 284, 285
   chemically modified, 109
   chitosan, and, 180
   coatings, 282–285, 290
   deacetylation, 240
   degradation, 110, 340
   drug delivery matrix, 333
   encapsulation, 289, 290
   extraction, 104–106
   films, 294
   gelatin, and, 294, 295
   gelation, 107–110
      polyphosphates, and, 111